BUILDING 2000

Volume 1

Schools, Laboratories and Universities,
Sports and Educational Centres

Commission of the European Communities

BUILDING 2000

Volume 1

Schools, Laboratories and Universities, Sports and Educational Centres

Edited by

C. DEN OUDEN

EGM engineering bv, Dordrecht, The Netherlands
Technical University, Delft, The Netherlands

and

T.C. STEEMERS

Commission of the European Communities, Brussels, Belgium

Springer Science+Business Media, B.V.

ISBN 978-94-010-5129-3 ISBN 978-94-011-2558-1 (eBook)
DOI 10.1007/978-94-011-2558-1

Publication arrangements by
Commission of the European Communities
Directorate-General Telecommunications, Information Industries and Innovation, Scientific and Technical Communications Service, Luxembourg

EUR 13858

Originally published by **ECSC, EEC, EAEC, Brussels and Luxembourg** in 1992.

PREFACE

This is the first volume of BUILDING 2000, a pilot project of the Commission's R&D-programme 'Solar Energy Applications to Buildings' with the purpose of encouraging the adoption of solar architecture in large buildings.

In this first volume the results of the design studies illustrating passive solar architecture in buildings in the European Community are presented in particular for the building categories: SCHOOLS, LABORATORIES AND UNIVERSITIES, and SPORTS AND EDUCATIONAL CENTRES.

In a second volume, a similar series of studies is presented for the building categories: OFFICE BUILDINGS, PUBLIC BUILDINGS and HOTELS AND HOLIDAY COMPLEXES.

There was an enthousiastic response from project teams responsible for the design of 32 large buildings with a total construction budget of more than 140 million ECU.

The willingness to improve their building concepts by collaborating with R&D-experts was encouraging to the Commission's action in this field.

These two books reflect the results of the exchange of information between the actual design practitioners and the European R&D-community.

Within the BUILDING 2000 programme 'Science and Technology at the Service of Architecture' became reality. This was not only realised by the various support activities initiated by BUILDING 2000, but also by the active exchange of ideas by architects and design team members with R&D-workers during the various workshops held within the BUILDING 2000 programme.

I highly recommend architects and engineers interested in passive solar architecture and modern daylighting approaches to study these final products of the BUILDING 2000 programme.

Dr. W. Palz
Directorate General XII for Science, Research and Development
Commission of the European Communities.

Contents

Volume 1

Contents

Volume 1

Commission of the European Communities

BUILDING 2000

SCHOOLS

S1
High School, Gavrion, Greece
S2
Elementary School, Barcelona, Spain
S3
Combined School, Newport Pagnell, United Kingdom
S4
Primary School, Methilhill, Scotland, UK
S5
Secondary School Valongo do Vouga, Agueda, Portugal
S6
Downsview County Primary School (infants), Swanley, Kent, UK

LABORATORIES AND UNIVERSITIES

L1
High Tech, Laboratory, Leuven, Belgium
L2
Business Innovation Centre, Porto, Portugal
L3
Architecture Faculty Building, Lisbon, Portugal
L4
School of Medicine, Ioannina, Greece
L5
Geosciences Research Laboratory, Ampus, France

SPORTS AND EDUCATIONAL CENTRES

E1
Indoor Public Swimming Pool, Cesson, France
E2
Education Centre, Madrid, Spain
E3
Study Centre, Machynllet, Wales, UK
E4
Research and Educational Centre, Steyerberg, Germany
E5
Educational and Vocational Training Centre, Huntlosen, Germany

Building 2000 is a series of design studies illustrating passive solar architecture in buildings in the European Community.

VOLUME 1 / **KLUWER ACADEMIC PUBLISHERS**

Commission of the European Communities

BUILDING 2000

SCHOOLS

S1
High School, Gavrion, Greece
S2
Elementary School, Barcelona, Spain
S3
Combined School, Newport Pagnell, United Kingdom
S4
Primary School, Methilhill, Scotland, UK
S5
Secondary School Valongo do Vouga, Agueda, Portugal
S6
Downsview County Primary School (infants), Swanley, Kent, UK

Building 2000 is a series of design studies illustrating passive solar architecture in buildings in the European Community.

VOLUME 1 / **KLUWER ACADEMIC PUBLISHERS**

BUILDING 2000

Commission of the European Communities

- High school with ten classrooms
- Passive solar heating in winter using direct gain through south-facing glazing and Trombe walls
- Prevention of overheating on warm days by shading and cross ventilation
- Adequate daylighting of classrooms through south-facing windows and north-facing skylights
- Heating of auditorium in winter using active solar system with air collectors

HIGH SCHOOL

GAVRION, ANDROS ISLAND/GREECE

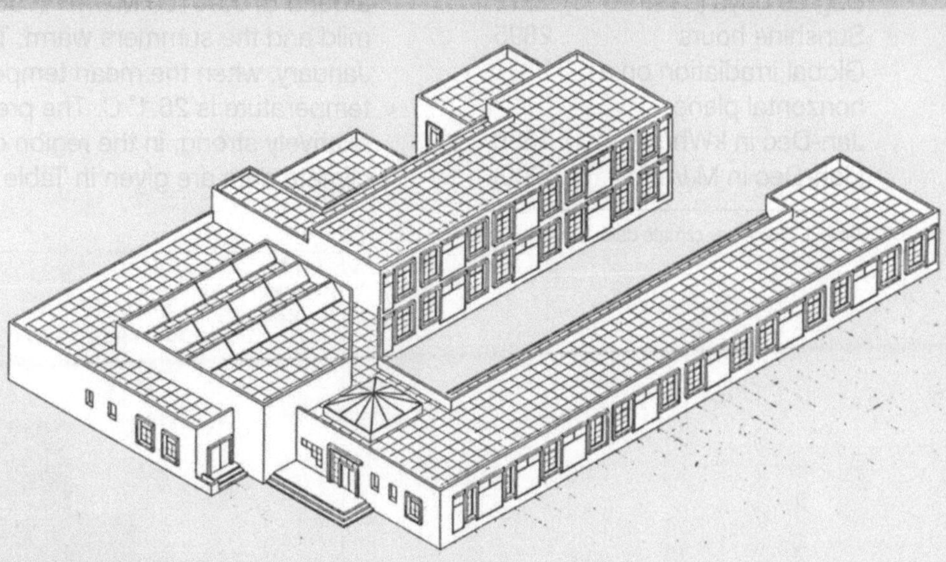

Building 2000 is a series of design studies illustrating passive solar architecture in buildings in the European Community.

ISSUE **S 1**

DEC 1990

PROJECT DESCRIPTION

Building Type

This is a high school with ten classrooms, a library, laboratory, auditorium, administration offices and the usual service areas such as storage spaces, warehouses, cloakrooms, corridors, etc.

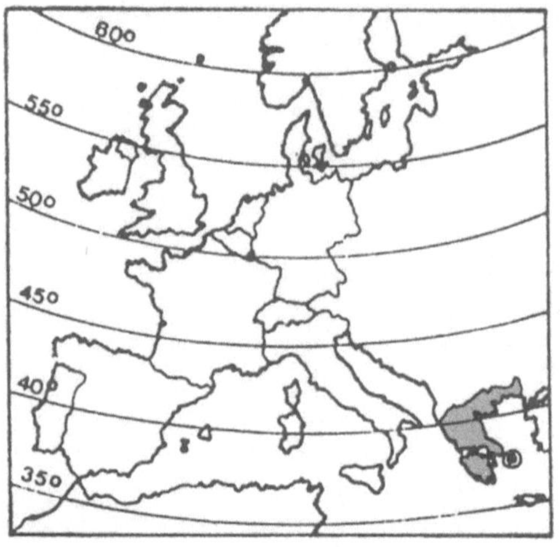

Figure 1. Location

Degree days (base 18° C)	1219
Sunshine hours	2895
Global irradiation on a horizontal plane	
Jan-Dec in kWh/m²	1639
(Jan-Dec in MJ/m²	5900)

Table 1. Some key climate data

Location

The building is at Gavrion on the island of Andros (latitude 37.3° N, longitude 24.8° E - see Figure 1), the most northerly island of the Cyclades archipelago. The 5 acre site is at sea level and perfectly flat which allows the school to be positioned away from shadows likely to be cast by other buildings constructed in the locality in future.

Site Microclimate

Andros is situated in the middle of the Aegean sea and the climate is typical of the Mediterranean marine area - temperate and humid. The annual amount of solar radiation falling on a horizontal plane is high - around 5700-6100 MJ/m² (1583-1694 kWh/m²). The winters are mild and the summers warm. The coldest month of the year is January, when the mean temperature is 10.3° C. In July the average temperature is 26.1° C. The prevailing winds are from the north and relatively strong, in the region of 3.5 m/s to 10.5 m/s. Some key climate data are given in Table 1 and Figure 2.

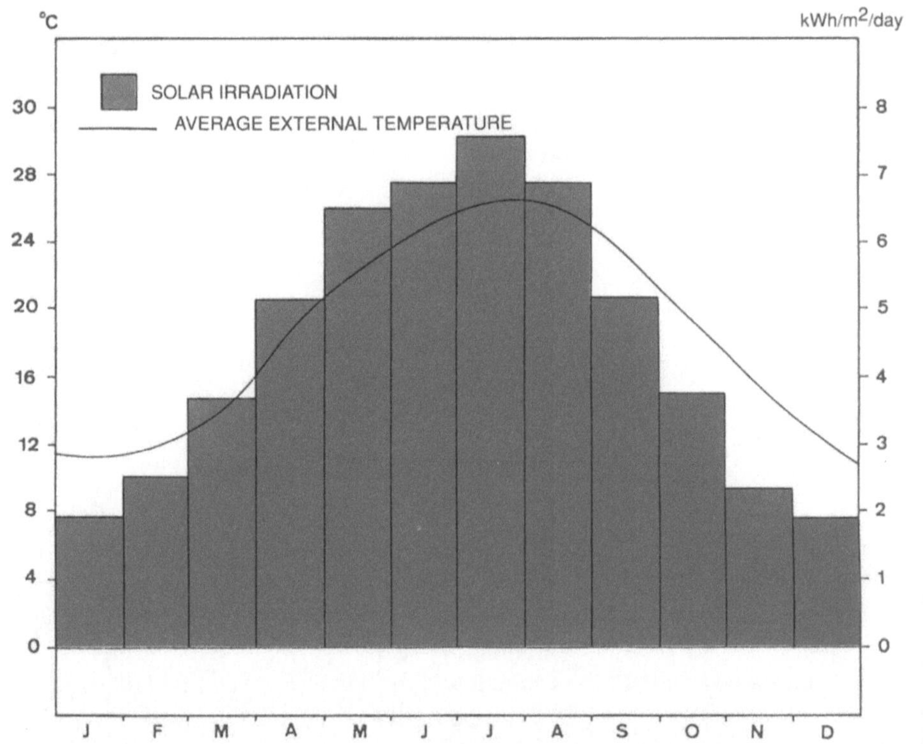

Figure 2. Solar irradiation and ambient temperature data (average values for 1960-75)

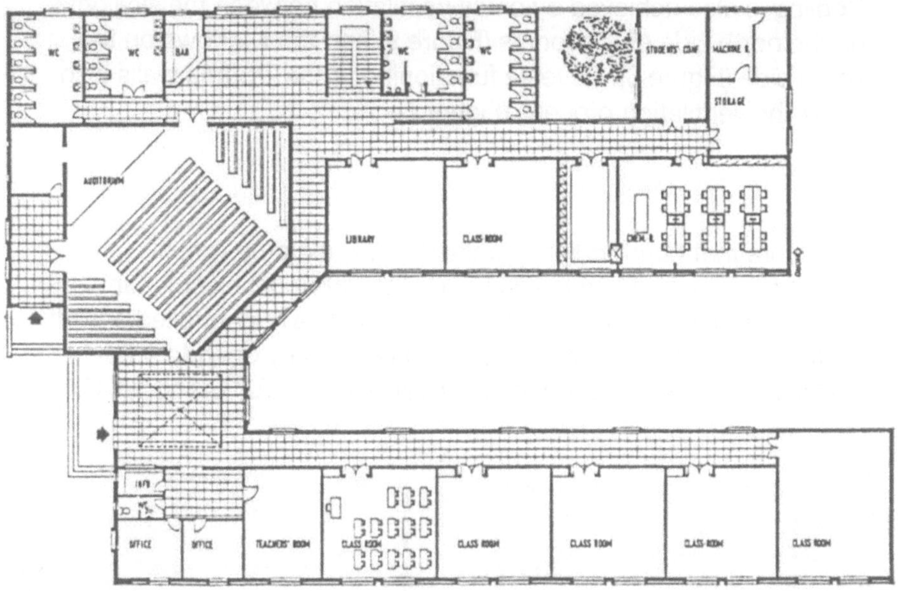

Figure 3. Ground floor plan

Design and Construction Details

The 1600 m² building has been designed to be in keeping with the traditional architecture of the island and yet to take into account bioclimatic principles and the needs of a school. It is arranged along an east-west axis with a slight veer towards the east and consists of two wings joined by the auditorium. It is on two floors. Six classrooms, the library, laboratory, auditorium, administrative offices and services are on the ground floor. The other four classrooms are on the upper floor. The classrooms, library, laboratory and offices are on the south side of the building. The storage areas, warehouses, cloakrooms and corridors, etc., are on the north side. The ground floor plan is shown in Figure 3 and the south facade in Figure 4.

The entrance hall has been turned into a sunspace by addition of a glass pyramidal roof (Figure 11). The building is well insulated, the insulation being placed on the interior of the structure. Polystyrene panels have been used on the roof and external walls. Aerated perlitic concrete has been used for the floor. There is double glazing on all north-facing windows. The U-values of some of the building elements are given in Table 2.

external walls	0.44
roof	0.31
floor	0.65
glazing	
single	5.80
double	3.70

Table 2. U-values of some building elements (W/m² K)

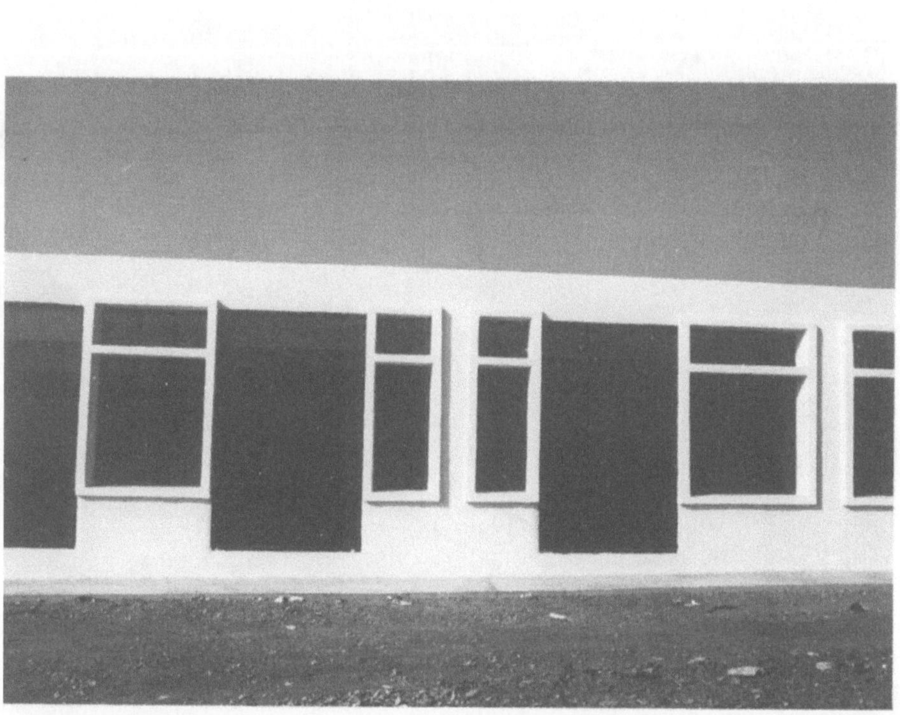

Figure 4. South facade

The school has been designed to use passive solar and other energy-saving features to provide thermal comfort and good daylighting during the period of occupation, which is from September to June.

Heating

Wintertime heating requirements are met in a number of ways. The glazed roof has turned the entrance hall into a sunspace. Glazing covers 85 m² (30%) of the south facade and there is direct solar gain through all of this (see Figures 4 and 7). The unheated spaces on the north side of the building form a thermal barrier and reduce heat loss. Trombe walls occupy 70 m² (25%) of the south side of the classrooms (see Figures 8 and 17). The auditorium (which is large but has a small glazing area) is heated by an array of approximately 15 m² solar air collectors on the roof. The warm air from these is circulated through ducts using electric fans (see Figure 6). To increase the contribution which passive solar systems make to the heating requirements of the school, classes will begin an hour later than usual.

Shading

On warm spring and autumn days, when the solar altitude is low and solar radiation high, overheating from direct solar gain is prevented by the use of lightcoloured internal venetian blinds which can be adjusted to maintain adequate but glare-free daylighting. Because of the high levels of direct solar gain through south facing windows, the total area of Trombe wall is only needed in extreme winter conditions; to prevent overheating at other times, the area of Trombe wall exposed to the sun can be reduced by using external roller blinds.

Ventilation

Cooling is also achieved by cross ventilation between the skylights on the north side of the rooms (Figure 9) and the windows on the south side (Figure 6). Reverse functioning of the Trombe walls also helps the ventilation process (Figure 7).

Auxiliary Heating

To ensure that comfort is maintained under extreme winter conditions, the building is equipped with individual electric fan convector units (3 kW each) with thermostatically-controlled on- off switches. When the heaters are switched on they operate automatically between upper and lower set points of 22° C and 18° C.

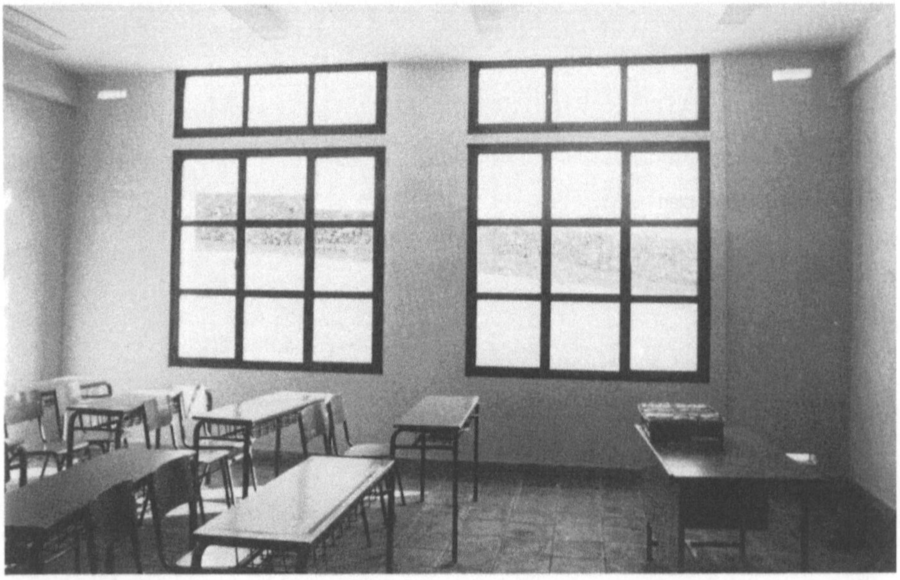

Figure 5. Internal view of Trombe walls and south-facing windows in classroom

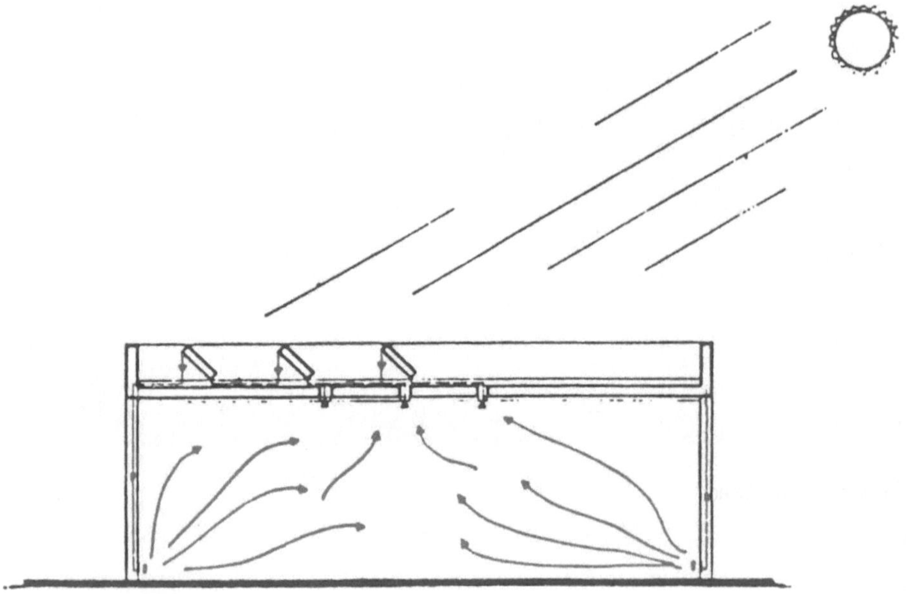

Figure 6. Mode of operation of air collectors

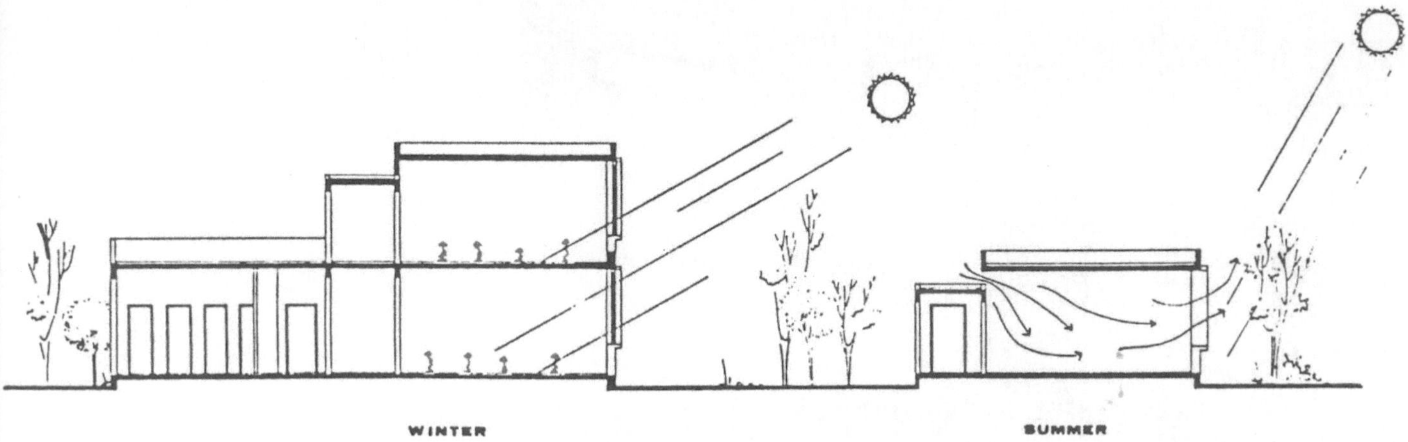

Figure 7. Direct gain: winter and summer operating modes

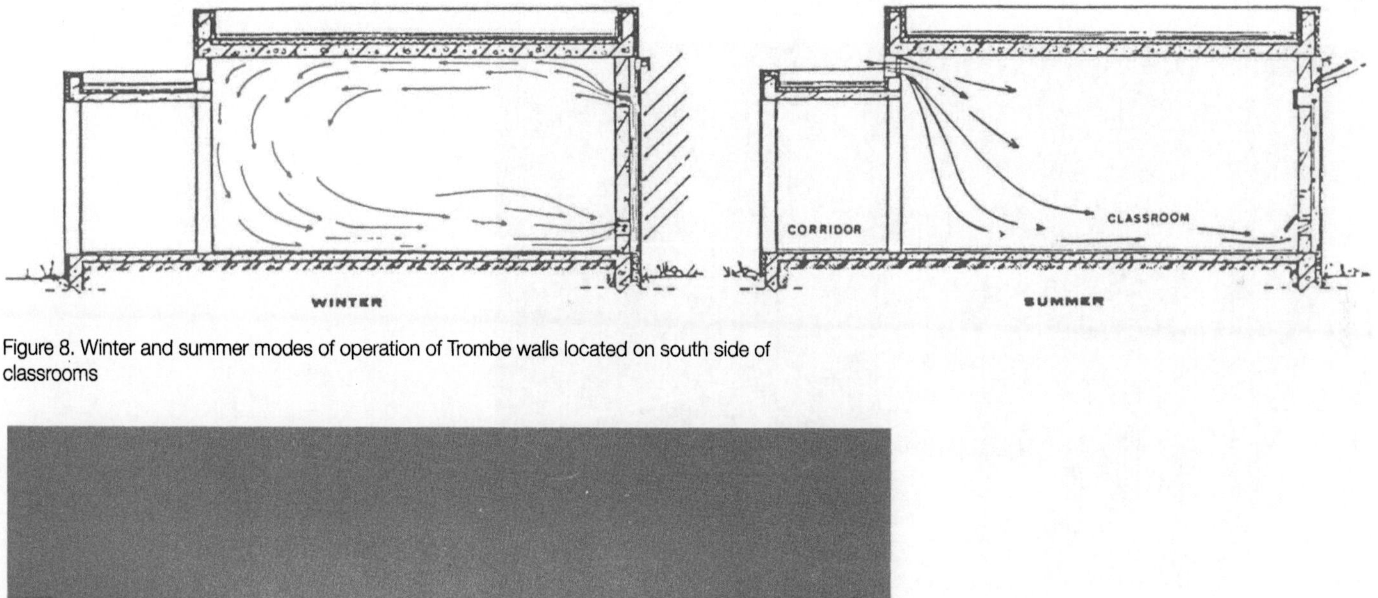

Figure 8. Winter and summer modes of operation of Trombe walls located on south side of classrooms

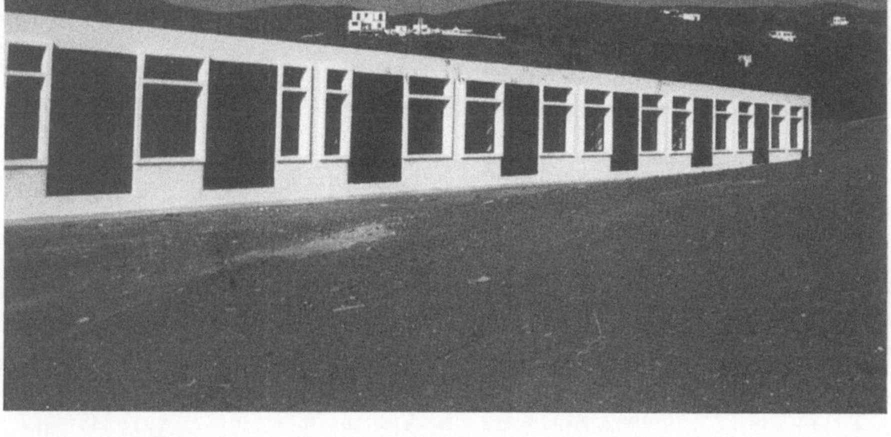

Figure 9. South facade

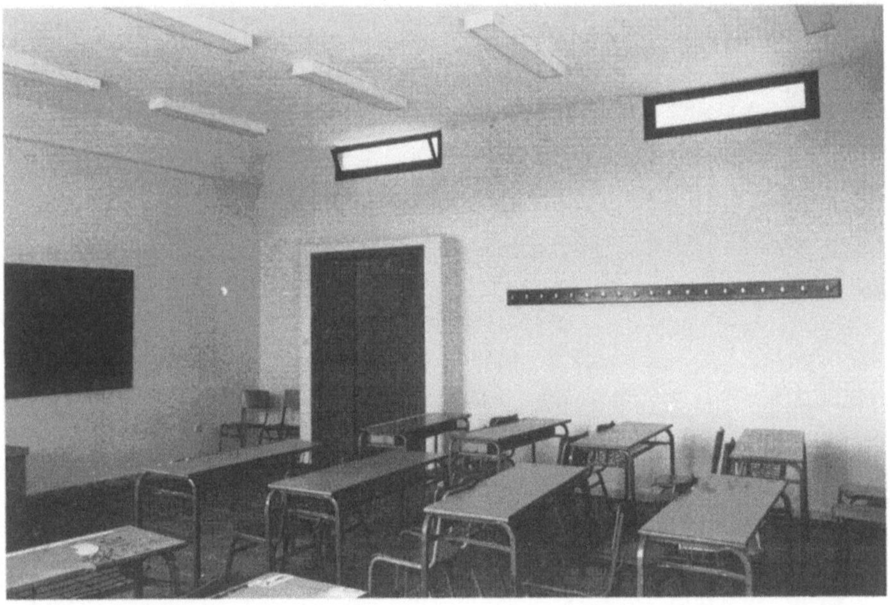

Figure 10. Internal view of skylights on north side of classrooms

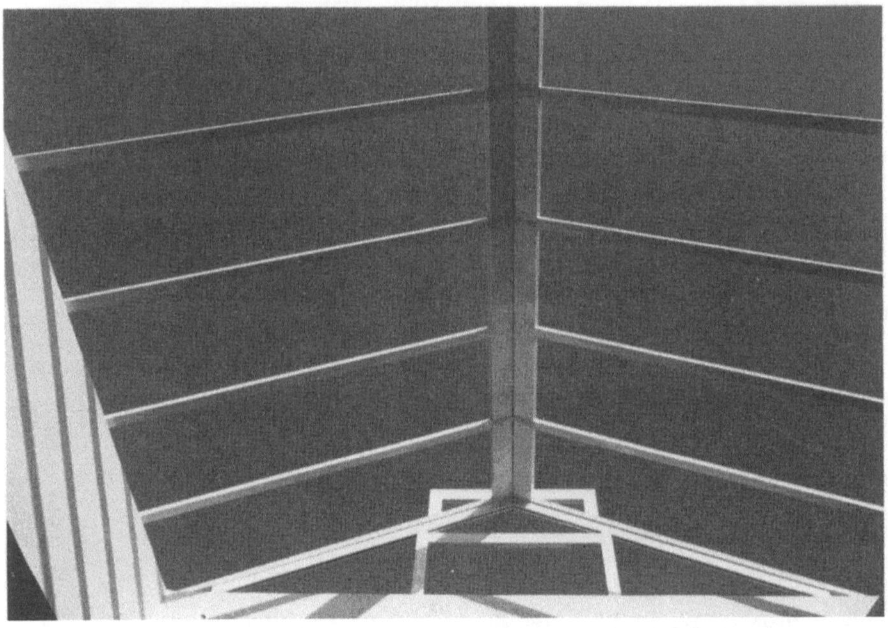

Figure 11. Internal view of glass roof in entrance hall

ENERGY CALCULATIONS PERFORMED AND DESIGN TOOLS USED

Design Tools

The heating, shading, ventilation and daylighting systems were designed with the help of a number of design tools.

Thermal Performance

The thermal performance was evaluated using the PASTEMP computer simulation model. This is based on a simplified thermal network method developed by Building Research Establishment in the UK. Daily internal temperature variations, heating or cooling loads and the passive solar contribution to heating or cooling are estimated using admittance factors. These indicate the extent to which heat enters the building materials over a 24-hour cycle of temperature variation. (Dense materials take up more heat and have higher admittance values than lightweight materials.) Various input parameters describe weather conditions, building structure and openings, etc., occupancy patterns and desired internal temperatures and ventilation rates.

Thermal Comfort

The thermal comfort of the building was estimated using the COMFORT 2.10 program developed by Bruel and Kjaer of the Technical University of Denmark. This calculates the predicted mean vote (PMV) index which indicates how close to comfort a space is taking into account air temperature, humidity and speed and the occupants' clothing, metabolic rate and activity levels. It also estimates the percentage of dissatisfied occupants (PPD).

Daylighting

Daylight factors and illuminances inside the building were determined from room, window and site characteristics using the MICROLITE 1.0 computer model developed in the USA.

	kWh	%
Energy losses		
fabric	17.1	56
infiltration and ventilation	13.6	44 +
Total losses	30.7	100
Energy gains		
casual	16.9	54
passive solar (direct gain and Trombe wall)	12.7	42
active solar (air collectors)	1.1	4 +
Total gains	30.7	100

Table 3. Annual energy balance

Results of Thermal Performance Analysis

The calculations showed that, because of the mild climate and relatively short occupancy period (9 am to 4 pm, September to June), the passive solar systems will supply almost all the heating needs of the building. They also indicated that the natural ventilation and shading systems would be sufficient to cool the building in the warmer months.

The energy balance of the building for a typical year is shown in Table 3. Figures 12 and 13 show the classroom temperatures over a cold sunny and a cold cloudy day in January and over a typical January and a typical June day. The curves show the effect of shading the Trombe wall.

Evaluation of the active solar system showed that the 15 m^2 array of solar collectors would provide around 90% of the auditorium heating require-ments. Use of the collectors is justified because the auditorium is large and has a low glazing area.

The building's high insulation levels and the low thermal mass of the floor lead to a rapid response to solar gains. Thermal comfort is highly dependent on ventilation rates. Because of the large number of openings in the building, the basic infiltration rate was assumed to be 2 air changes per hour (ACH). The required rate, which depends on the number of pupils in the building, is set at 3 ACH in winter and 5 ACH in the warmer months when cross ventilation can be induced in daytime through the Trombe walls. Ventilation is increased at night to cool the building structure.

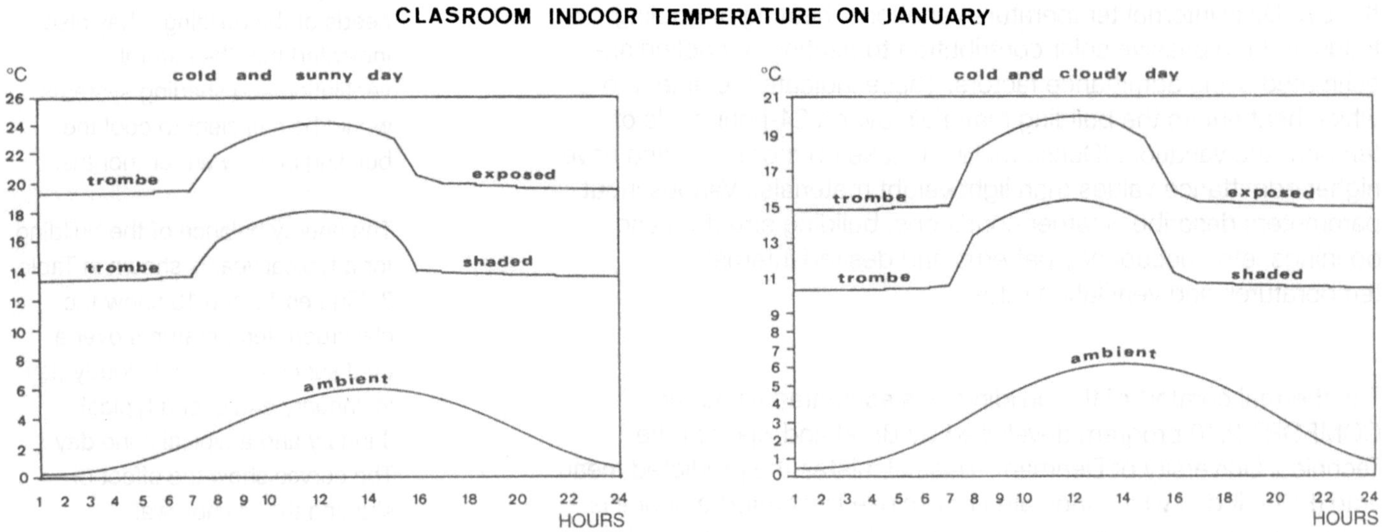

Figure 12. Variation in classroom temperature on a cold sunny and a cold cloudy January day. The curves show the effect of shading the Trombe wall. The occupancy period is from 8 am to 4 pm.

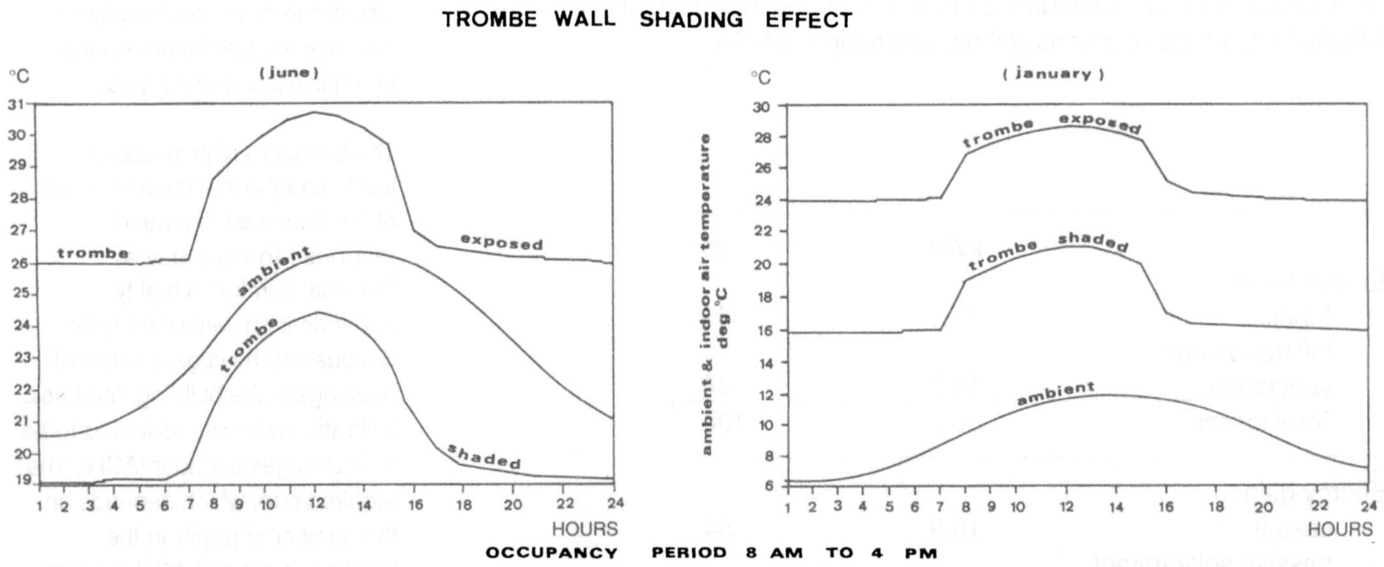

Figure 13. Variation in classroom temperature on a typical June and a typical January day. The curves show the effect of shading the Trombe wall. The occupancy period is from 8 am to 4 pm.

Results of Thermal Comfort Analysis

The comfort analysis showed that the PMV and PPD values vary considerably not only from month-to-month but over a single day. This is because thermal comfort depends on variables such as air temperature, mean radiant temperature, relative humidity and ventilation rate (i.e. air speed) as well as on clothing and activity levels. The results showed that the building will be comfortable without the use of auxiliary heating or cooling systems both in winter and in the warmer months of occupation: the PMV was always in the range - 0.5 to + 0.5 and the PPD in the range 5% to 10%. The actual figure depended on the ventilation rate and the use of the Trombe walls. The daily variation of the PMV in a classroom on a typical June and January day is shown in Figure 14.

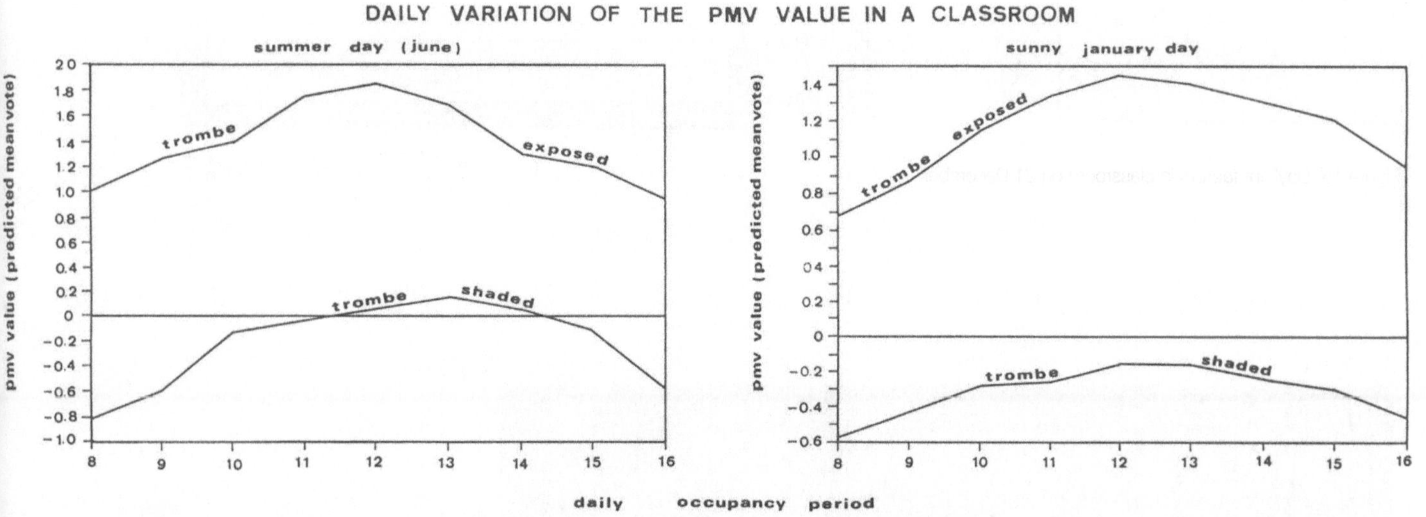

Figure 14. Variation of the PMV value in a classroom on a sunny June and a sunny January day over the period of occupancy

Results of Daylighting Analysis

The analysis was carried out for a classroom at a height of 1 m from the floor for different hours of the day on 21 December with an overcast sky. The results are given in Figure 15. The building has extensive areas of glazing and, as would be expected, produced average daylight factors of between 7% and 9%. At 12 noon, with a sky illumination of 24,000 lux, the minimum daylight factor was 2%. This is above the 300 lux required on average for a classroom. Glare arising from reflective surfaces (on, say, student desks) is prevented by the use of venetian blinds.

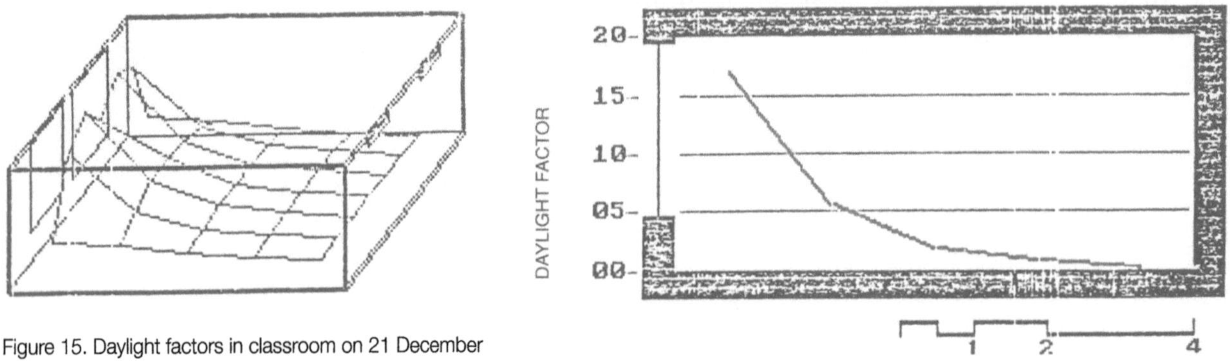

Figure 15. Daylight factors in classroom on 21 December

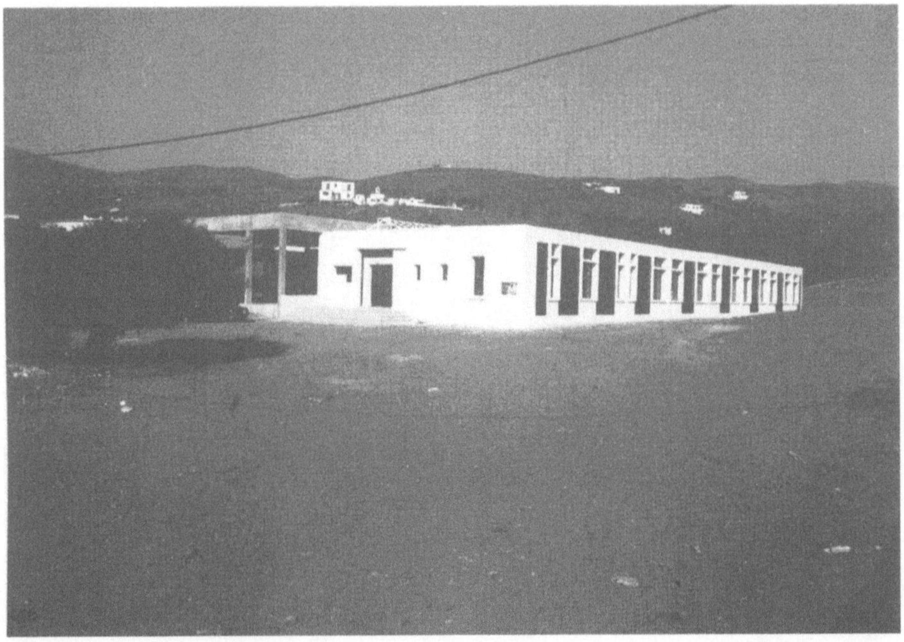

Figure 16. View from the south west

GENERAL DESIGN GUIDELINES/POINTS OF INTEREST RESULTING FROM THIS PROJECT

Bioclimatic design, in its broadest sense, was known to the island's early architects. Modern buildings on Andros, on the other hand, have ignored the climate factor. This project aimed to demonstrate that bioclimatic principles can be used effectively in modern buildings in the Cyclades region. A school building was chosen for the demonstration because, with its high internal gains and lack of summer use, it is particularly suitable for passive solar systems. Because the project is a demonstration, the systems used in the school needed to be simple to operate and cheap to install. Andros' regulations concerning building shape led to the choice of direct solar gain, Trombe walls and solar air collectors. The project has showed that these systems can fit into the local architectural environment. The studies carried out as part of the Building 2000 programme proved that:

- the chosen heating systems can provide assured thermal comfort conditions and supply the majority of the heating demands of buildings in the region;

- cross ventilation is a simple and effective means of providing cooling in this geographical area because there are cool northerly winds most of the year;

- high levels of insulation installed in a way which takes into account the building use are a decisive factor in the success of bioclimatic design.

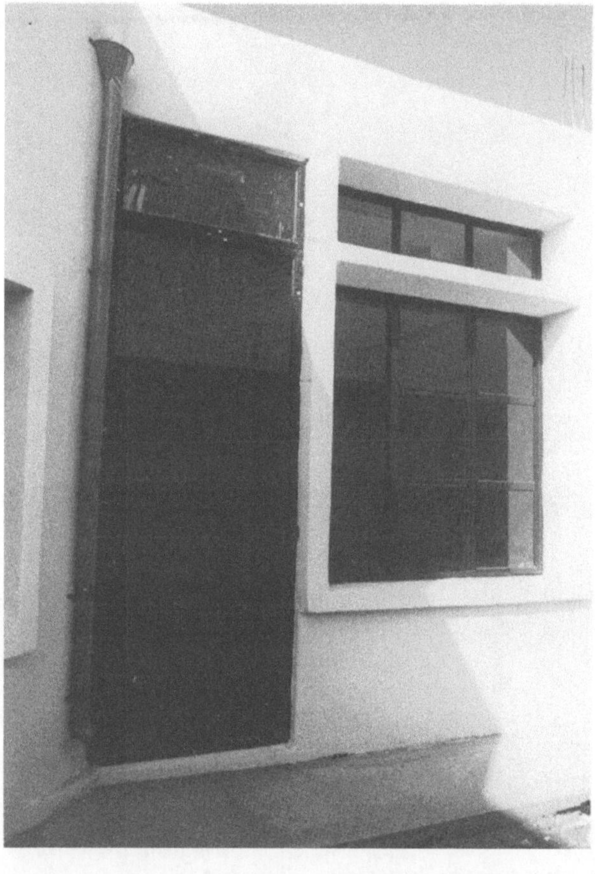

Figure 17. Detail of Trombe wall

BUILDING 2000 Building 2000 brochures are published by Directorate General XII of the Commission of the European Communities to show how design studies can help architects and other building designers use passive solar principles to the best effect to produce attractive energy efficient buildings. Each brochure describes studies carried out with the support of the Commission during the design phase of one or thirty-six non-domestic buildings in the EC Member States. The studies were on such topics as daylighting, heating, cooling, ventilation, comfort, control systems and urban design. They were carried out with the help of acknowledged European experts in these fields and drew heavily on lessons learned and techniques developed through the Commission's research and development programme on solar energy applications to buildings.

Commission of the European Communities/Directorate-General for Science, Research and Development

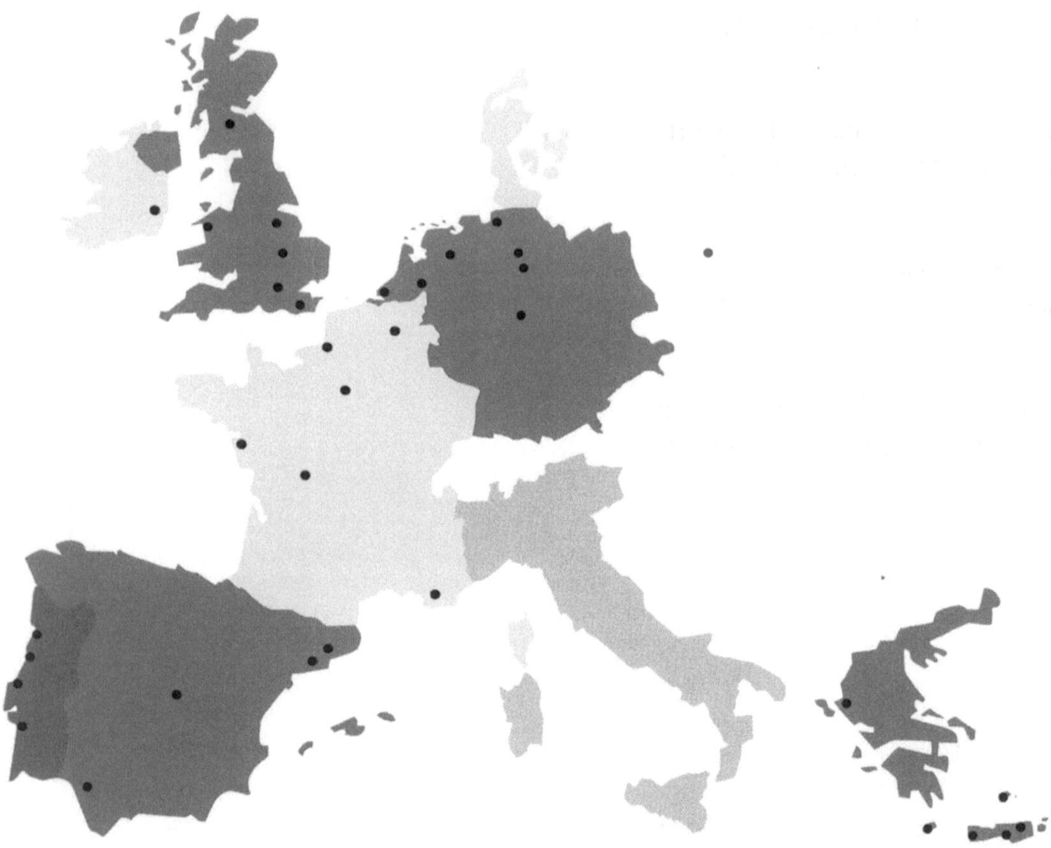

List of Design Team Participants and Advisers

Client
Prefecture of Cyclades
Greece

Architect
Alexander Kapon
Charilaou Tricoupi Str 146
Athens 11472
Greece

Energy Consultant
George Zannis

Structural Engineer
Savas Varthalitis

This set of **Building 2000** brochures illustrates how architects and other building designers can successfully apply passive solar principles to produce energy-efficient buildings.

BUILDING 2000 Participants

Project Director
Theo C. Steemers

Coordinator
Cees den Ouden

Technical Steering Committee
Dean Hawkes
Nick Baker
Alex Lohr
Jean P. Lepoivre

Regional Liaison Agents
(D) Jörn Behnsen
(E) Vicente Sifre
(F) Michel Raoust
(GB) Alan Hildon
(GR) Matheus Santamouris
(P) Eduardo Maldonado

Further information or copies of the brochures can be obtained from prof. ir. Cees den Ouden, EGM Engineering BV, P.O. Box 1042, 3300 BA Dordrecht, The Netherlands.

BUILDING 2000

Commission of the European Communities

- Elementary school for 640 pupils
- Improvement in daylighting by use of sun-ducts, light shelves and south-facing clerestories
- Passive heating using walls containing water store
- Natural cross ventilation

ELEMENTARY SCHOOL

BARCELONA/SPAIN

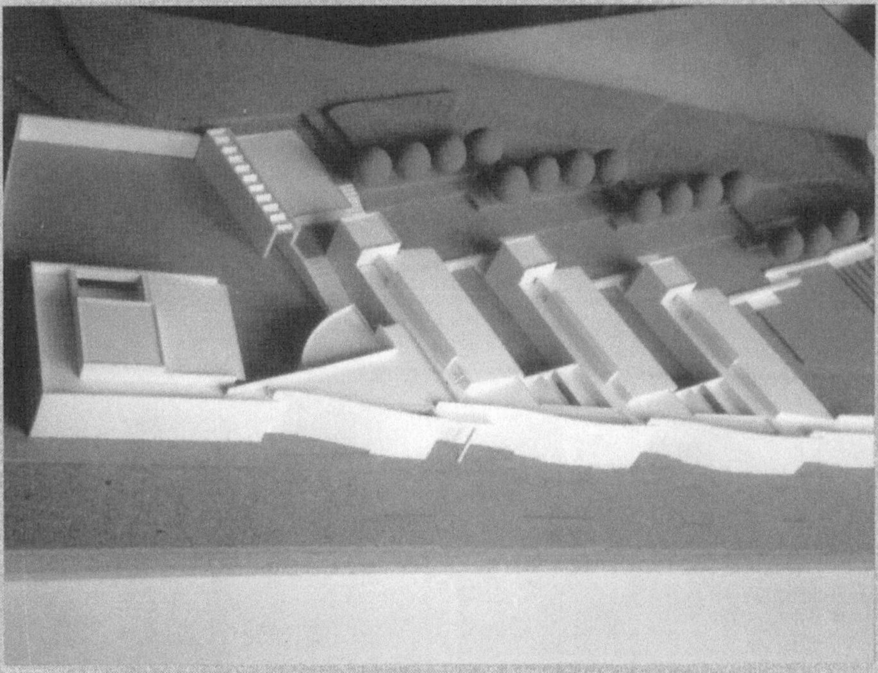

Building 2000 is a series of design studies illustrating passive solar architecture in buildings in the European Community.

ISSUE

S 2

DEC 1990

PROJECT DESCRIPTION

Building Type

This is a 640-pupil elementary school with 20 classrooms, a library, laboratory, administration offices, gymnasium and playground. It is an extension of an existing hall (now used as a nursery). It was designed in 1987-89 and is being built during 1990-91.

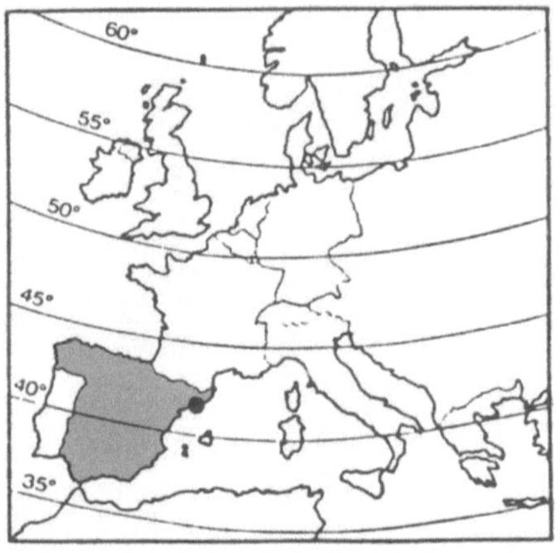

Figure 1. Location of Barcelona

Location

The site, which is 350 m above sea level, is in "Les Franqueses del Valles" (latitude 41° 38'N, longitude 2 51'E) - an industrial zone to the north of Barcelona. It is not overshadowed and has a slope of 12% from east to west.

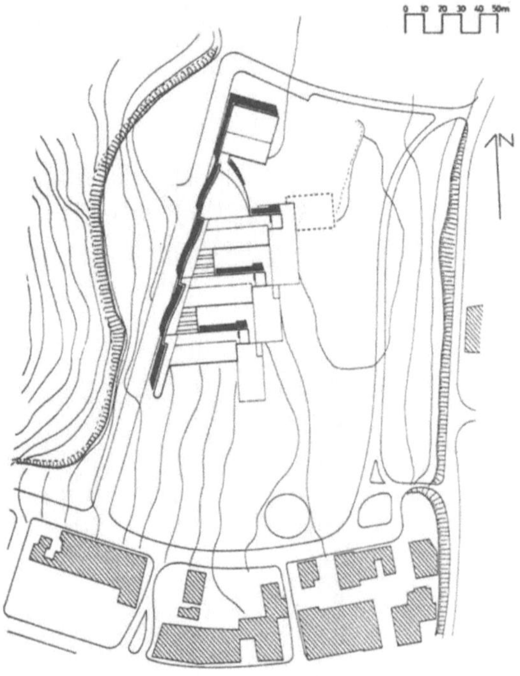

Figure 3. Site plan

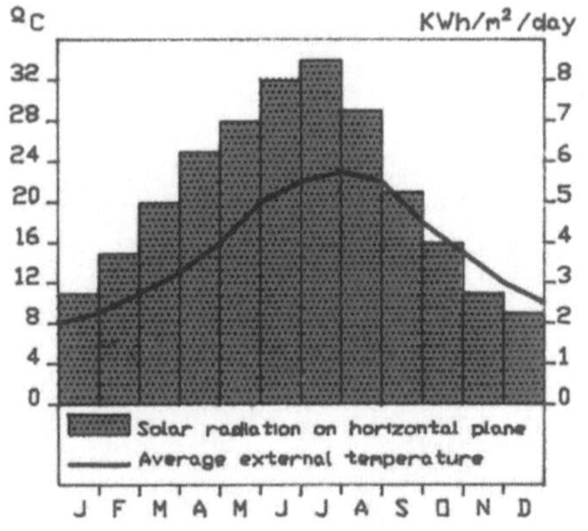

Figure 2. Mean daily solar irradiation and ambient temperature data

Site Microclimate

The climate is typically Mediterranean with moderate temperatures. There is fog in autumn and winter; the summers are very humid; and the winters are cool. The winds are from the NE. There are also breezes from the SE and NW. The sun shines 60% of the daytime throughout the year. Solar irradiation and temperature data are given in Figure 2.

Design and Construction Details

The school, which has two storeys, is arranged in blocks, all the classrooms being accessed from the main walkway. All teaching areas face south; only the corridors and service areas are on the north side.

The ground floor plan, section and views of the building from the south are shown in Figures 4-7. Views of a scale model are given in Figures 8 and 9.

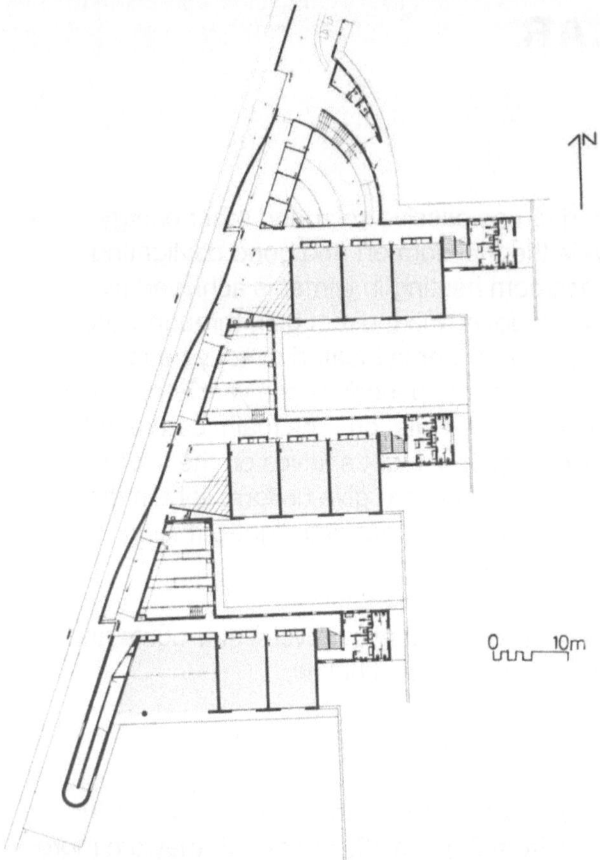

Figure 4. Ground floor plan

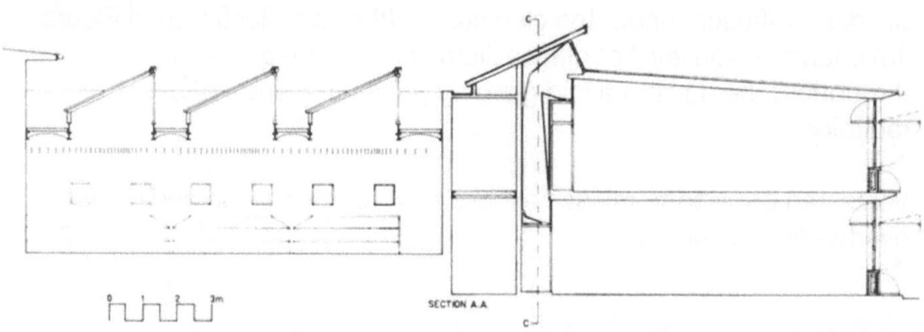

Figure 5. Section

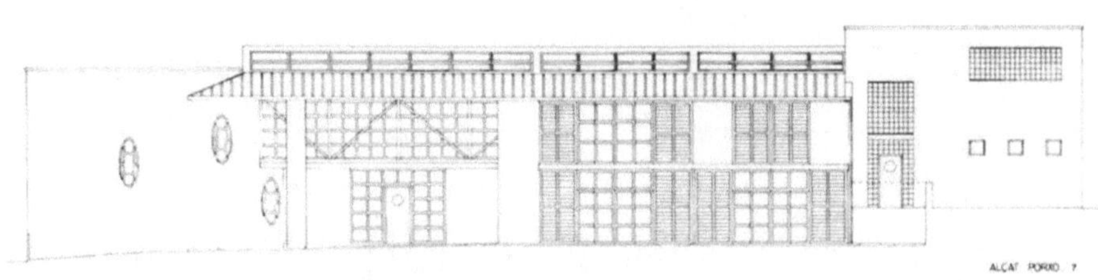

Figure 6. South facade

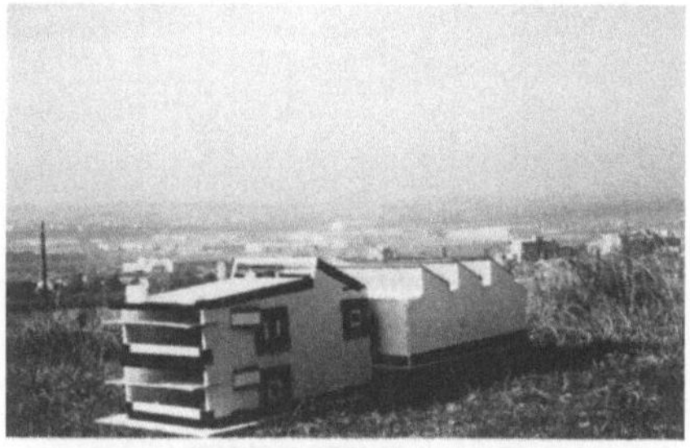

Figure 8. Scale model

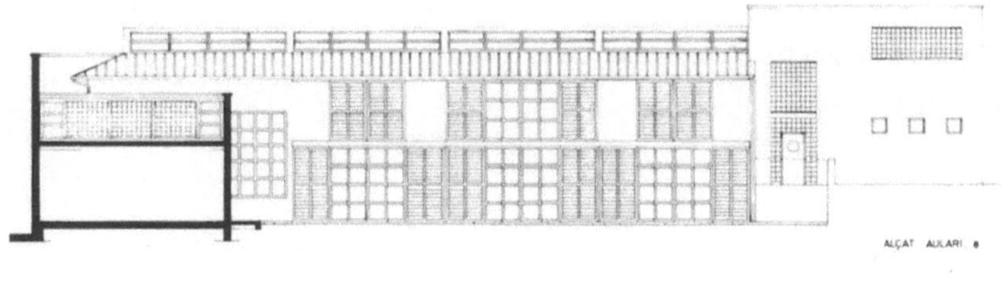

Figure 7. South facade

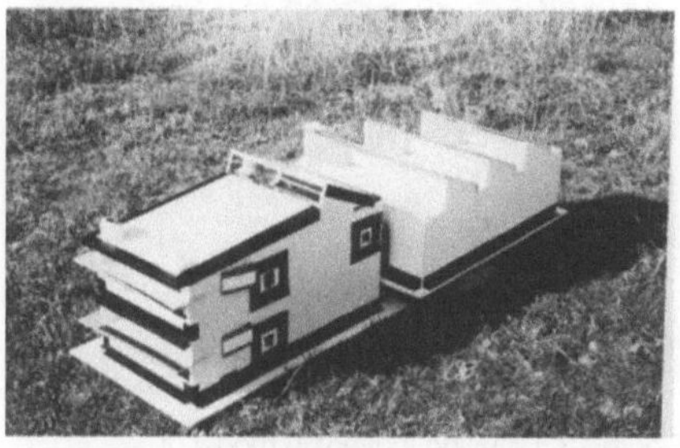

Figure 9. Scale model

DESCRIPTION OF PASSIVE SOLAR FEATURES/COMPONENTS

The school was designed to use passive solar and other energy-saving features to provide thermal comfort and good daylighting throughout the year. Classroom heating in winter is achieved by direct solar gain through windows and indirect gain through water-storage walls with a high thermal inertia located directly beneath the window sills. Comfort in the warmer periods of the year is produced by means of high levels of cross ventilation. The main feature of the building, however, is the daylighting. Light shelves on the south-facing facades and sun-ducts at the rear give uniform light in the classrooms while south-facing clerestories with light-diffusing louvers achieve a similar effect in the laboratory.

The south-facing clerestories and diffusing louvers, sun-ducts and light shelves are described in more detail below.

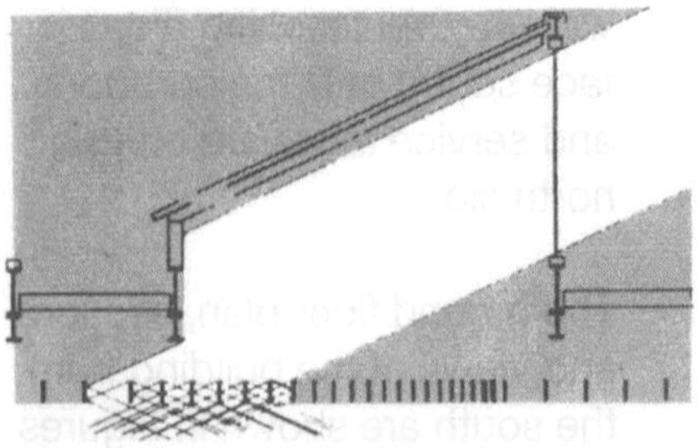

Figure 10. Entry of daylight through clerestories on 21 December (angle of sun 25°)

South-Facing Clerestories and Diffusing Louvers

The south-facing clerestories used in the laboratory are shown in Figure 14. As can be seen from Figures 10, 11 and 12, they are more effective in winter than in summer in bringing sunlight into the building. Diffusers under the clerestories (they are illustrated in Figure 13) enable a high level of diffuse illumination to be achieved throughout the day but avoid glare by preventing entry of direct radiation.

Vents can be used to release excess heat to the outside and prevent overheating in summer.

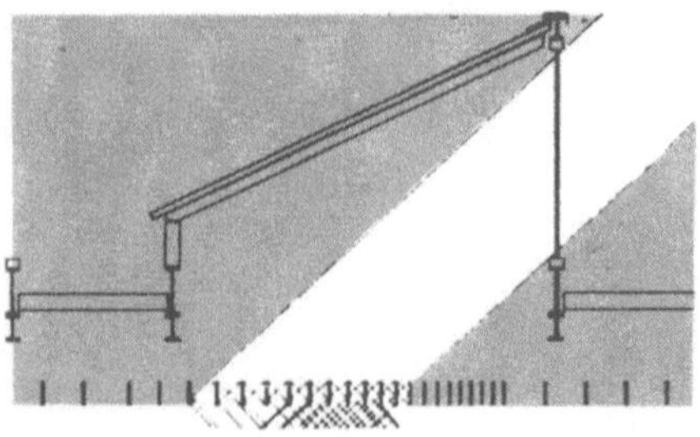

Figure 11. Entry of daylight through clerestories on 21 March (angle of sun 47°)

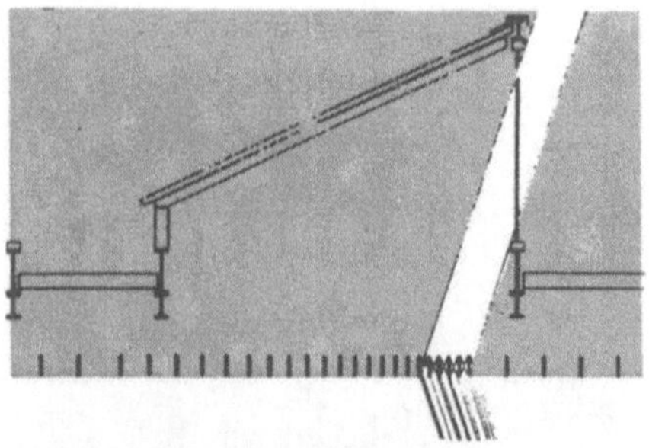

Figure 12. Entry of daylight through clerestories on 21 June (angle of sun 72°)

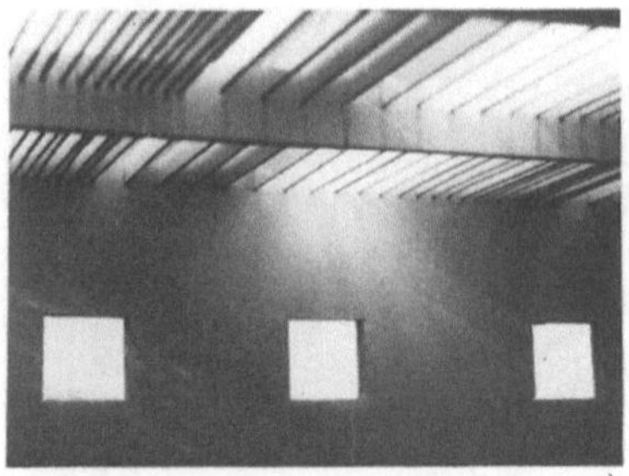

Figure 13. Diffusing louvers in model of laboratory

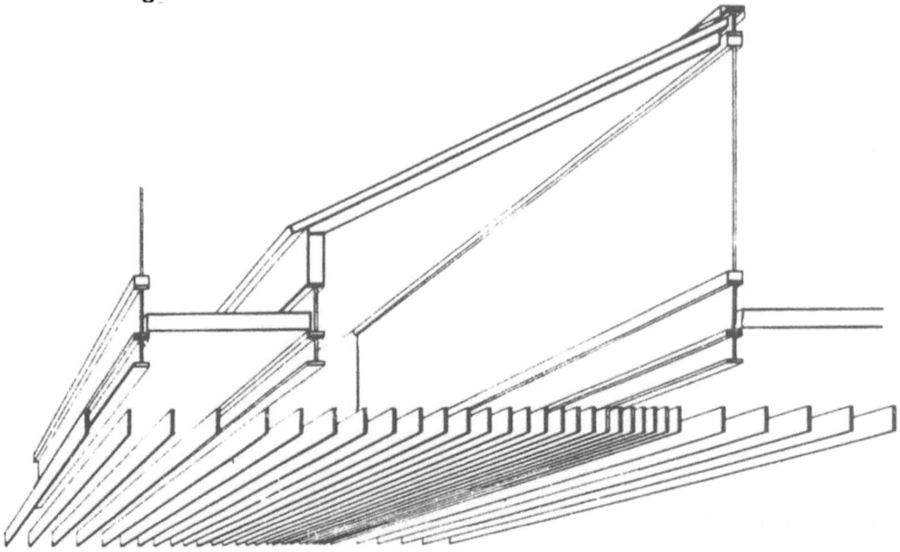

Figure 14. Perspective of south-facing clerestories

Sun-Ducts

The sun-ducts are illustrated in Figure 15. Their top entrances are oriented to the south in order to catch the sun's rays and direct the light to the north ends of the ground- and first-floor classrooms and the north-facing corridors (see Figures 18 and 19). The penetration in December, March and June is shown in Figures 16, 17 and 20.

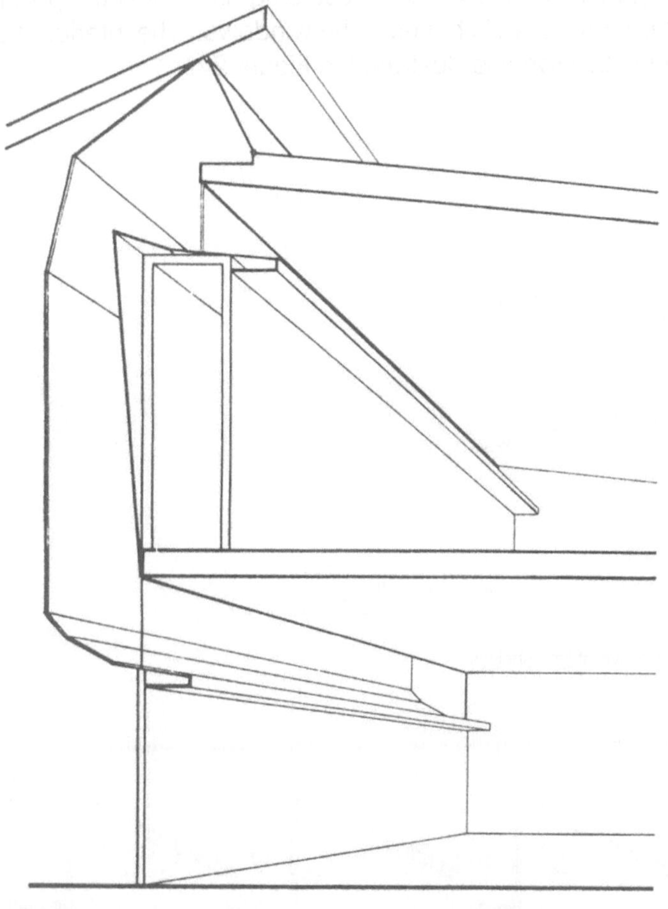

Figure 15. Perspective of sun-ducts

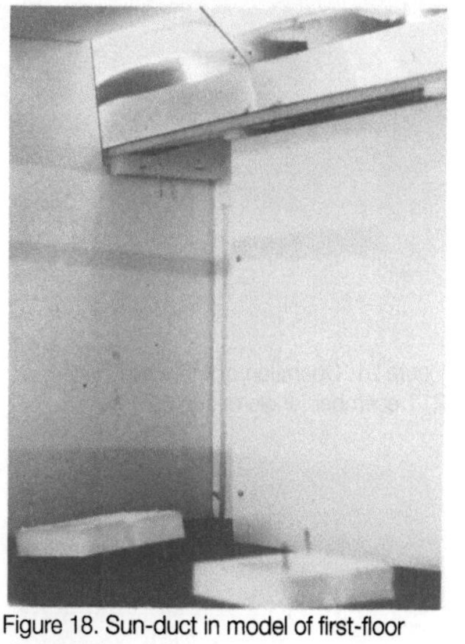

Figure 18. Sun-duct in model of first-floor classroom

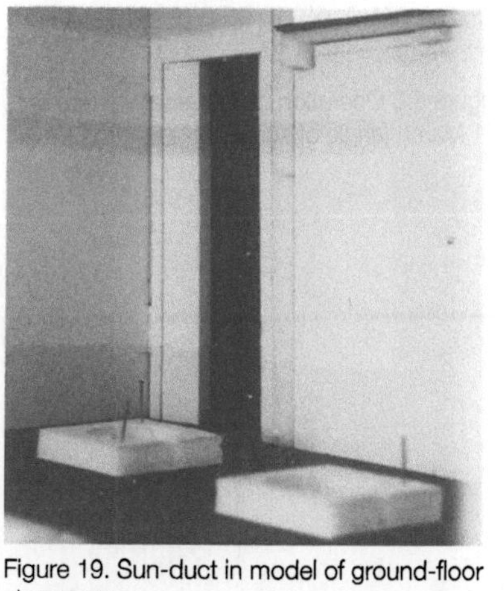

Figure 19. Sun-duct in model of ground-floor classroom

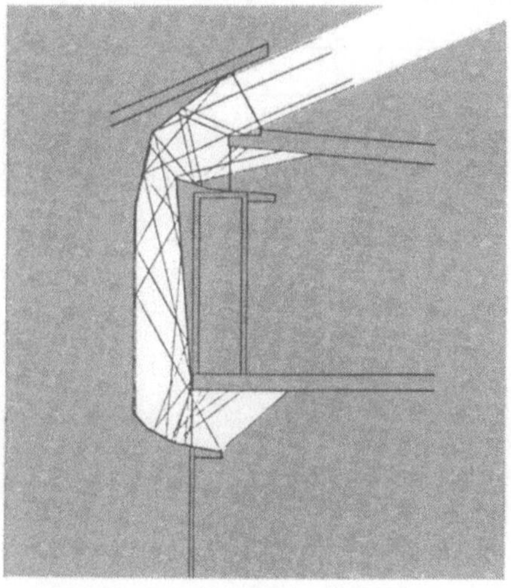

Figure 16. Entry of daylight through sun-ducts on 21 December (angle of sun 25°)

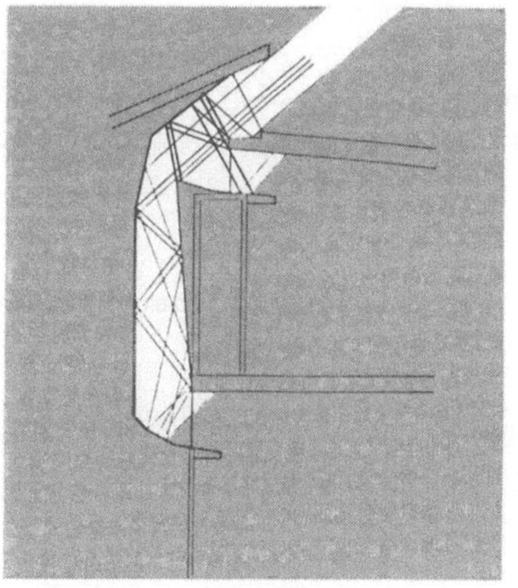

Figure 17. Entry of daylight through sun-ducts on 21 March (angle of sun 47°)

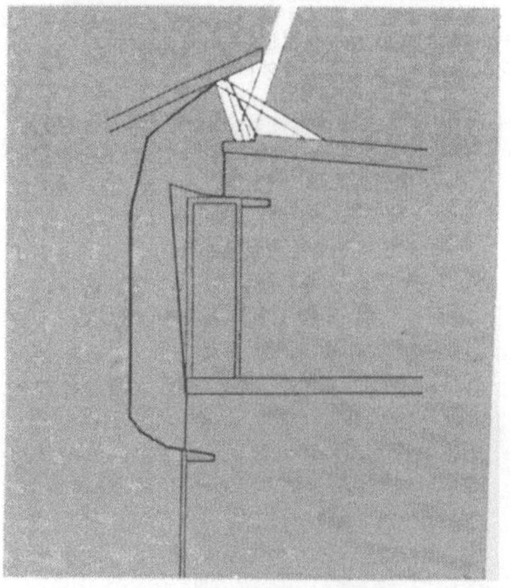

Figure 20. Entry of daylight through sun-ducts on 21 June (angle of sun 72°)

Light Shelves

The light-shelves on the south-facing facade are shown in Figure 25. They improve the quality of the daylighting by protecting the south windows of the classrooms from direct solar radiation and redirecting the light to the central areas of the building. The operation of the light shelf system in December, March and June is shown in Figures 21, 22 and 23. (These diagrams show the position of the water-storage walls beneath the windows.) The interior of a south-facing classroom is illustrated in Figure 24.

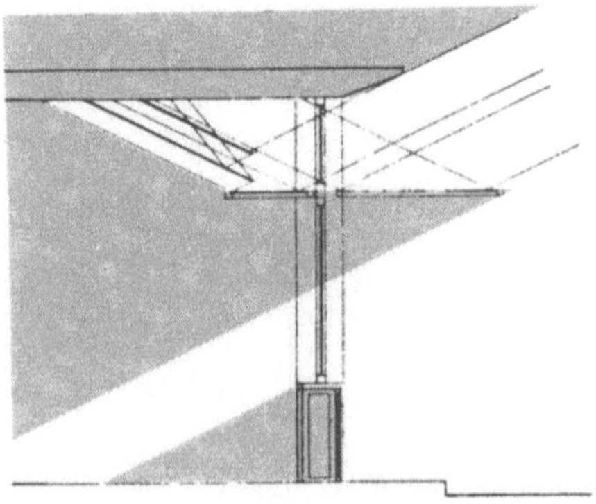

Figure 21. Operation of light-shelf system on 21 December (angle of sun 25°)

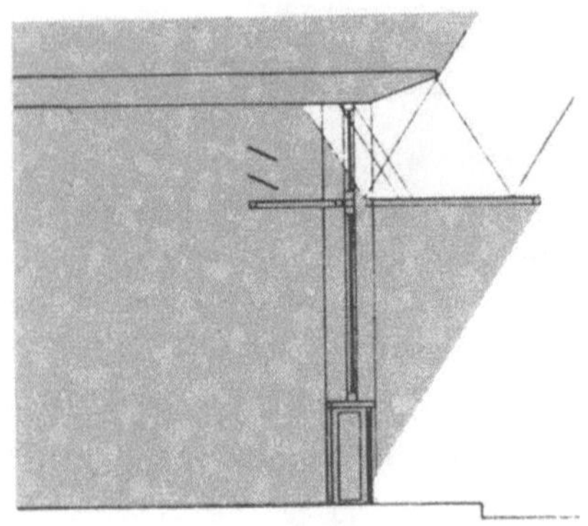

Figure 22. Operation of light-shelf system on 21 March (angle of sun 47°)

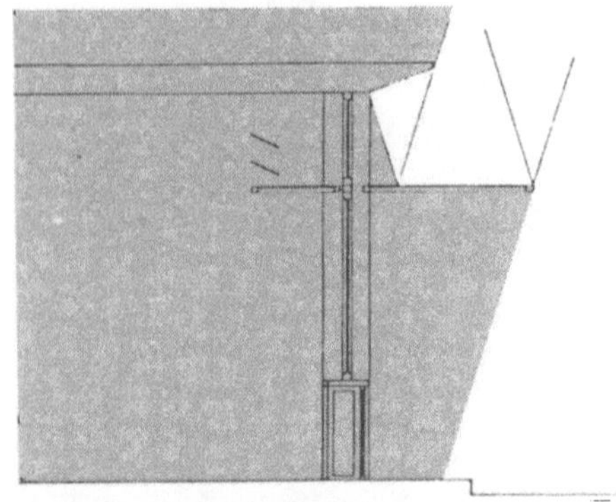

Figure 23. Operation of light-shelf system on 21 June (angle of sun 72°)

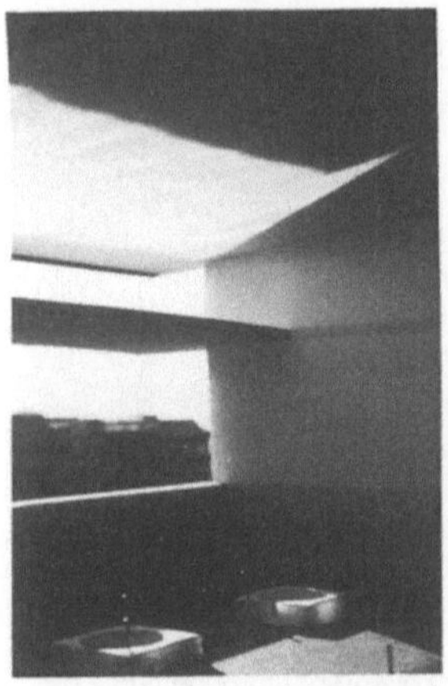

Figure 24. Interior of model of south-facing classroom

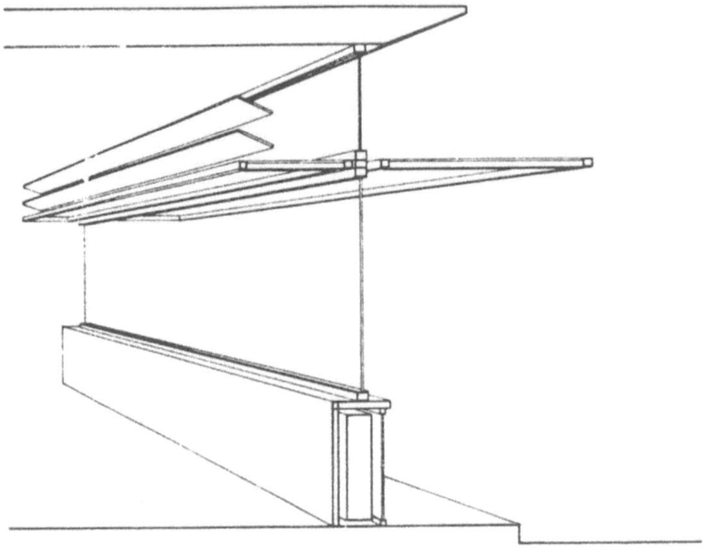

Figure 25. Perspective of light shelves

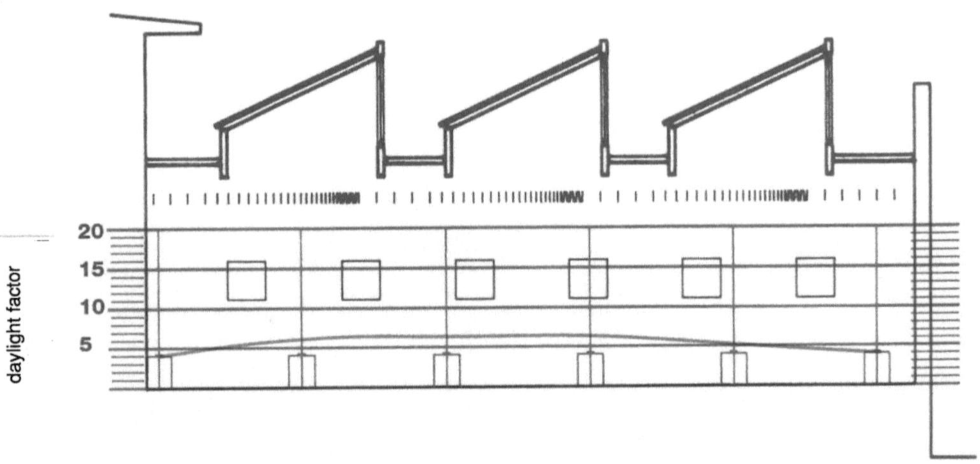

Figure 26. Lighting levels in laboratory with an overcast sky

ENERGY CALCULATIONS PERFORMED AND DESIGN TOOLS USED

Performance of Daylighting Systems

Lighting Levels in Laboratory from the South-Facing Clerestories
The measurements described opposite showed the illumination levels in the laboratory to be satisfactory throughout the year. With an overcast sky, the daylight factor varies between 4% and 7% (Figure 26). With a clear sky, the relationship between the external conditions and the lighting levels inside suggests that the correct solar radiation penetration had been selected (Figure 27-29). There was no glare on a horizontal plane at any time of year.

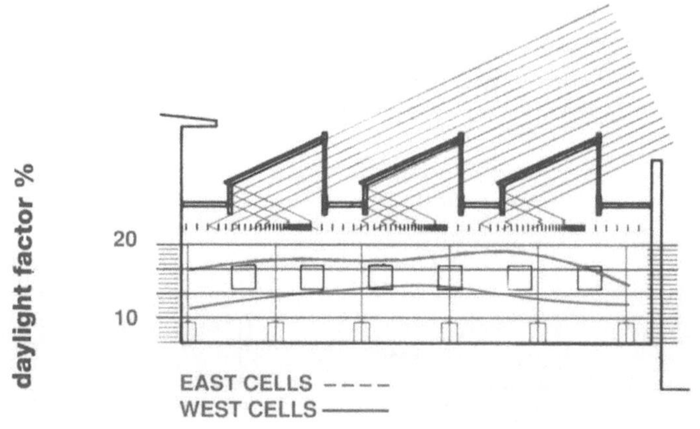

Figure 27. Lighting levels in laboratory with a clear sky on 21 December (sun angle 25°)

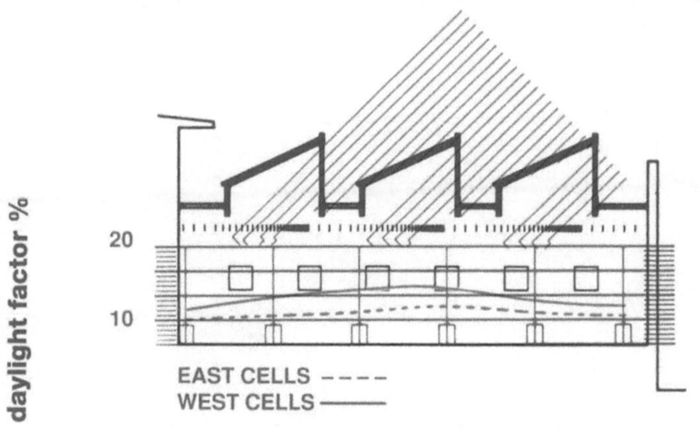

Figure 28. Lighting levels in laboratory with a clear sky on 21 March (sun angle 47°)

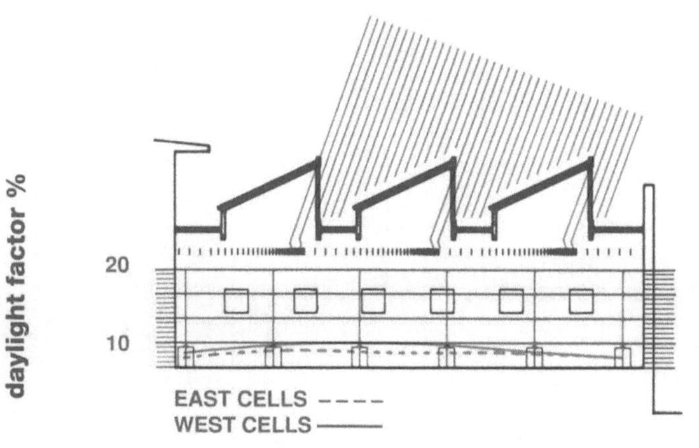

Figure 29. Lighting levels in laboratory with a clear sky on 21 June (sun angle 72°)

Design Tools

The daylighting systems were designed with the help of a number of design tools. The preliminary work was carried out using graphical methods. Then studies were carried out on 1:20 scale models of the classrooms and laboratory under natural light conditions, measurements of light levels in different parts of the building being made using photometric cells (see Figure 30). For an overcast sky, the measurements were made outdoors. For clear sky conditions, the measurements were carried out with the aid of a heliodon device of the LASH laboratories of Ecole Nationale des TPE, France. This enables the sun's position at a particular latitude to be simulated at a specific hour and day. In these measurements only the direct component of the radiation was taken into account. Ray-tracing tests were also carried out on 1:10 scale models of the sun-ducts in the Escola Tecnica Superior d'Arquitectura laboratory in Barcelona. Computer simulation studies of the daylighting systems were conducted using the GENELUX model.

The thermal performances of the classrooms, laboratory and library in winter were calculated using a simplified thermal network method.

Computer studies of the ventilation of the complex were carried out in the LASH laboratories.

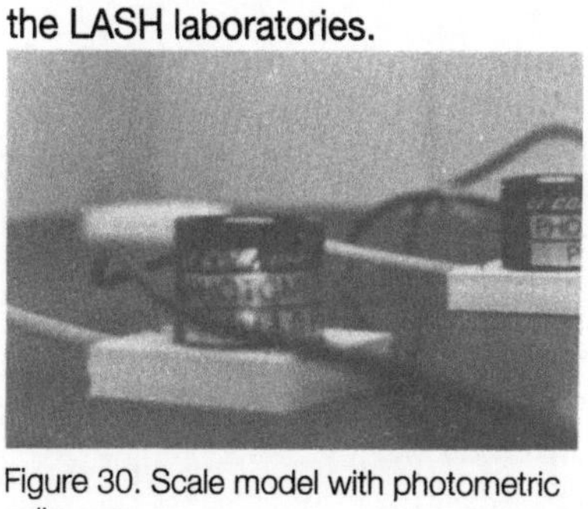

Figure 30. Scale model with photometric cells

Lighting Levels in Classrooms from Light Shelves

The south facade receives a considerable amount of light - more than enough for requirements. The problem, therefore, is to provide adequate control of the direct solar radiation while achieving good distribution of the light which does enter. The light shelves are positioned to reduce considerably the amount of direct solar radiation which enters through the bottom part of the window - especially in summer time, when excess solar gains could cause discomfort. The penetration of solar radiation is confined to areas and times when it can cause least discomfort, i.e. outside each pupil's immediate work area and at the beginning and end of the working day. To reduce glare, the louver at the top of each window is fitted with a reflector at the top and a diffuser at the bottom.

Lighting Levels in Classrooms from Sun-Ducts

The tests showed that the sun-ducts bring a certain amount of light to the north ends of the classrooms (see Figures 31, 32 and 33). This light is not capable of illuminating the rooms totally. It should be regarded as being complementary to the light entering through the south windows. The contribution varies according to season. There is a 5% contribution on average in December. This decreases to 3% in March. The daylighting contribution in June is insignificant.

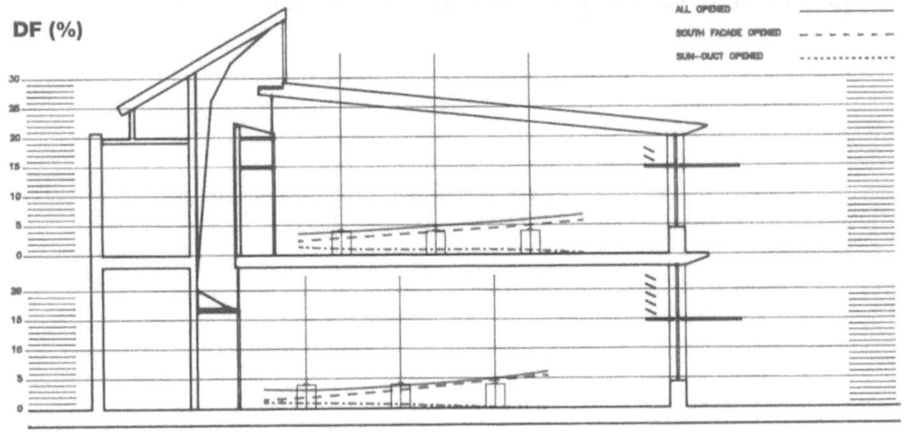

Figure 31. Lighting levels from sun-ducts with an overcast sky

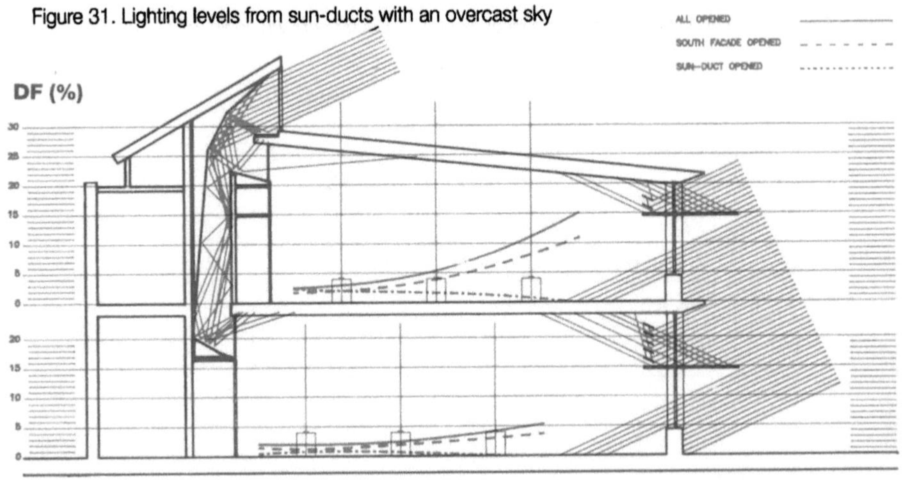

Figure 32. Lighting levels from sun-ducts with a clear sky on 21 December (sun angle 25°)

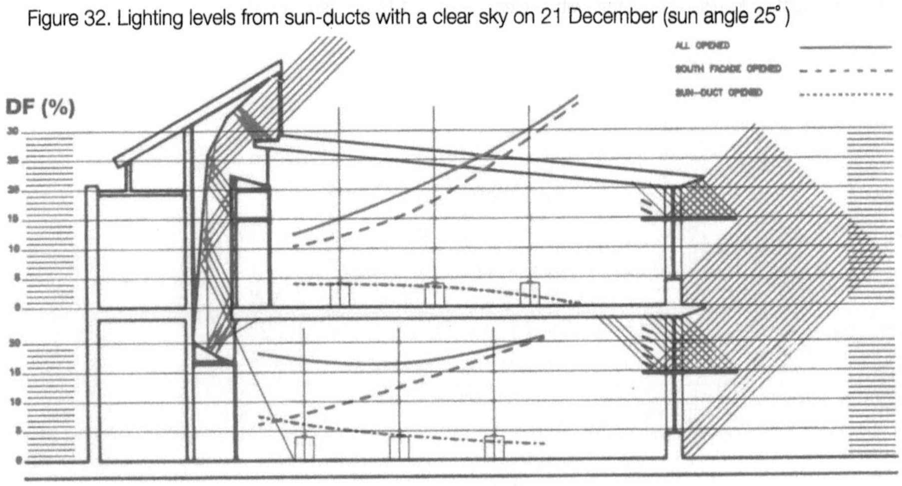

Figure 33. Lighting levels from sun-ducts with a clear sky on 21 March (sun angle 47°)

Thermal Performance

The school has a high level of occupancy but is used intermittently. It is not used in summer. The thermal performance analysis, therefore, was carried out for January, the coldest month of the year. The calculations were performed for two central classrooms in a block, one on top of the other, and for two end classrooms (again, one on top of the other) at the east end of a block. The calculations were carried out using the following equation:

$$Ti = Te + \frac{I + D}{G}$$

where Ti is the mean indoor temperature in °C, Te is the mean external temperature in °C, I is the mean solar gains (W/m³), D is the mean internal gains from occupants, equipment, etc., (W/m³) and G is the heat exchange coefficient (W/m³ °C).

The results of the calculations when the mean external temperature is 6° C, internal gains are 3 W/m³ and the heat exchange coefficient due to ventilation is 0.53 W/m³ °C are shown in Figures 34 and 35. It can be seen that the mean internal temperature (Ti) is around 15° C in both cases - some 3-4° C above what is regarded as a comfortable temperature. This would give a 75% reduction in heating bills.

The indoor conditions will, of course, vary with time and to find the extent of these changes the following equation was applied:

$$\Delta Ti = (Ti - T'e - \frac{I'}{G'})\,(1 - e^{-\frac{h^*G'}{Md}})$$

where Ti, T'e, I' and G' are the same parameters as before but for night-time conditions. I' takes into account the inertia effect of the storage system. Ti is the variation in the mean indoor temperature (°C), h is the duration of the night-time period and Md is the thermal mass by unit volume operating during period h (Wh/m³ °C).

The results of the calculations when the mean night-time temperature is 4° C are shown in Figures 36 and 37. It can be seen that the indoor temperature drop at night is very small - less than 3° C. Therefore the temperature of the classrooms will rise to an acceptable level early in the working day.

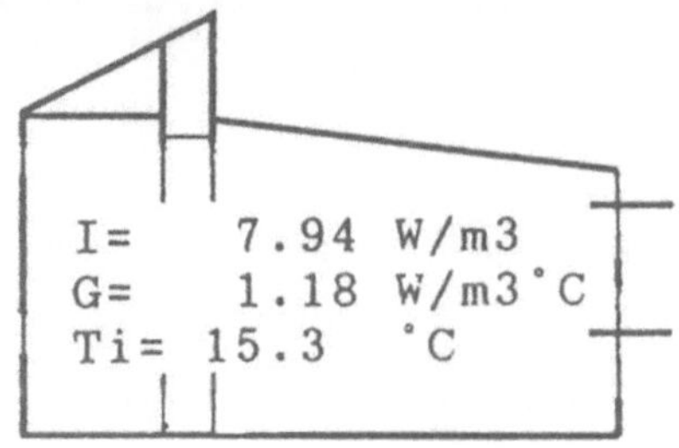

Figure 34. Thermal performance of central classrooms on an average January day

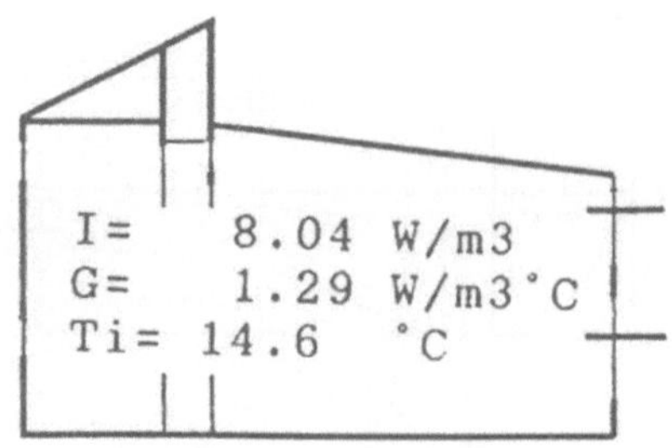

Figure 35. Thermal performance of end classrooms on an average January day

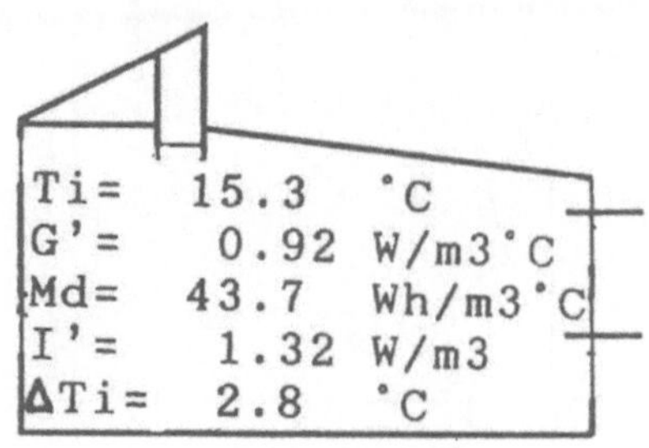

Figure 36. Thermal performance of central classrooms on a January night

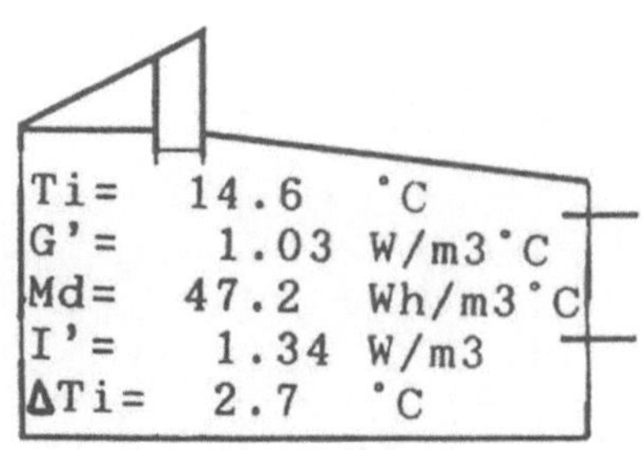

Figure 37. Thermal performance of end classrooms on a January night

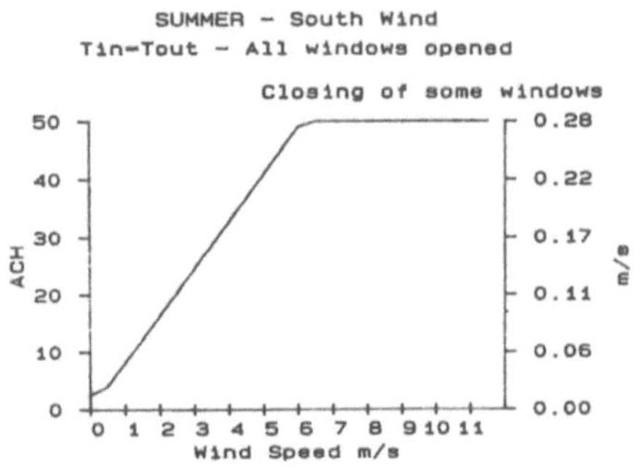

Figure 38. Cross ventilation for summer cooling

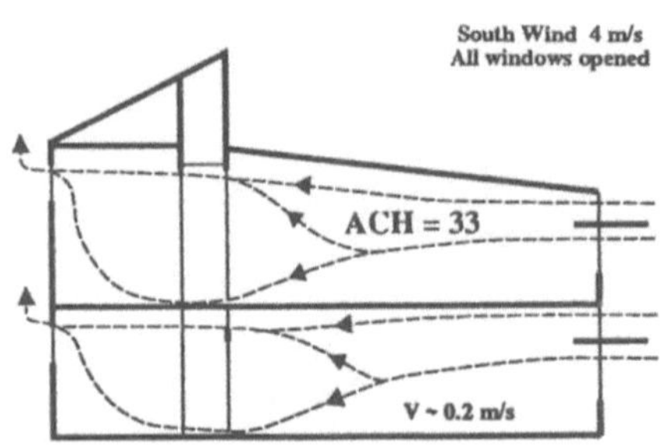

Figure 39. Daytime ventilation in summer

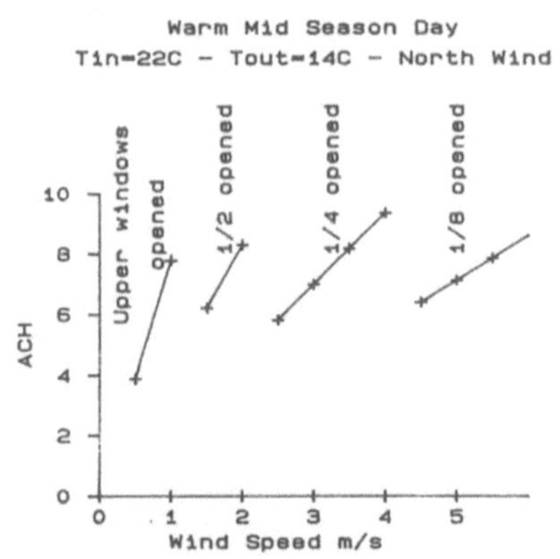

Figure 40. Mid-season ventilation when there is a north wind

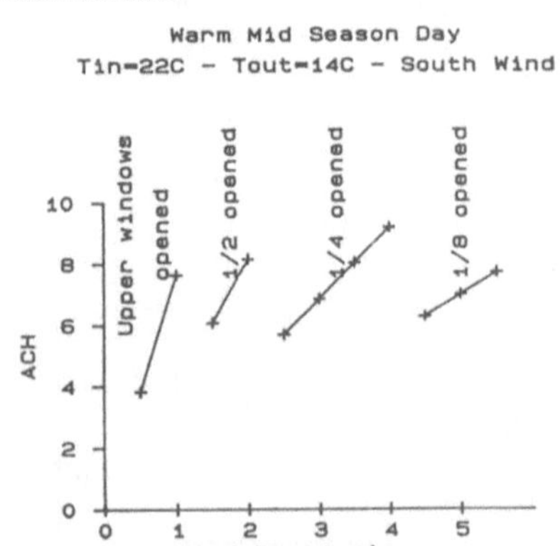

Figure 41. Mid-season ventilation when there is a south wind

Performance of Natural Ventilation Systems

The classroom ventilation systems are illustrated in Figure 42. The aim is to achieve:

- a minimum air change rate during cold winter days while maintaining a good air quality in the rooms;
- a minimum air change rate in mid-season to remove excess internal gains quickly;
- air velocities of 0.2-0.4 m/s in the classrooms to maintain thermal comfort and keep pupils cool on hot days;
- flushing of the building at night during the hot season.

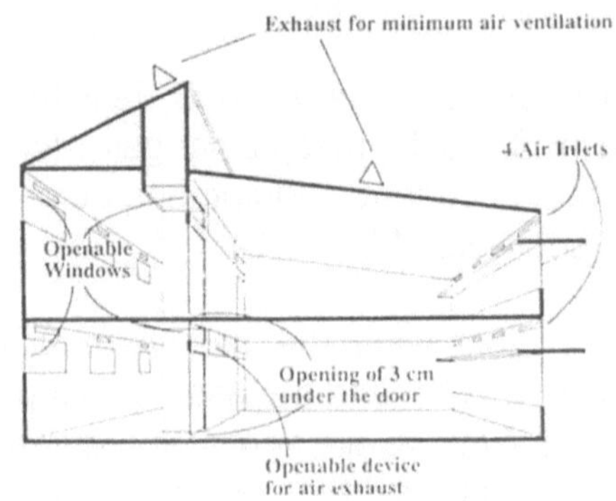

Figure 42. The classroom ventilation systems

In summer, the strategy is to store as much cold at night without damaging the building and achieve good cross ventilation during the daytime occupancy hours (see Figures 38 and 39).

In mid-season, when internal gains exceed heat losses, the strategy is to adjust the upper window openings to give the amount of cooling required. Analysis has showed that this is effective regardless of whether the winds are from the south or north (see Figures 40 and 41). For a typical mid-season day when the external temperature is 15° C and the temperature inside is 22° C, an air change rate of 7 air changes per hour (ACH) is required to overcome the unwanted internal gains (see Figure 43).

In winter, it was found that a minimum rate of 2.2 air changes per hour (i.e. 13 m^3 per hour per pupil) is required to maintain air quality (see Figure 44). The air inlets and outlets have been designed to achieve this.

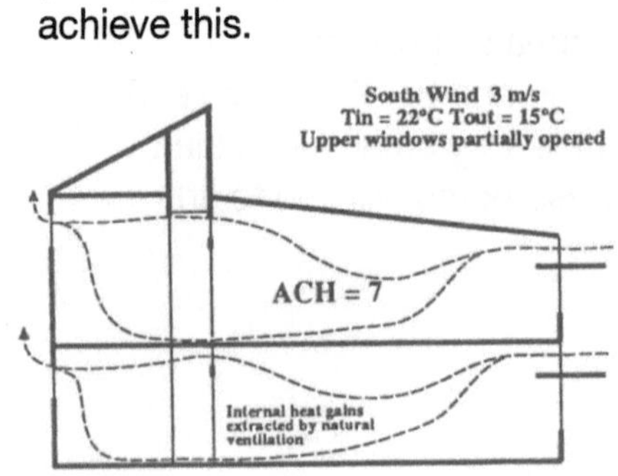

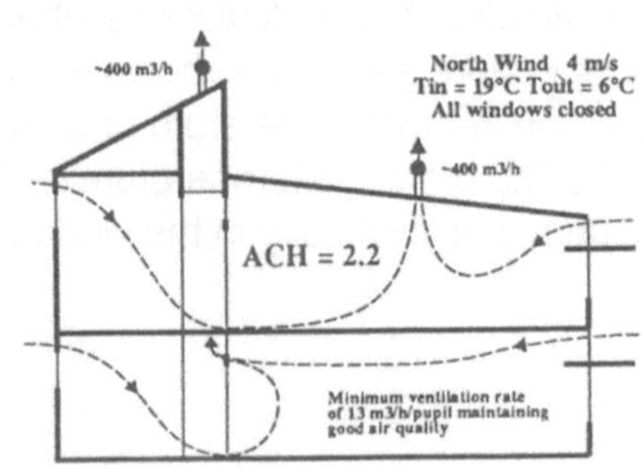

Figure 43. Daytime ventilation in mid-season Figure 44. Daytime ventilation in winter

GENERAL DESIGN GUIDELINES/POINTS OF INTEREST RESULTING FROM THIS PROJECT

The achievement of a good relationship between classroom length and width is a common problem for a school building architect. It is generally held that, to achieve good lighting, it is necessary to place all the rooms so that their length runs along an outer wall (Figure 45). This produces a building which is expensive to construct because it has a long facade and long corridors. The design developed in this project (Figure 46) overcomes this problem. Here, the classroom width (which is between 6 m and 8 m) runs along the facade, which keeps building costs low, but there is no drop in lighting levels. The sun-ducts provide light to zones which would otherwise be too dark. They also permit natural ventilation in each classroom.

The project proved that performance evaluation of daylighting systems using scale models is very easy. The scale itself does not create any problems. Once an accurate model has been prepared reliable results are assured. In general, a good model can be achieved with a scale of 1:20 if materials are used which have reflectances similar to those found in the end building. The measurements in this project were carried out using photometric cells connected to a daylight data acquisition system. A similar system with a minimum of two cells could be used by any architect.

Evaluation of the building's thermal performance and ventilation systems was not found to be so easy: more sophisticated techniques with computer models were required.

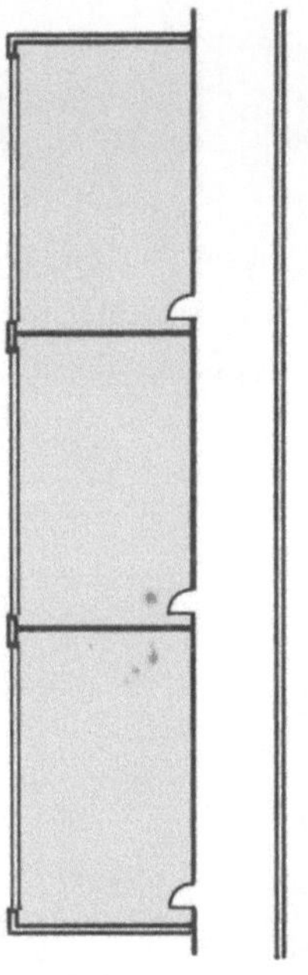

Figure 45. A conventional school layout

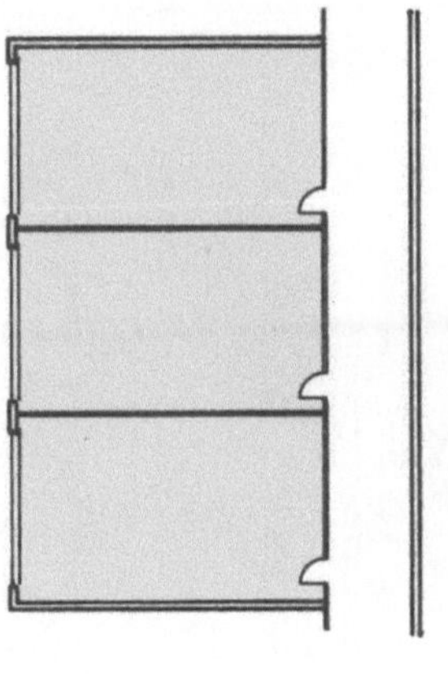

Figure 46. Layout in this project

BUILDING 2000

Building 2000 brochures are published by Directorate General XII of the Commission of the European Communities to show how design studies can help architects and other building designers use passive solar principles to the best effect to produce attractive energy-efficient buildings. Each brochure describes studies carried out with the support of the Commission during the design phase of one of thirty-six non-domestic buildings in the EC Member States. The studies were on such topics as daylighting, heating, cooling, ventilation, comfort, control systems and urban design. They were carried out with the help of acknowledged European experts in these fields and drew heavily on lessons learned and techniques developed through the Commision's research and development programme on solar energy applications to buildings.

Commission of the European Communities/Directorate-General for Science, Research and Development

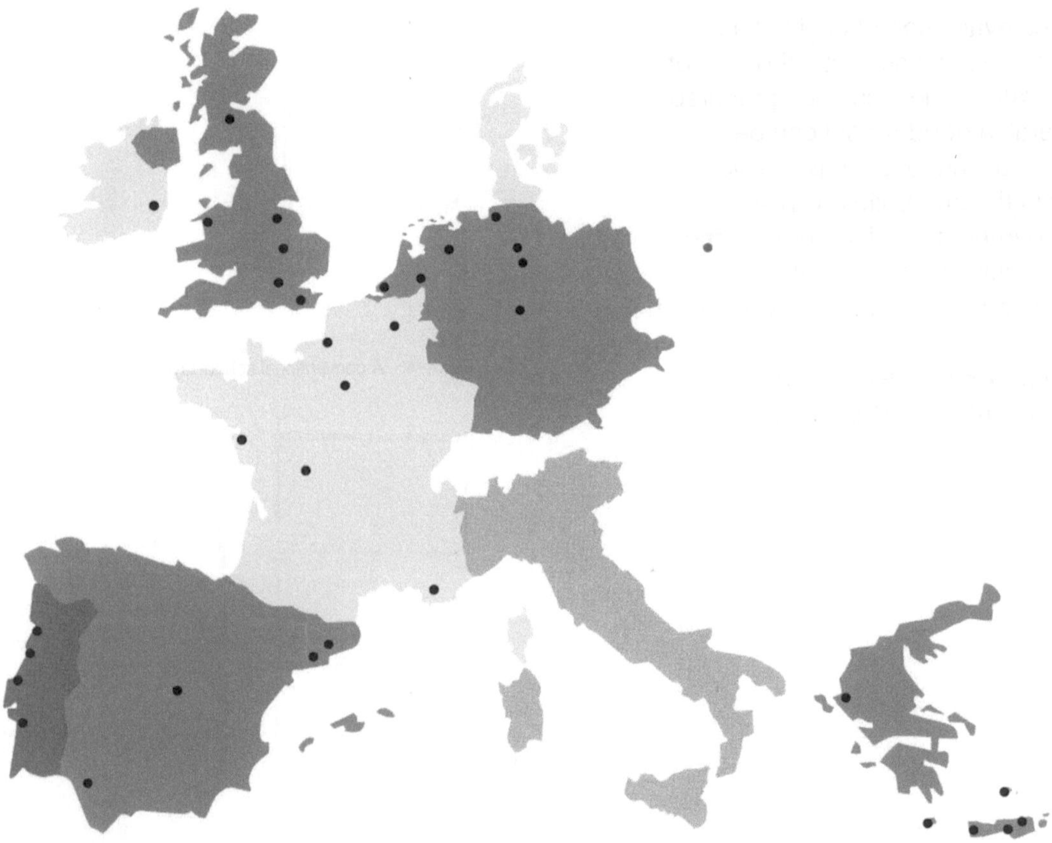

List of Design Team Participants and Advisers

Client

Ayuntamiento "Les Franqueses del Valle

Architects

J Sanmarti
Carrer Bailen 25 bis
09010 Barcelona
Spain

Prof. Rafael Serra Florensa
Escola Tècnica Superior d'Arquitectura d
Barcelona
Universitat Politecnica de Catalunya
Av. Diagonal 649
08028 Barcelona
Spain

Daylighting and Ventilation Consultants

Bernard Fleury and Marc Fontoynont
LASH,
Ecole Nationale des TPE,
rue M. Audin
F-69120 Vaulx en Velin
France

This set of **Building 2000** brochures illustrates how architects and other building designers can successfully apply passive solar principles to produce energy-efficient buildings.

BUILDING 2000 Participants

Project Director
Theo C. Steemers

Coordinator
Cees den Ouden

Technical Steering Committee
Dean Hawkes
Nick Baker
Alex Lohr
Jean P. Lepoivre

Regional Liaison Agents
(D) Jörn Behnsen
(E) Vicente Sifre
(F) Michel Raoust
(GB) Alan Hildon
(GR) Matheus Santamouris
(P) Eduardo Maldonado

Further information or copies of the brochures can be obtained from prof. ir. Cees den Ouden, EGM Engineering BV, P.O. Box 1042, 3300 BA Dordrecht, The Netherlands.

BUILDING 2000

Commission of the European Communities

- School for 5-12 year olds.
- Roofspace collector systems for preheating input air to warm air auxiliary heating system.
- Central courtyard to provide good levels of daylight and natural ventilation.

GREEN PARK COMBINED SCHOOL

NEWPORT PAGNELL/UNITED KINGDOM

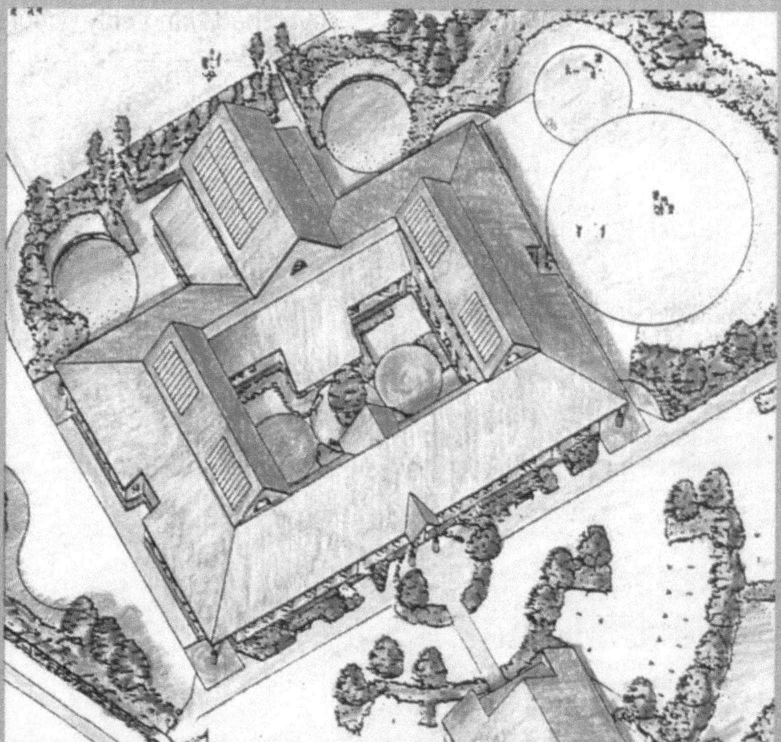

Building 2000 is a series of design studies illustrating passive solar architecture in buildings in the European Community.

ISSUE

S 3

DEC 1990

PROJECT DESCRIPTION

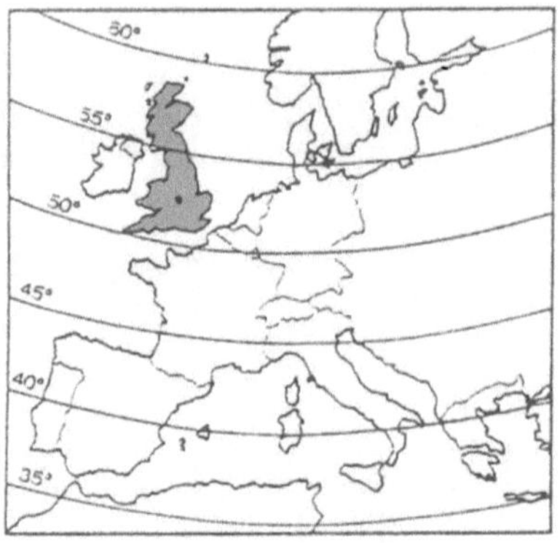

Figure 1. Location of Newport Pagnell

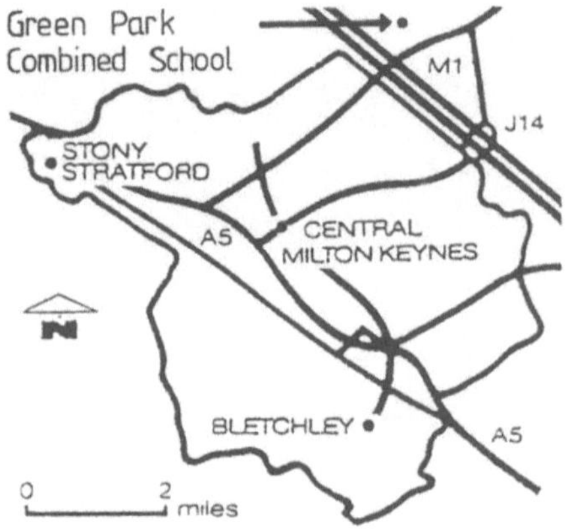

Figure 2. Location of school

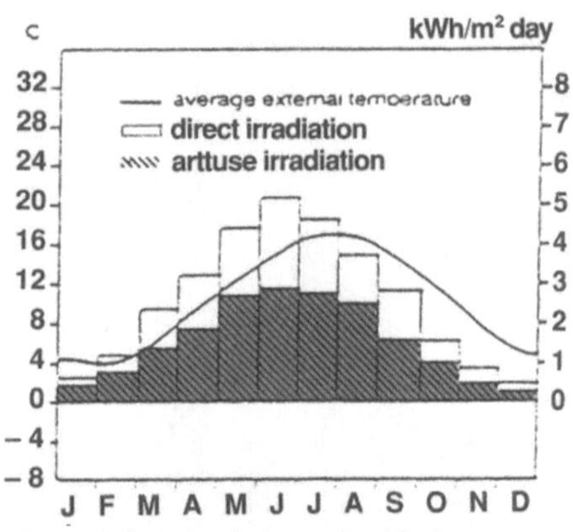

Figure 3. Solar irradiation and ambient temperature data

Degree days (base 15.5 C)	2308
Average temperature	
Jan	3.7 °C
Jul	17.5 °C
Sunshine hours	1460
Global irradiation on a horizontal plane	
Jan-Dec in kWh/m^2	945
(Jan-Dec in MJ/m^2	3402)

Table 1. Some key climate data

Building Type

This is a school for 360 pupils aged 5-12 years containing classrooms, a hall, activity room and library and the usual administrative offices, cloakrooms, changing rooms and stores, etc.

Location

The site, which is 100 m above sea level, is located in Newport Pagnell (latitude 52 ° 05', longitude 0 ° 44'), Buckinghamshire, England. Newport Pagnell is some 100 km north-west of London, immediately to the north of Milton Keynes (see Figures 1 and 2).

A site plan is shown in Figure 4. The site slopes gently from east to west. There are modern two-storey houses on three sides but overshadowing only occurs in winter, when the sun is very low in the sky.

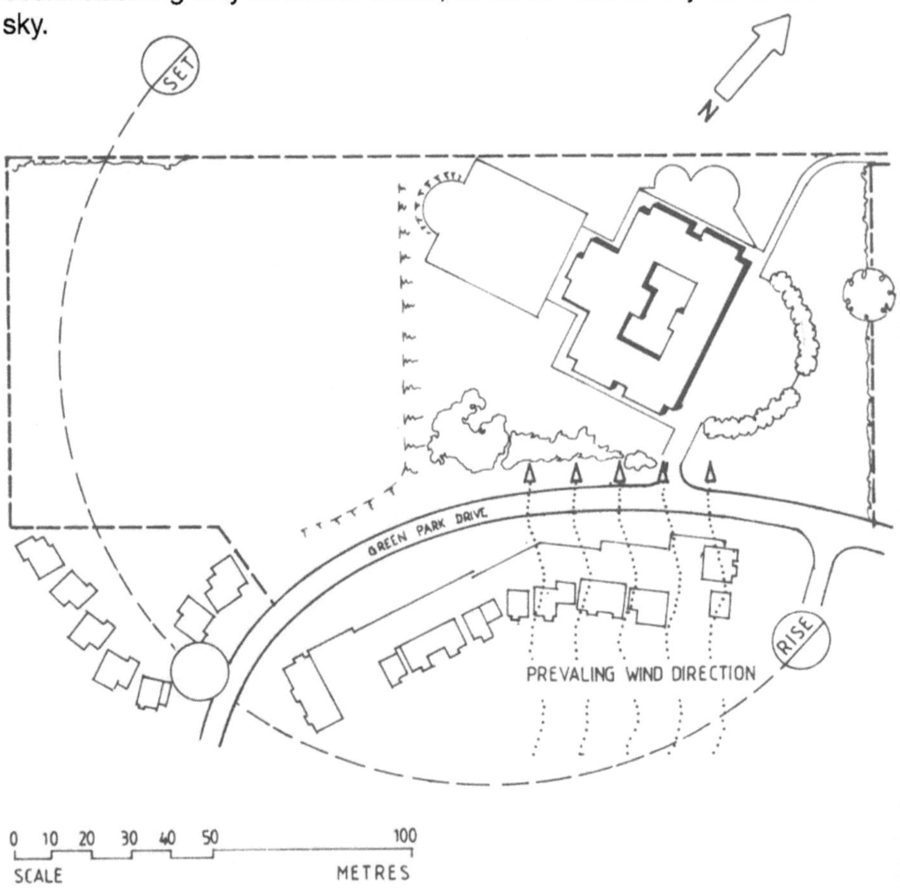

Figure 4. Site plan

Site Microclimate

Climate conditions are typical of the mild, temperate weather found in much of northern Europe. The minimum average daily temperature is 1.6 °C and the maximum (which occurs in July) is 21.8 °C. The prevailing wind is from the south-east. Some additional climate data are found in Figure 3 and Table 1.

Design and Construction Details

The school, which was commissioned by the Buckinghamshire County Council Education Committee, has been designed to a brief supplied by the Chief Education Officer. This stipulates that central government requirements outlined in the Department of Education and Science Design Note 17 entitled 'Guidelines for environmental design and fuel conservation in buildings' be met and normal comfort conditions maintained.

The school is a single-storey structure arranged around an internal courtyard. The total floor area is 1400 m^2 and the total volume 4643 m^3. The floor area of the heated part of the building is 1380 m and the volume 4593 m^3.

The communal facilities (shared teaching areas, school hall, library, activity room, etc.) are immediately adjacent to the courtyard and the individual teaching spaces grouped into four areas on the perimeter (see Figure 5). There is easy access from the teaching areas to the courtyard, which is used for play and learning. The building is positioned so that external services do not cross the site.

The school is built on a reinforced concrete strip foundation with brick and insulating 150 mm blockwork external cavity walls. These walls are further insulated with 25 mm extruded polystyrene. All internal walls are constructed of dense concrete block or loadbearing brickwork. The ground floor slab is 125 mm reinforced concrete topped with a 75 mm screed.

The windows are double glazed and weatherstripped throughout. The main external doors have a draught lobby.

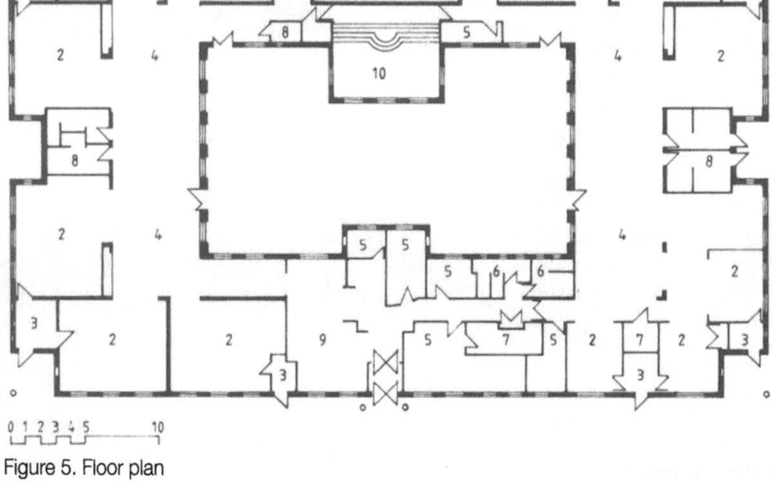

Figure 5. Floor plan

The lower roofs have 22.5 ° pitch trussed rafters with concrete interlocking tiles on felt and battens. The higher level roofs have a 35 ° pitch to give improved solar radiation collection. The south-facing portions are glazed with twin-walled polycarbonate, selected for its weight and cost advantages compared with double layers of glass.

U-values of some of the building elements are given in Table 2. Auxiliary heating is by means of a gas-fired warm air heating system. The design temperatures are as follows: external temperature -1 °C; frost protection 5 °C minimum; classrooms 18 °C; hall 14 °C.

external walls	0.42
roof	0.28
floor	0.38
windows	3.00

Table 2. U-values of some building elements (W/m^2 K)

DESCRIPTION OF PASSIVE SOLAR FEATURES/COMPONENTS

Introduction

The passive solar features incorporated in the building design aim to improve thermal comfort and achieve good daylighting throughout the year. Direct solar gains through the south-facing openings make a contribution to winter-time heating. In addition, the glazing on the south-facing roofs has the effect of converting the roofspaces (which are ventilated in the normal way) into solar collectors. The air warmed in them forms a source of pre-heated fresh air for the gas-fired warm air auxiliary heating system. Summer-time comfort is improved by the incorporation of high levels of internal mass in the building to help even out peak temperatures. The aim has been to produce a building where 27 °C is exceeded in the habitable spaces no more than 10 days a year.

Further information on the roofspace collector system is given opposite.

Roofspace Collector System

Figure 6 shows the position of the high level roofs containing the Rohm Makrolon Longlife 16 mm double skin (CLEA 283) polycarbonate glazing. This material has a daylight transmittance of 80% and is expected to degrade by 5% over twenty years.

The air in the roofspaces is mixed with air recirculated from the building and used as a source of incoming air for the auxiliary heating system. On cold sunny days, the air from the roofspaces will have been warmed by the incident solar radiation. On fully overcast winter days, it will be no colder than the outside air. (The possibility of radiative cooling to the cold night sky creating lower temperatures in the roofspaces in the very early morning is being monitored following the building's completion.)

In summer, the heat collected in the roofspaces is not, of course, needed. To a certain extent, overheating in these areas does not matter because these spaces are isolated from the rest of the building and are not habitable. However, the heat could cause decomposition of the building fabric. In addition, there could be unwanted conductive gains to the room below. To reduce this, the height of the spaces below the solar roofs has been increased and the ceilings have been well-insulated. The possibility of using external blinds or other techniques on the glazing to prevent entry of direct solar radiation into the roofspace in summer was rejected as being too complicated and costly. Instead, dampered louvered vents with heat-activated controls have been added to the gable ends of the roofs to vent unwanted hot air mechanically when a pre-determined temperature is reached.

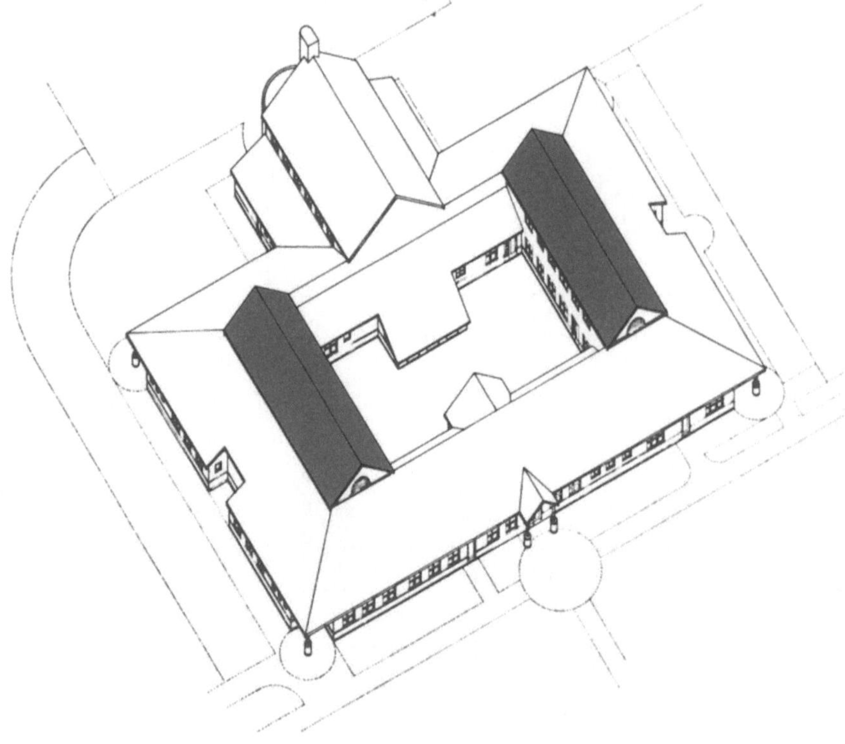

Figure 6. View from south-east

ENERGY CALCULATIONS PERFORMED AND DESIGN TOOLS USED

Introduction

The design was developed with the aid of computer studies carried out using the SERI-RES 1.2 dynamic thermal simulation program. These were carried out at the end of the sketch plan stage. Two models were taken, with and without the roofspace collector system (see Figures 7 and 8). At the beginning, windows were assumed to be single glazed and U-values were taken as being the minimum allowed by current building regulations. As the design developed, the U-values were improved and the windows double glazed and all the results of the analyses reported below are based on models incorporating these improved figures.

The simulations were carried out as follows. Annual runs were conducted to give an idea of the overall savings which might be achieved with the basic roofspace collector system described above and an improved version where the internal surface of the collectors is painted black and the ceilings between the classrooms and roof collectors are insulated. The annual performance of the daylighting system was also investigated.

Monthly simulations gave a breakdown of the solar gains, lighting energy use, incidental (casual) gains from appliances, lighting equipment, occupants, etc., and auxiliary heating requirements and the energy inputs and losses over the heating season.

Simulations over a few characteristic days enabled particular aspects of the building performance to be determined.

Finally, a number of sensitivity analyses were carried out to investigate the effect on energy (and in some cases daylighting) performance of changing building orientation, the area of glazing in the roof collectors, the proportion of fresh air which is preheated by the roof collectors, the area of glazing on the external facades of the classrooms, the insulation in the ceilings between the classrooms and roof collectors and the thermal mass in the floor slabs.

Ordinary roofspace

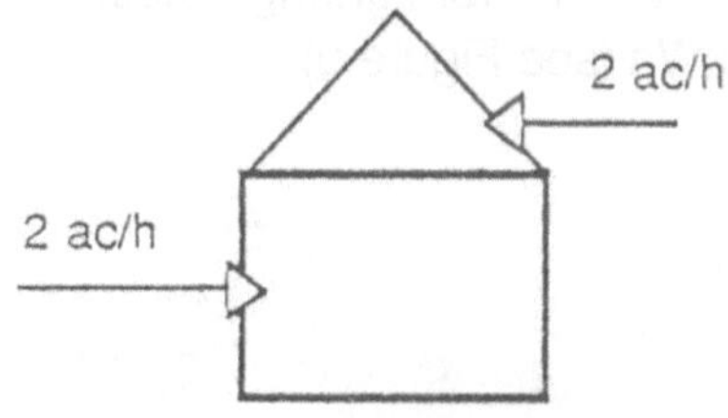

Figure 7. Model of school without roofspace collector system

Roofspace collector

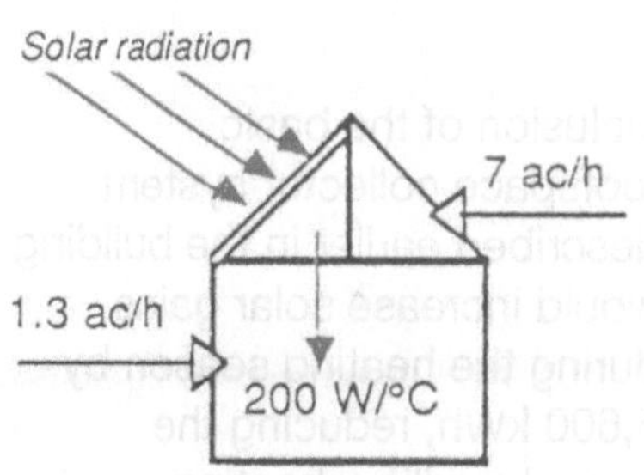

Figure 8. Model of school with roofspace collector system

Roofspace Collector System

It was calculated that, without the roofspace collector system, the annual auxiliary heating requirements of the building would be 97,000 kWh and the consumption of electricity for lighting 19,900 kWh (see Figure 9).

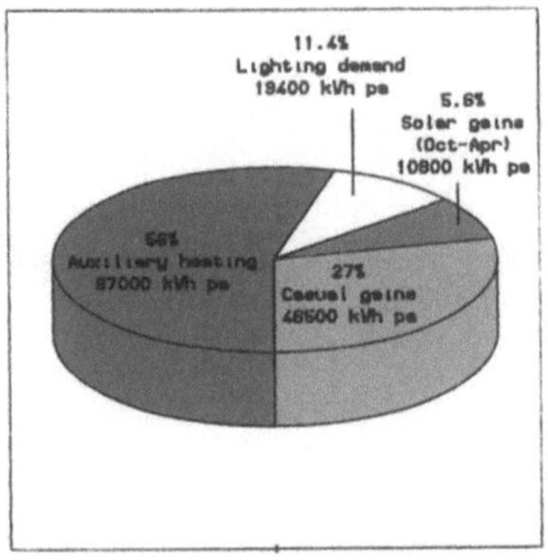

Figure 9. Energy consumption of original design without roofspace collector system

Inclusion of the basic roofspace collector system described earlier in the building would increase solar gains during the heating season by 5,600 kWh, reducing the annual auxiliary heating requirements to 91,400 kWh (see Figure 10 and Table 3). The consumption of primary energy is significantly less than the targets set by the Department of Education and Science in Design Note 17.

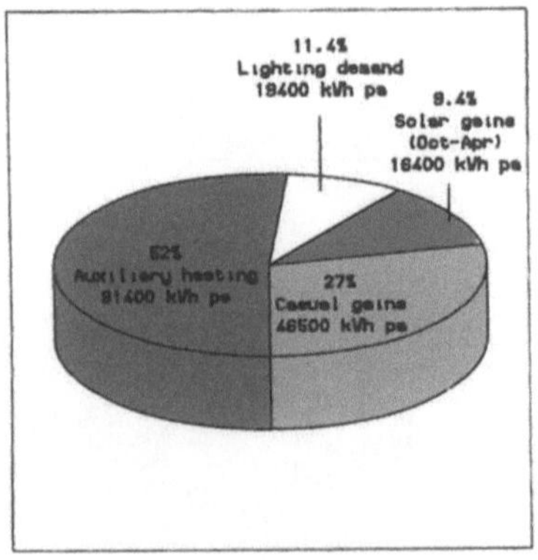

Figure 10. Energy consumption of original design with basic roofspace collector system

Results of Annual Simulations

	delivered kWh pa	delivered kWh/m^2 pa	primary kWh/m^2 pa	DN17 target kWh/m^2 pa
Auxiliary heating	91,400	64.1	107.7	182
Lighting demand	19,900	13.9	51.8	54
Solar gains (Oct-Apr)	16,400	11.5		
Casual gains	46,500	32.6		

Notes: The figures for primary energy assume that the ratio of primary energy to delivered energy is 1.68 for gas and 3.73 for electricity.
The DN17 target is the target for primary energy consumption set by the Department of Education and Science Design Note 17.

Table 3. Primary energy consumption of original design with basic roofspace collector system.

It was also found that this basic roofspace collector system could be improved at little extra cost by painting the internal surfaces of the roofspaces black and increasing the insulation between ceilings and collectors in order to increase the temperature within the roofspaces. Annual simulations showed that this improved system could give savings of auxiliary energy of 9,700 kWh a year.

Artificial Lighting

The initial simulations showed that, setting aside certain areas where the lighting was always on during the period the building was occupied, use of artificial lighting was 573 hours a year for the south classrooms and 692 hours a year for the north classrooms. These figures are high compared with a UK Building Research Establishment (BRE) survey of lighting use (presumably in term-time only) in teaching spaces designed to be daylit which put an average use of artificial lighting at 400 hours a year, of which 300 were between 9 am and 4 pm. The high result for Green Park Combined School was attributable to the fact that:

- Christmas, Easter and half-term holidays were not accounted for in the simulations. In 1988, these totalled 36 days. Cutting them out would (bearing in mind that lighting use would be high during these periods) reduce artificial lighting use by about 140 hours a year.

- The simulation predicts that about 80 hours of artificial lighting will be required from May to September. In practice, the figure could be much less because occupants are less likely to switch lighting on as a matter of course at this time of year.

- The switching model assumes that, if lighting is switched on during the morning or afternoon, it will remain on until lunch-time or evening even if the daylighting levels improve. If a more optimistic view is taken of occupants' energy-awareness and, say, lighting is switched off one hour after daylight levels improve, then the simulation model predicts a further savings of 10%.

If all these factors are taken into account, the predicted artificial lighting use in Green Park Combined School can be reduced to 320 hours for the south classrooms and 420 hours for the north classrooms. These figures are more consistent with the BRE survey results.

Results of Monthly Simulations

Breakdown of Energy Inputs

A month-by-month breakdown of the solar gains (including those from the basic roofspace collector system), lighting energy use, incidental (casual) gains from appliances and occupants, etc., and auxiliary heating requirements for the classrooms and hall during the weekday heating periods from September to June is given in Figure 11.

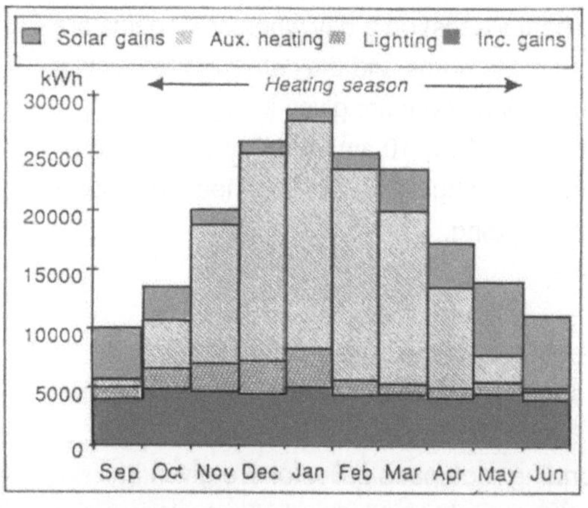

Figure 11. Monthly energy inputs for design with basic roofspace collector system

Energy Balance

The energy balance from the heated zones of the building during the heating season is given in Table 4. It shows that over a third of the total heat input during the day is stored in the fabric of the building and lost overnight. This suggested that the thermal mass in, say, the floor slab might be usefully reduced - an issue which is discussed in further detail in the sensitivity analyses below.

Gains	kWh pa ('000)	%
Auxiliary heating	86.8	56
Solar gains (direct)	16.4	11
Preheating	5.2	3
Lighting	16.1	11
Casual gains	29.2	19 +
Total	153.7	100

Losses	kWh pa ('000)	%
Windows	59.2	39
Walls	8.2	5
Roof	16.5	11
Ground	32.8	21
Infiltration	37.0	24 +
Total	153.7	100

Table 4. Energy gains and losses in heated zones, October to April

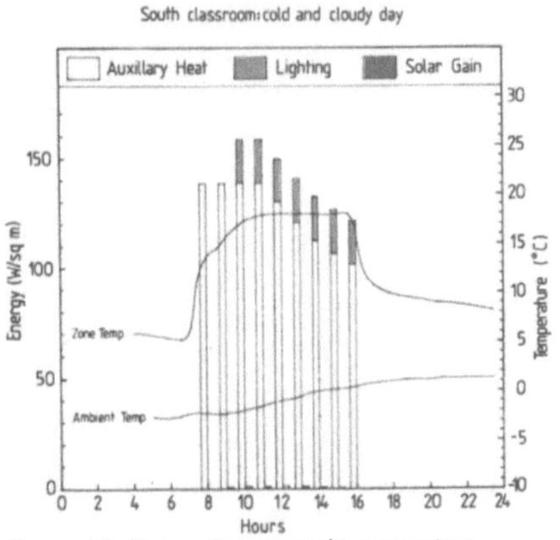

Figure 12. Energy inputs and temperature for south classroom on a cold and cloudy day

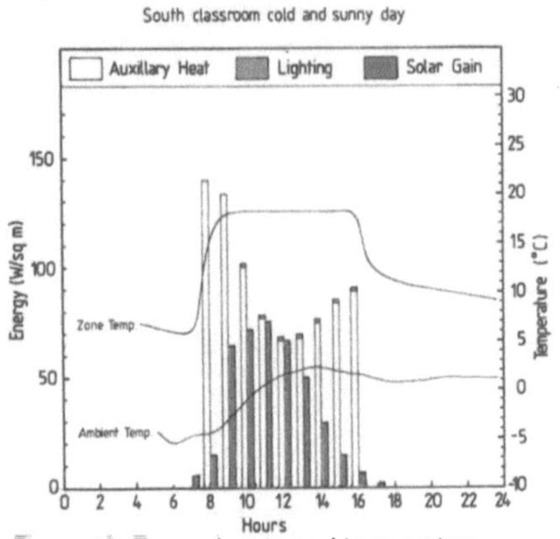

Figure 13. Energy inputs and temperature for south classroom on a cold and sunny day

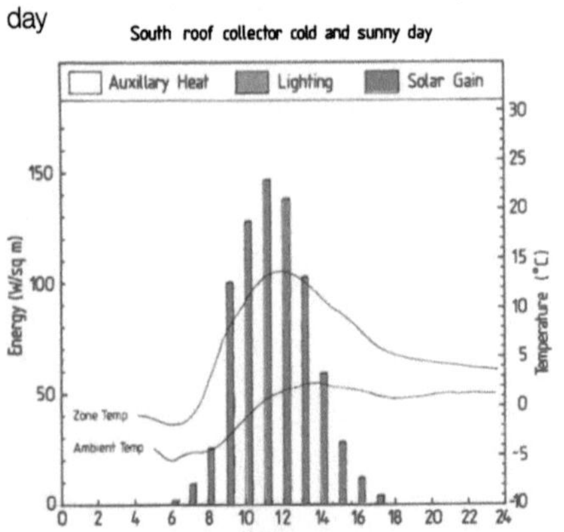

Figure 14. Energy gains and temperature in south roofspace on cold and sunny day

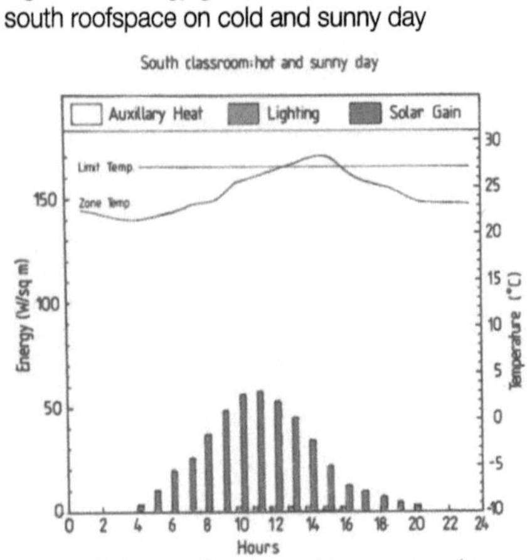

Figure 15. Energy inputs and temperature in south classroom on hot and sunny day

Results of Daily Simulations

In order to illustrate particular aspects of the performance of the building with the basic roofspace system, a number of simulations were carried out to establish the pattern of energy inputs and temperatures on cold and cloudy, cold and sunny and hot and sunny days. The results are as follows.

Cold and Cloudy Days

The aim here was to check the sizing of the auxiliary heating system (which had originally been sized for steady-state heat losses) on a day when solar gains are at their lowest. It was found that all the classrooms and the hall exhibited similar behaviour. The results for a south-facing classroom are given in Figure 12. It was found that the temperature reached 17 °C by 10 am. A higher temperature could be achieved either by increasing the sizing of the auxiliary heating system or by reducing the thermal mass in the building.

Cold and Sunny Days

The aim was to show the maximum potential of the passive solar features. It was found that all the heated zones showed similar behaviour. The results for a south-facing classroom are given in Figure 13. The energy gains and temperature reached in the corresponding roofspace collector are given in Figure 14. The increased solar gains compared to a cold and cloudy day reduced the auxiliary heating requirements by up to 50%. The temperature in the roofspace rose to 14 °C when the ambient temperature was around zero.

Hot and Sunny Days

To show the likelihood and extent of summer-time overheating a simulation was carried out for one of the hottest days in the school year when the ambient temperature was 25 °C and there were high solar gains. The results for a south-facing classroom are shown in Figure 15. It was found that the temperature only reached 28 °C because of the high thermal mass of the building. This suggests that there is scope for reducing the mass without causing overheating.

Results of Sensitivity Analyses

Introduction

To explore opportunities for further improvements in performance, a number of simulations were carried out using a model of the school with the improved roofspace collector system where the internal surfaces of the roofspaces have been painted black and the insulation between the classrooms and the roofspace collector has been increased. These simulations investigated the effect on energy (and, sometimes, daylighting) performance of changing building orientation, the area of glazing of the roof collector, the proportion of fresh air which is preheated by the roof collectors, the area of glazing on the external facades of the classrooms, the insulation in the ceilings between the classrooms and roof collectors and the thermal mass in the floor slab. The results are discussed below.

Building Orientation

The original orientation was 20 ° east of south (see Figure 16). Simulations were also run for orientations 30 ° on either side of this. The results are given in Table 5. They show that the original orientation is close to the optimum.

	Auxiliary heating (kWh pa)	Lighting demand (kWh pa)	Change Auxiliary heating (kWh pa)	Change Lighting demand (kWh pa)
Original orientation (20 ° east)	87,300	19,000		
10 ° west	87,450	20,000	up 150	up 1000
50 ° east	87,700	19,950	up 400	up 950

Table 5. Effect on auxiliary heating and lighting demand of changing building orientation

Area of Glazing in Roofspace Collectors

The results of simulations of the original design with and without the improved roofspace collectors have already been discussed. They show that the auxiliary heating requirements of the original design without collectors were 97,000 kWh a year and that this would be reduced by 9,700 kWh a year by inclusion of the improved roofspace collectors. Two further simulations were carried out where the glazing of the roofspace collector is 70% and 50% of the original value. The results are given in Table 6. They show that reducing the glazing area reduces the savings in auxiliary heating - but not very much.

	Auxiliary heating (kWh pa)	Saving auxiliary heating (kWh pa)
Original design without collectors	97,000	
Original design with collectors	87,300	9,700
70% glazing area	88,400	8,600
50% glazing area	89,300	7,700

Table 6. Effect on auxiliary heating of reducing glazing area of roofspace collectors

Proportion of Fresh Air Intake Being Preheated

In the original system, about a third of the fresh air was preheated in the roofspace collectors. Two additional simulations were run where this is increased to one half and to two-thirds. In both cases, the direct infiltration to the classrooms is decreased by a corresponding amount. The results are shown in Table 7. When two-thirds of the air is preheated, the savings in auxiliary heating were around 14,750 kWh a year. This is the maximum which could be achieved by the roofspace collector system although there will also be a certain amount of direct heating from the warm roofspaces. When half the fresh air is preheated, the savings in auxiliary heating are around 77% of this 'maximum' figure. With one-third, the savings are 66% of the 'maximum'. Thus, significant further savings were obtained by increasing the transport of the preheating system above the originally specified level.

	Auxiliary heating (kWh pa)	Savings auxiliary heating (kWh pa)	Savings as % of 'maximum'
Original design without collectors	97,000		
10 m^3 per child per hour	87,300	9,700	66%
15 m^3 per child per hour	85,700	11,300	77%
20 m^3 per child per hour	82,250	14,750	100%

Table 7. Effect on auxiliary heating of increasing proportion of auxiliary heating system input air which is preheated by roofspace collectors .pa

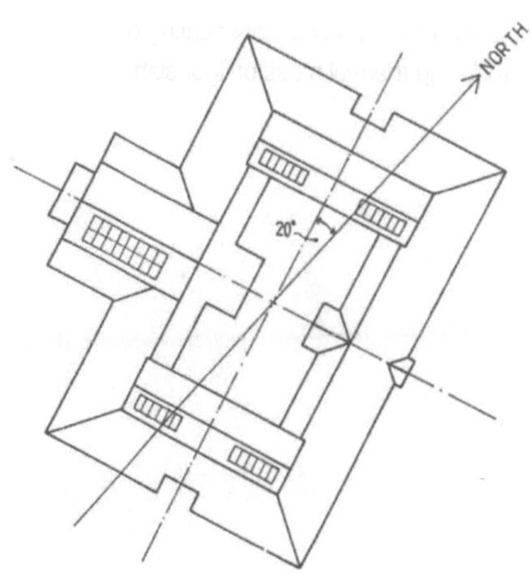

Figure 16. Building orientation

	Auxiliary heating (kWh pa)	Change (kWh pa)
Original	87,300	
Insulation removed	87,600	up 300

Table 8. Effect on auxiliary heating of removing insulation between classrooms and collectors

	Auxiliary heating (kWh pa)	Change (kWh pa)
Original floor slab	87,300	
Reduced weight floor slab	84,900	down 2,400

Table 9. Effect on auxiliary heating of reducing thermal mass of floor slab

Area of Glazing on the External Facades

Simulations were run where the glazing area on the external facades was increased or decreased by 15%, leaving the glazing area on the inner courtyard unchanged. The percent change in glazing area gave rise to the same percent change in minimum daylight factor, i.e. from 1% to 1.15% or 0.85% respectively. The marginal effects on auxiliary heating and lighting are shown in Table 10. Generally, the effects are small but suggest that lighting demand increases sharply as the daylight factor is reduced below 1%. However, increasing the daylight factor to 1.15% does not produce large savings, which suggests that the area of glazing specified originally is close to the optimum, except for the north zone where it could be reduced.

	Zone Auxiliary heating (kWh pa)	Change Lighting demand (kWh pa)	Auxiliary heating (kWh pa)	Lighting demand (kWh pa)
South zone: 15% more glazing	20,980	3,500	up 150	down 150
South zone: 15% less glazing	20,510	4,000	down 320	up 350
North zone: 15% less glazing	21,490	4,680	down 400	up 260
East zone: 15% less glazing	19,790	5,190	down 490	up 440
West zone: 15% less glazing	18,850	6,240	down 480	up 500

Table 10. Effect on auxiliary heating and lighting demand of altering glazing area on external facades

Insulation Between Classrooms and Collectors

A single simulation was run where the ceilings between the classrooms and roof collectors were left uninsulated. The aim was to see if the insulation was necessary. The results, which are shown in Table 8, showed a small increase in heating demand. The architect did not feel that removal of the insulation was a realistic option.

Thermal Mass of Floor Slab

The results of the heating season energy balance and of the daily simulations suggested that reducing the thermal mass of the building might lead to reductions in auxiliary heating demand without leading to overheating. A simulation was, therefore, run for a typical hot and sunny day where the layer of dense concrete in the floor slab was replaced by a layer of lightweight concrete. The results are shown in Table 9. For the lightweight slab, the peak temperature was about 0.2 °C higher - indicating that this measure would not lead to excessive overheating. Further reductions in the mass in the floor slab led to an increase in auxiliary heating demand.

Actions Following Sensitivity Analyses

If all the recommendations from the sensitivity analyses had been incorporated in the design, the energy savings would be 14,000 kWh a year - or about 15% of the heating demand for the original design without roofspace collectors. In practice, the following actions were taken: the orientation was left unchanged; the area of roof glazing was reduced to curb costs without unduly affecting the saving in auxiliary heating; the recirculation and fresh air intake rates were optimized; the area of glazing on the external facades was left unchanged because of the lead time for manufacture of new windows and the marginal energy savings which would be involved; the insulation between classrooms and collectors was left unchanged; the floor slab was left as dense concrete because it is punctuated by floor ducts for the auxiliary heating system and is finished with a variety of materials (including carpet) which may have the effect of reducing the mass of the slab.

GENERAL DESIGN GUIDELINES/POINTS OF INTEREST RESULTING FROM THIS PROJECT

Overall Results from Energy Performance Analyses

Table 11 summarizes the energy savings which would result from the recommendations for improvements to the overall design which resulted from all the energy performance analyses.

			Change from original case	
	Auxiliary heating (kWh pa)	Lighting demand (kWh pa)	Auxiliary heating (kWh pa)	Lighting demand (kWh pa)
1. Original case with no collectors	97,000	19,900		
2. Original case with collectors	91,400	19,900	down 5,600	
3. Collectors painted black and roof insulated	87,300	19,900	down 9,700	
4. as 3 + increased pre-heating (15 m per hour per child)	85,700	19,900	down 11,300	
5. as 3 + 15% less glazing on north facade	87,300	20,160	down 9,700	up 260
6. as 3 + reduced weight floor slab	84,900	19,900	down 12,100	

Table 11. Savings in auxiliary heating and lighting demand at various stages of design development

Advantages of Roofspace Collectors

Roofspace collectors have a low capital cost because their physical construction may not differ greatly from that of a conventional pitched roof. A further reduction in initial cost can result from use of components such as fans and controls which would already be present in an auxiliary air heating system. In contrast to a conservatory, the roofspace collector can (because it is not occupied) be designed for optimal heat collection and attainment of high temperatures.

In addition, it is often easier to ensure that roofspace collectors are unshaded at all times than it is to avoid overshadowing of ground-level passive solar features. In urban locations with high building densities the amount of overshadowing at low sun angles can be such that passive solar features are rendered ineffective. Roof levels are less frequently overshadowed.

BUILDING 2000 Building 2000 brochures are published by Directorate General XII of the Commission of the European Communities to show how design studies can help architects and other building designers use passive solar principles to the best effect to produce attractive energy-efficient buildings. Each brochure describes studies carried out with the support of the Commission during the design phase of one of thirty-six non-domestic buildings in the EC Member States. The studies were on such topics as daylighting, heating, cooling, ventilation, comfort, control systems and urban design. They were carried out with the help of acknowledged European experts in these fields and drew heavily on lessons learned and techniques developed through the Commision's research and development programme on solar energy applications to buildings.

Commission of the European Communities/Directorate-General for Science, Research and Development

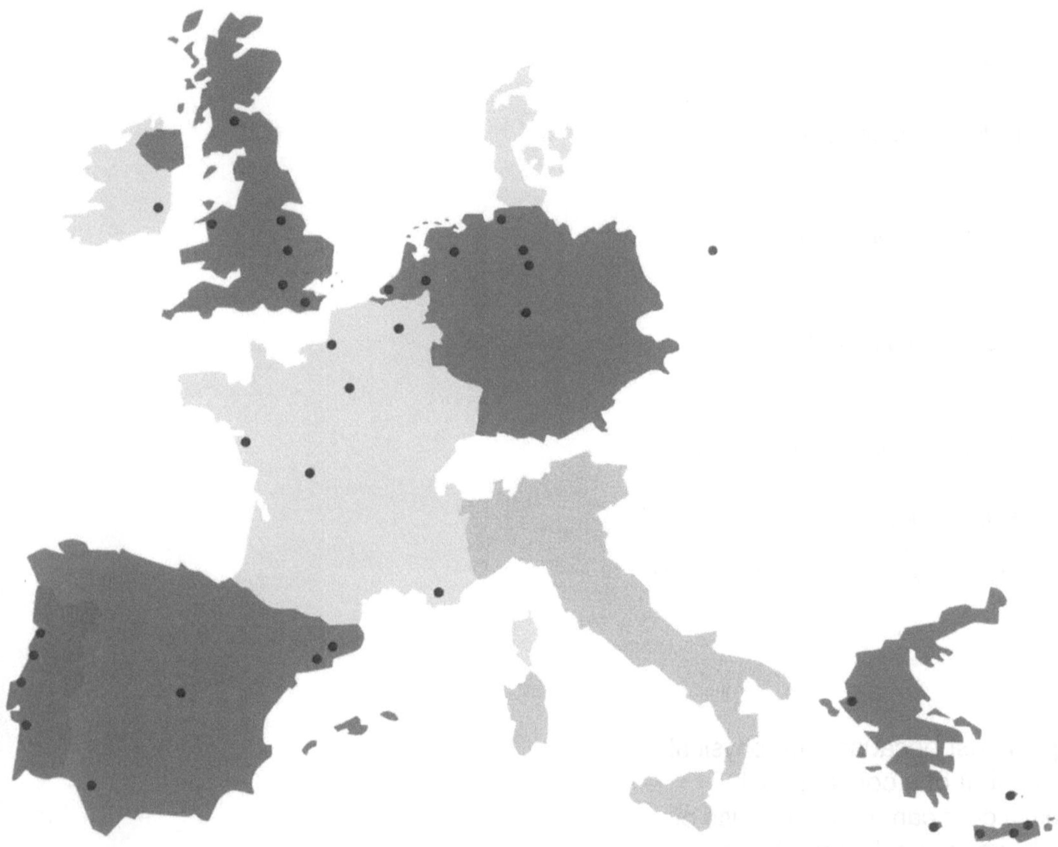

List of Design Team Participants and Advisers

Client
Buckinghamshire County Council
Education Committee
County Hall
Aylesbury
Buckinghamshire HP20 1UA
United Kingdom

Architect
Paul Markcrow, County Architect
County Hall
Aylesbury
Buckinghamshire HP20 1UA
United Kingdom

Contractor
Llewellyn Construction Limited
Bleak Hall
Milton Keynes MK6 1LA
United Kingdom

Energy Consultant
Professor Brian Norton
Department of Building
University of Ulster
Newtonabbey BT37 0QB
Northern Ireland

Thermal Analysis
Principia Mechanica Limited
233 High Holborn
London WC1V 7DJ
United Kingdom

This set of **Building 2000** brochures illustrates how architects and other building designers can successfully apply passive solar principles to produce energy-efficient buildings.

BUILDING 2000
Participants

Project Director
Theo C. Steemers

Coordinator
Cees den Ouden

Technical Steering Committee
Dean Hawkes
Nick Baker
Alex Lohr
Jean P. Lepoivre

Regional Liaison Agents
(D) Jörn Behnsen
(E) Vicente Sifre
(F) Michel Raoust
(GB) Alan Hildon
(GR) Matheus Santamouris
(P) Eduardo Maldonado

Further information or copies of the brochures can be obtained from prof. ir. Cees den Ouden, EGM Engineering BV, P.O. Box 1042, 3300 BA Dordrecht, The Netherlands.

BUILDING 2000

Commission of the European Communities

- Primary school for 644 pupils.
- Two covered courtyards for use througout the year.
- Daylighting through windows, clerestory glazing and atrium roof to reduce consumption of electricity for artificial lighting.
- Solar preheating of ventilation air for all internal rooms.

METHILHILL PRIMARY SCHOOL

METHILHILL/SCOTLAND/

UNITED KINGDOM

Building 2000 is a series of design studies illustrating passive solar architecture in buildings in the European Community.

ISSUE S 4

DEC 1990

PROJECT DESCRIPTION

Building Type

This is a primary school for 464 pupils containing 16 class areas, withdrawal rooms, a general purpose hall, staff rooms and the usual service areas. It is being built to replace an existing building on an adjacent site. Apart from being used as a school, the building will also serve as a social and recreational centre for the local community, during and after school hours. Construction began in November 1989 and the building is due for completion early in 1991.

Location

The site (formerly a farmstead) is in the centre of the town of Methilhill near the east coast of the region of Fife in Scotland (see Figure 1).

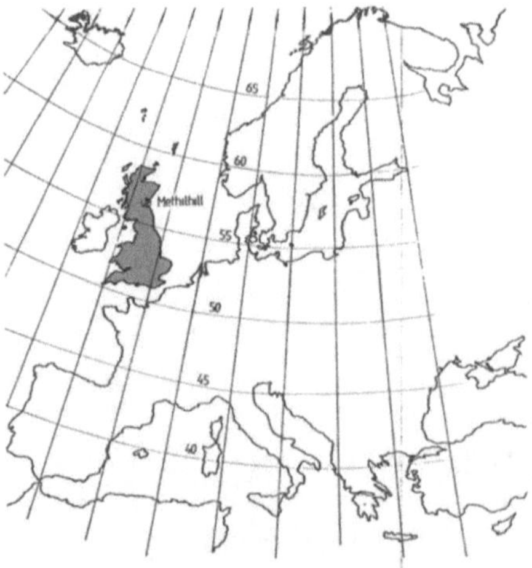

Figure 1. Location of Methilhill

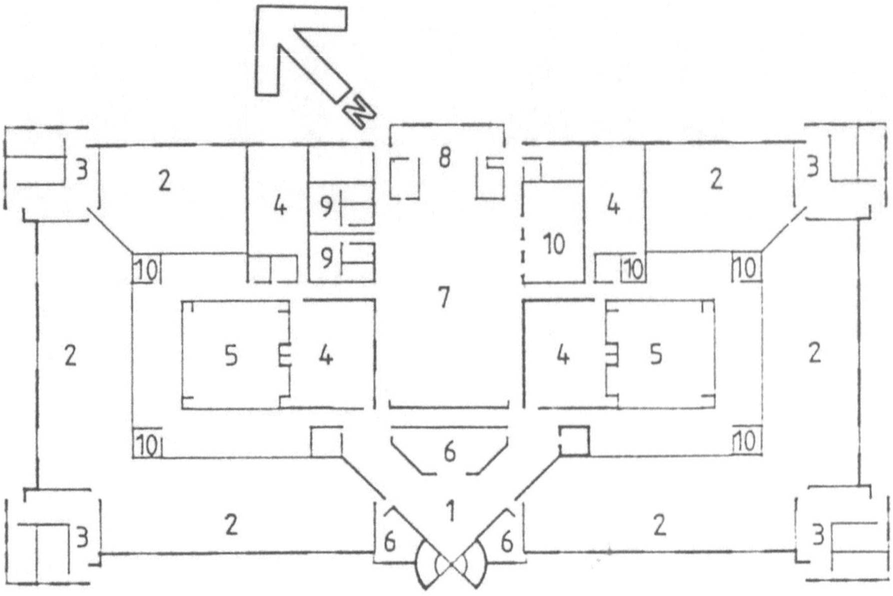

Figure 2. Plan of ground floor

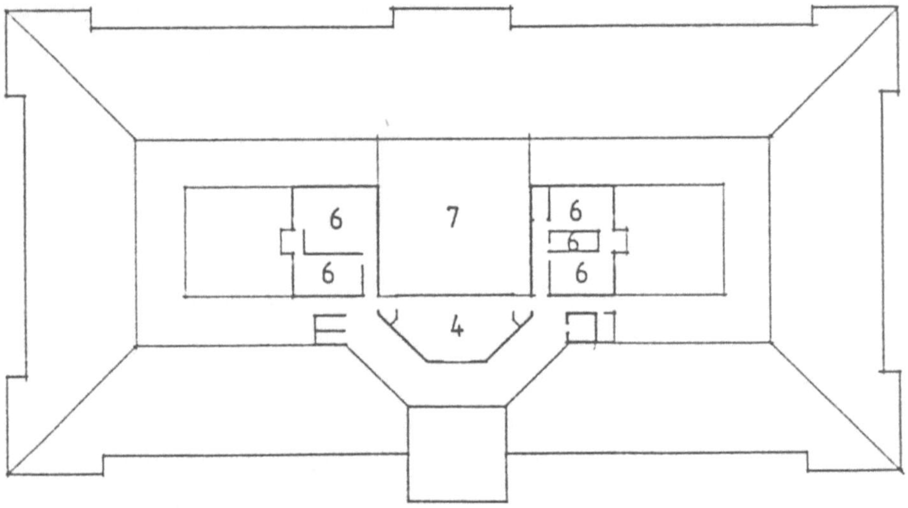

Figure 3. Plan of first floor

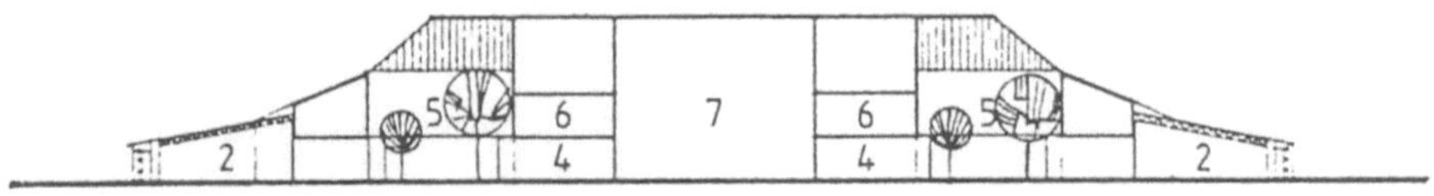

Figure 4. Long section

Key to Figures 2, 3 and 4:

1. Main entrance
2. Class areas
3. Toilets/cloakrooms
4. Withdrawal rooms
5. Courtyards
6. Staff areas
7. General purpose hall
8. Kitchen
9. Changing rooms
10. Storage

Design and Construction Details

In accordance with a long-established policy of the regional Education Department, the building layout is open plan with class areas grouped around two central courtyards. In previous schools constructed in the region, these courtyards remained exposed to the weather. In the present building, they are covered so that they can be used as teaching spaces for most of the year and benefit from available solar radiation. The building has two storeys and the plans of the two floors are given in Figures 2 and 3.

The building is arranged along an axis 35 ° east of south (see Figure 4) so that it receives the early morning sun and also benefits from the available solar radiation throughout the building occupation period. The longitudinal axis is aligned with the prevailing wind. This will enable adequate ventilation to take place and reduce the winter chill factor to the minimum. The south west and north east elevations are shown in Figures 5 and 6.

The external walls are of a traditional cavity construction with a 102 mm facing brick external leaf, a plastered common blockwork inner leaf and a 100 mm cavity containing a Styrofoam insulation board. The U-value (see Table 1) easily satisfies current and proposed Building Regulation requirements.

The ground floor consists of a 150 mm in-situ concrete slab.

The roof is constructed of a variety of elements and materials ranging from natural slating to stainless steel sheet cladding. A composite sarking/insulation decking is used throughout, giving U-values which are never greater than 0.4 W/m^2 K. Again, this satisfies current Building Regulations and those proposed for 1990. Sealed double glazing units are used throughout.

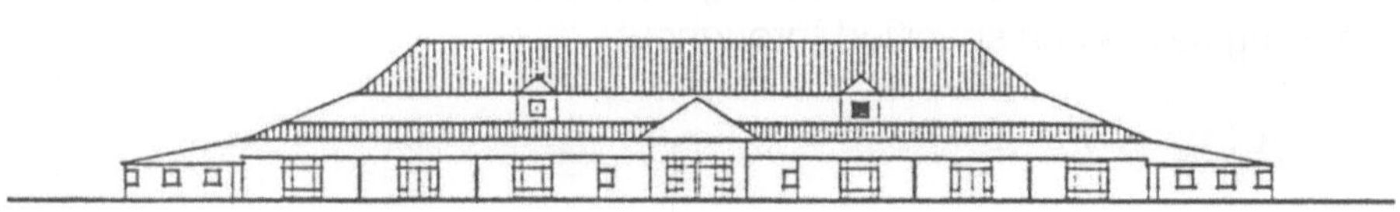

Figure 5. South-west facade

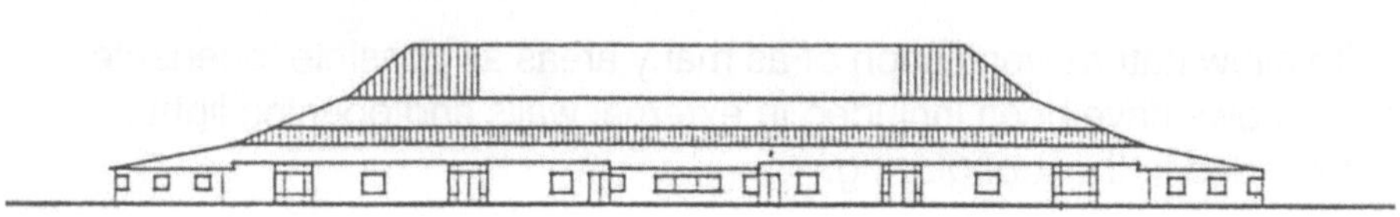

Figure 6. North-east facade

Site Microclimate

The site is at the top of a shallow rise and is not overshadowed. The prevailing wind is from the south east.

external wal	0.39
ground floor	0.37
roofs	<0.4
double glazing units	2.8

Table 1. U-values of some building elements in W/m^2 K

DESCRIPTION OF PASSIVE SOLAR FEATURES/COMPONENTS

Introduction

A number of passive solar and other features have been included in the building to save heating and lighting energy and help provide thermal and visual comfort throughout the year.

Initially, the intention was to obtain optimum solar gains for space heating. As the design developed, however, the priorities changed so that in the final design emphasis has been given to obtaining good levels of daylight and using solar gains to temper the ventilation air supplied to the internal rooms. Auxiliary heating is provided by a gas-fired boiler, radiators in the internal rooms and underfloor pipework in the courtyards.

	kWh/year
Barrel vault	
Single-glazed	4958
Double-glazed	3850
Triangular	
Single-glazed	4864
Double-glazed	3821

Table 2. Collection of solar energy by barrel vault and triangular section roof structures

Daylighting

The windows on the facade, the clerestory glazing and the atrium roof give high levels of daylight in the occupied parts of the building thus reducing the need for artificial light.

Solar Gains

The above glazed areas also allow direct and reflected solar radiation to contribute to heating the class areas and hall in winter-time.

Roof Geometry

Special care was taken over design of the roof. A comparsion of two different geometries (barrel vault and triangular section structures) was conducted to establish which, if any, was more efficient in terms of solar collection. The barrel vault had a radius of 5 m. The triangular section had an overall height of 5 m and sides pitched at 45 . In each case, both single- and double-glazed constructions were tested. The results, which are shown in Table 2, indicated that the thermal performance of the two designs is very similar. The triangular section, however, was felt to be more 'buildable' and therefore likely to be cheaper to construct. It was this design which was chosen for the final building.

Preheating Ventilation Air

The air in the roofspace, warmed by direct solar radiation entering through the atrium roof, acts as a source of input air for the mechanical ventilation system serving various internal rooms. Under optimum conditions it may make a significant contribution to space heating.

Insulation

To prevent unwanted heat loss, high levels of insulation and double glazing have been specified throughout.

Shading

To avoid summer-time overheating, the building has been fitted with various permanent and movable shading devices such as eave overhangs, movable/retractable blinds and venetian blinds.

Natural Ventilation

To allow natural ventilation of as many areas as possible, openable windows have been included in external walls and opening lights provided in the roof glazing.

Mode of Operation of Passive Systems in Winter

The mode of operation of the passive solar systems in winter are shown in Figure 7. In the courtyards, the solar blind/reflector directs solar radiation to the floor slab where it is stored for re-radiation later. In the class areas, reflected diffuse radiation from the clerestory and direct gain from solar radiation entering through windows in the external wall is absorbed by the floor slab for later emission, when required. The air in the roofspace is warmed by direct solar radiation and the warm air is used as input for the mechanical ventilation system.

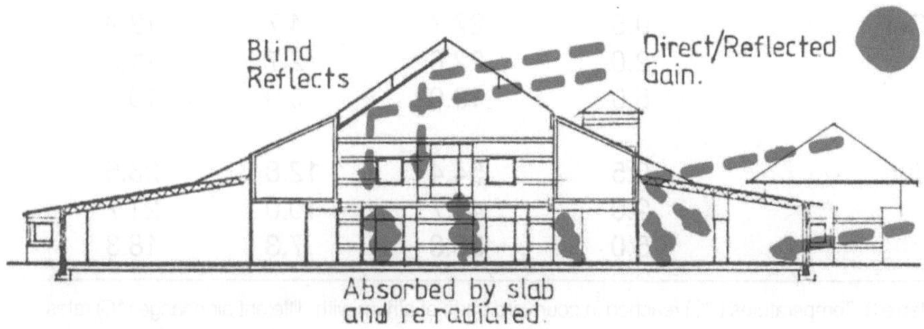

Figure 7. Mode of operation in winter

Mode of Operation of Passive Systems in Summer

The summer-time mode of operation is illustrated in Figure 8. The solar blind/reflector prohibits excessive solar gain in the courtyards. Surplus heat is ejected through electronically-operated opening vents. Reflected diffuse radiation enters the class areas via the clerestory glazing. The blinds are used when necessary. The entry of direct radiation through the windows on the external facade is reduced by the overhangs; additional screening is provided by venetian blinds. Fresh air is introduced through windows in the external wall and the building is ventilated naturally by means of the stack effect.

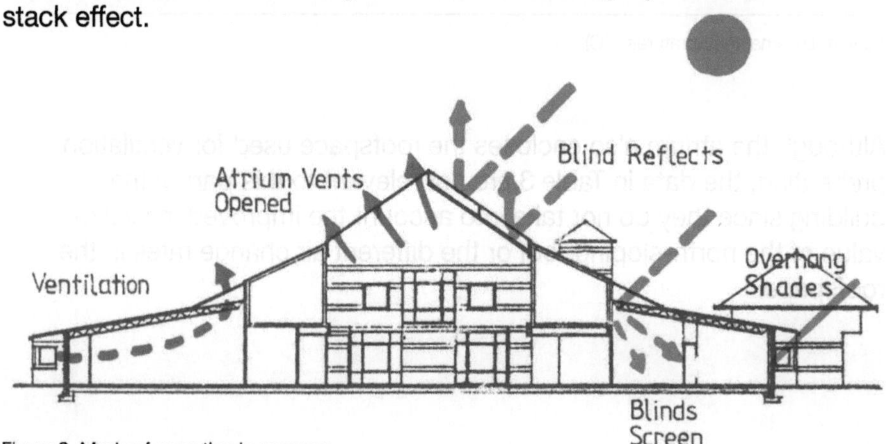

Figure 8. Mode of operation in summer

Auxiliary Heating

The majority of the space heating requirements are met by a conventional system consisting of a gas-fired central boiler with 'wet' distribution to radiators in the internal rooms and PVC underfloor pipework in the courtyards. Temperature control is effected by room thermostats so that any direct solar gains can reduce the demand on the auxiliary system. A manual override can be used when required.

ENERGY CALCULATIONS PERFORMED AND DESIGN TOOLS USED

Introduction

In addition to the comparison of the likely solar gains of two different roof geometries mentioned earlier, a number of thermal performance studies were carried out during the design stage, all of which made use of the computer simulation model ESP.

Temperatures in Courtyard Parts of Central Atrium

Table 3 gives the anticipated air temperatures for the central atrium (double glazed throughout with a pitch of 35 ° to the horizontal) where it encloses the two courtyards.

Month	AC/h	Maximum	Minimum	Mean
Jan	0.5	13.8	0.8	5.7
	2.0	11.1	-0.7	4.6
	5.0	9.9	-1.9	3.6
Oct	0.5	27.7	4.7	12.9
	2.0	22.6	2.7	11.5
	5.0	19.0	0.9	10.2
Jun	0.5	54.4	12.8	23.6
	2.0	44.7	10.0	21.7
	5.0	36.0	7.3	18.3

Table 3. Temperatures (°C) reached in courtyard parts of atrium with different air change (AC) rates

The mean temperatures in the courtyard atrium show a significant improvement on the external temperatures for the same months (Table 4). The extreme values shown (i.e. the June maximum and January minimum) would be moderated in practice by appropriate use of blinds and natural ventilation. The underfloor heating in the courtyard floor slab would improve the mean conditions in January and October.

	Maximum	Minimum	Mean
Jan	8.9	-3.8	2.0
Oct	15.6	-1.6	8.4
Jun	25.6	2.2	13.4

Table 4. External temperatures (°C)

Although the atrium also encloses the roofspace used for ventilation preheating, the data in Table 3 are not relevant to this part of the building since they do not take into account the improved insulation value of the north-sloping roof or the different air change rates in the roofspace.

GENERAL DESIGN GUIDELINES/POINTS OF INTEREST RESULTING FROM THIS PROJECT

As a result of the design study, the final building incorporated more passive solar features than had originally been anticipated. The end result is a building which should be capable of responding to users' needs and influencing its own energy use, rather than relying heavily on mechanical systems.

The designers' original idea of incorporating active systems such as those using heat pumps into the building was rejected because the simulation studies showed these features to be inefficient and economically non-viable in this particular situation. Additional simulation time was taken up examining various alternative roof structures and atrium geometries before the final choice was made. The simulation studies on atrium courtyard temperatures, however, encouraged the designers to incorporate these features in the design on the grounds that they could be used for much of the year.

In general, it was found that an early appreciation of the actual space demands of the building is essential in order to establish which systems/features should be included. In this project, the requirements for service space changed in the course of the design period and it became difficult to resolve certain sectional arrangements.

As this design has evolved from previous projects, so this school, too, should generate ideas for future designs. If it does nothing else, the completed building should serve as a working model from which lessons may be learned and future strategies formulated. A fundamental factor in the success or otherwise of the project will be the manner in which the occupants adapt to the new building and run it so that it responds to changes in the weather.

BUILDING 2000 Building 2000 brochures are published by Directorate General XII of the Commission of the European Communities to show how design studies can help architects and other building designers use passive solar principles to the best effect to produce attractive energy-efficient buildings. Each brochure describes studies carried out with the support of the Commission during the design phase of one of thirty-six non-domestic buildings in the EC Member States. The studies were on such topics as daylighting, heating, cooling, ventilation, comfort, control systems and urban design. They were carried out with the help of acknowledged European experts in these fields and drew heavily on lessons learned and techniques developed through the Commision's research and development programme on solar energy applications to buildings.

Commission of the European Communities/Directorate-General for Science, Research and Development

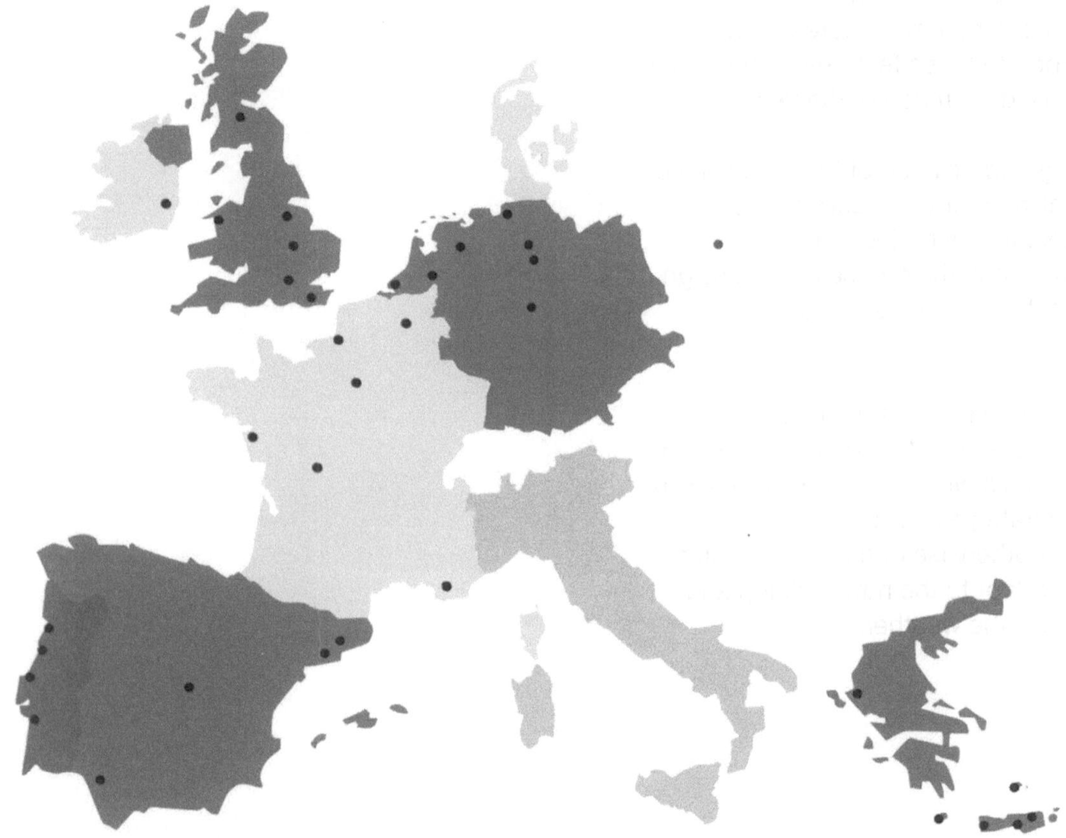

List of Design Team Participants and Advisers

Client
Department of Education
Fife Regional Council
Glenrothes
Fife
United Kingdom

Architect
Department of Architectural Services
Fife Regional Council
Glenrothes
Fife
United Kingdom
Director: J.D.T. Cowling
Architect: G. Gibson
Structural Engineer: S. Dryburgh
Quantity Surveyor: F. McDonald
Electrical Engineer: H. Stuart

Services Consultant
T. Harley Haow & Partners
Edinburgh
United Kingdom

Simulations
Abacus Simulations Limited
Glasgow
United Kingdom

This set of **Building 2000** brochures illustrates how architects and other building designers can successfully apply passive solar principles to produce energy-efficient buildings.

BUILDING 2000 Participants

Project Director
Theo C. Steemers

Coordinator
Cees den Ouden

Technical Steering Committee
Dean Hawkes
Nick Baker
Alex Lohr
Jean P. Lepoivre

Regional Liaison Agents
(D) Jörn Behnsen
(E) Vicente Sifre
(F) Michel Raoust
(GB) Alan Hildon
(GR) Matheus Santamouris
(P) Eduardo Maldonado

Further information or copies of the brochures can be obtained from prof. ir. Cees den Ouden, EGM Engineering BV, P.O. Box 1042, 3300 BA Dordrecht, The Netherlands.

BUILDING 2000

Commission of the European Communities

- Secondary school with 18 classrooms
- Daylighting using south-facing windows, light shelves, dual-mode rooflights and light ducts
- Passive solar heating using direct solar gain through south- facing windows and rooflights and heat storage in mass of floors and walls
- Prevention of overheating by external overhangs and draught- free cross ventilation
- Detailed monitoring

SECONDARY SCHOOL

VALDINGO DO VOUGA, AGUEDA/PORTUGAL

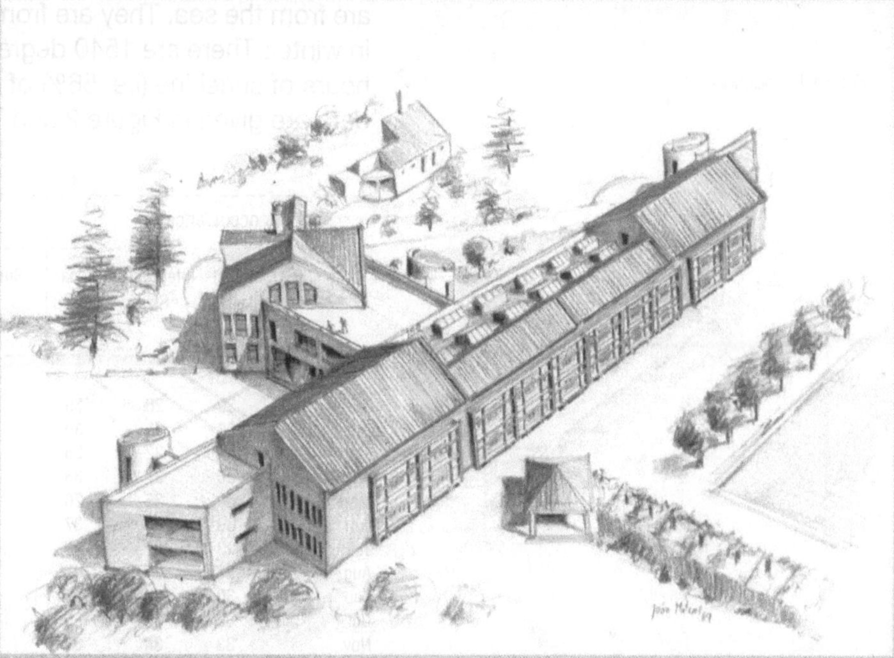

Building 2000 is a series of design studies illustrating passive solar architecture in buildings in the European Community.

ISSUE S 5

DEC 1990

PROJECT DESCRIPTION

Building Type

This is a secondary school with 18 classrooms, laboratories, seminar rooms, a theatre, library, administration offices, kitchen and refectory. It is one of several secondary schools being built under a joint programme of the Ministry of Education and the Agueda municipal council.

Location

The site, which is 26 m above sea level, is located some 50 km from the sea in the village of Valonga do Vouga (latitude 40° 35') in the municipality of Agueda in Portugal. The land slopes smoothly from north to south and is well exposed to the sun. The site is not overshadowed by other buildings.

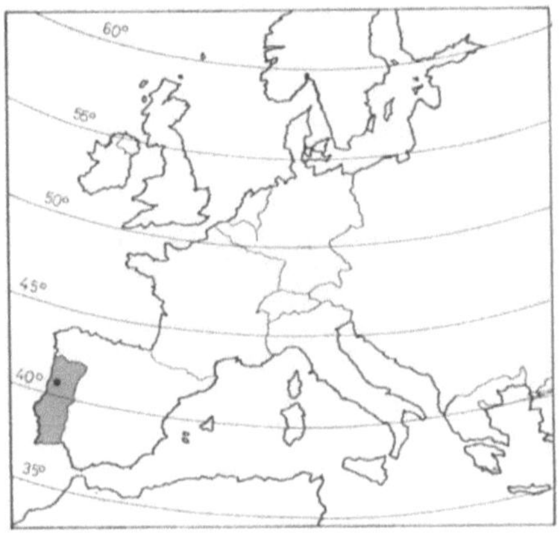

Figure 1. Location

Site Microclimate

The climate is moderate and typical of Portugal. The prevailing winds are from the sea. They are from the NW in summer and from the SW in winter. There are 1540 degree days a year on average and 2520 hours of sunshine (i.e. 56% of daylight hours). More detailed climate data are given in Figure 2 and Table 1.

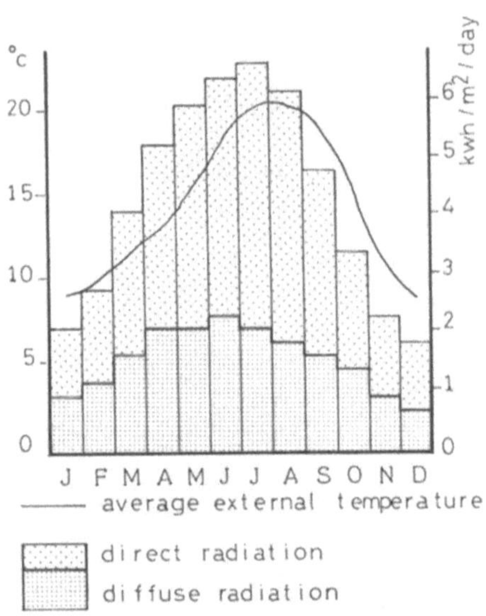

Figure 2. Mean daily irradiation and ambient temperature data

	% occurrence of:							
	Overcast Skies	Intermediate Skies	Clear Skies	Sunshine %	Solar irra-diation Horiz. pl. (kWh/m²)	External temperature (°C)		
						Min.	Average	Max
Jan	39	26	35	42	57.5	5	8.5	11.5
Feb.	36	25	39	47	81	5	8.5	11.5
Mar.	42	32	26	48	123	6	11.5	17
Apr.	30	37	33	55	166	7	13	19
May.	39	35	26	56	198	10	15	20
Jun.	30	33	37	56	210	11	16	21
Jul.	23	29	48	65	225	14	19	24
Aug.	23	29	48	68	200	15	19	23
Sep.	30	37	33	59	147	14	18	22
Oct.	32	33	35	56	106	11	16	21
Nov.	33	34	33	49	69	6	9	12
Dez.	39	29	32	44	56	5	8.5	11.5
ANNUAL AVERAGE	33	31	36	56	-	9	13.5	17.8

Table 1. Sunshine, global solar irradiation and temperature data

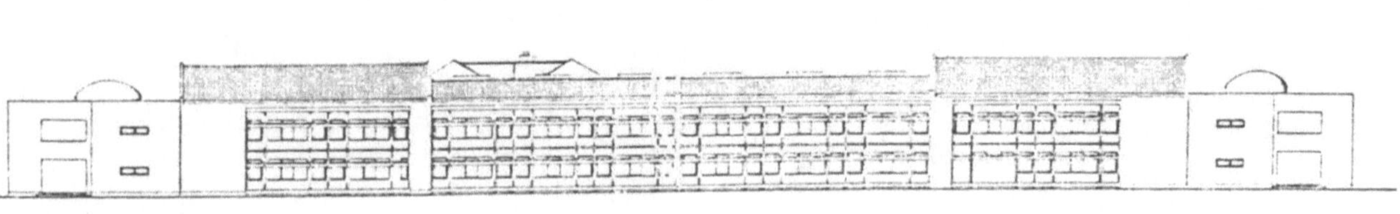

Figure 3. South elevation

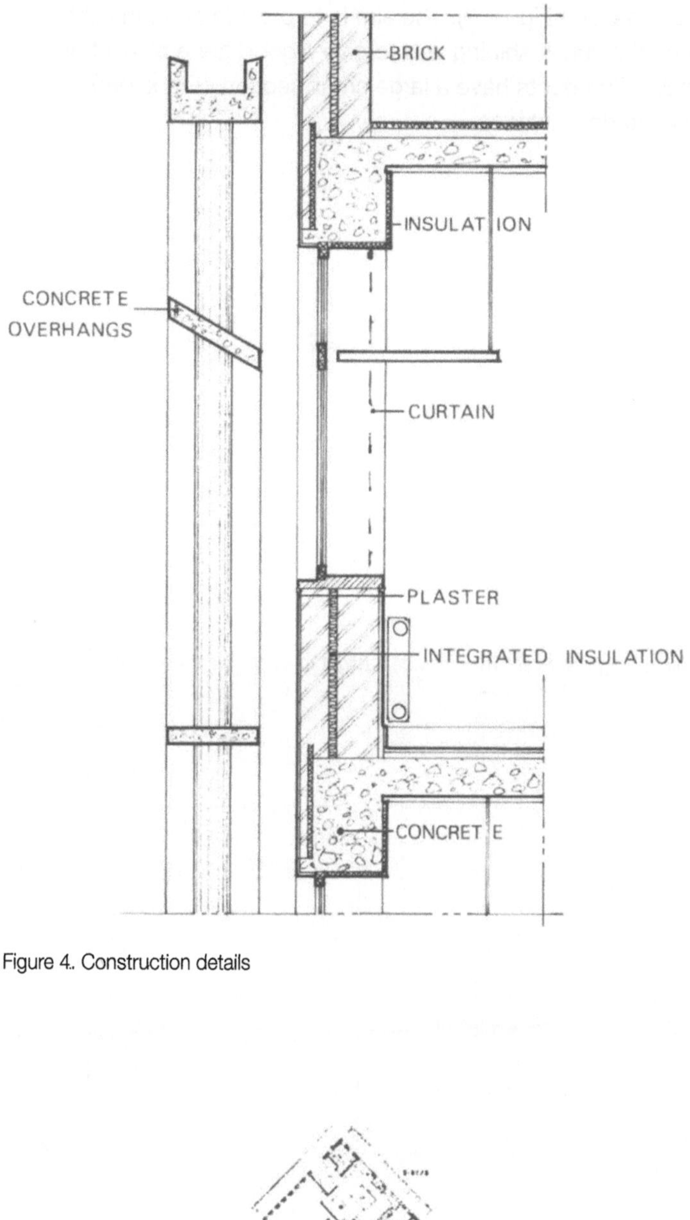

Figure 4. Construction details

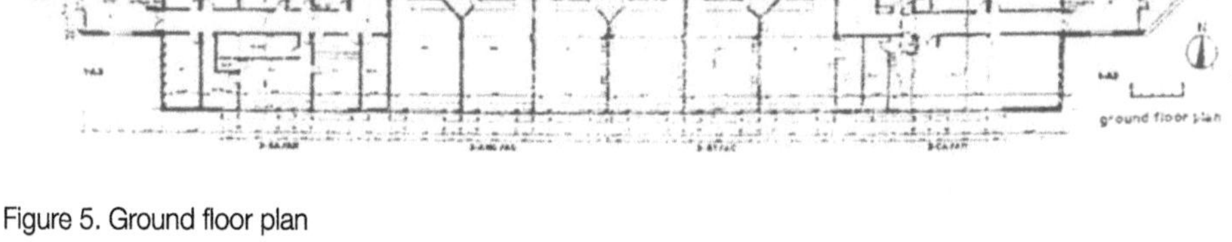

Figure 5. Ground floor plan

Design and Construction Details

The design is the result of joint efforts by the Ministry of Education and the University of Coimbra. During the design process, particular attention was paid to defining potential problems and testing new solutions with the idea of developing a design acceptable for other school buildings in the Ministry of Education/Agueda programme.

The building has a floor area of 2,000 m² and a volume of 18,000 m³. The cost is 2 M ECU.

The classrooms are arranged in two-storey modules along the south facade. The ground floor plan and south and north elevations are shown in Figures 3, 5, 6. Some construction details are given in Figure 4. U-values of some building elements are presented in Table 2.
Energy management, data logging for building performance evaluation and fire and security alarm systems are controlled by a central data acquisition and control system.

external walls	0.53
beams and columns	0.65
internal walls (room/room)	1.75
internal walls (room/corridor)	1.12
ground floor room floor	0.58
first floor room ceiling	0.52

Table 2. U-values of some building elements (W/m² K)

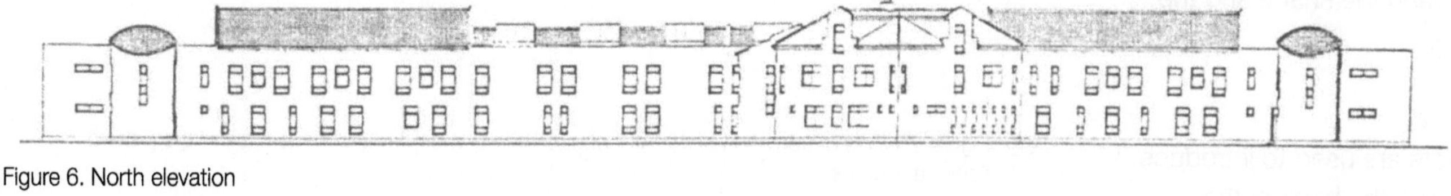

Figure 6. North elevation

DESCRIPTION OF PASSIVE SOLAR FEATURES/COMPONENTS

The design aims to use passive solar and other energy-saving features to provide lighting and thermal comfort in the building throughout the year. Good daylighting is produced by using light shelves, rooflights and light ducts. Thermal comfort in winter is achieved by direct solar gains, storage and a carefully controlled auxiliary heating system. Overheating in summer is prevented by shading and cross ventilation. The various systems are described in more detail below.

Daylighting Systems

Light Shelves

The light shelves were designed so that good quality daylight adequate for visual tasks is achieved in as much of the classrooms as possible when the sun shines. Located on the south facade, they reflect sunlight onto a white ceiling (which acts as a diffuser) to provide light to the south side of both the ground- and upper-floor classrooms.

Dual-Mode Rooflights

The aim of the rooflights is to achieve sufficient daylighting to enable visual tasks to be carried out at the back of the upper- floor classrooms most of the day. Each rooflight is provided with an internal white shade with good thermal insulation properties. The shade is movable so that it can be used to cover the north or south rooflight glazing (see Figure 8). When the shade is in its summer position on the south glazing, the room can be ventilated through a gap between the shade and the glazing.

Light Ducts

Light ducts are used to introduce daylight into the back of the ground-floor classrooms (see Figure 9). The aim is to provide enough light for visual tasks when the sun is shining and to give a good general level of light on overcast days. The ducts have a large cross section (3 m x 1 m) and highly reflective interior surfaces.

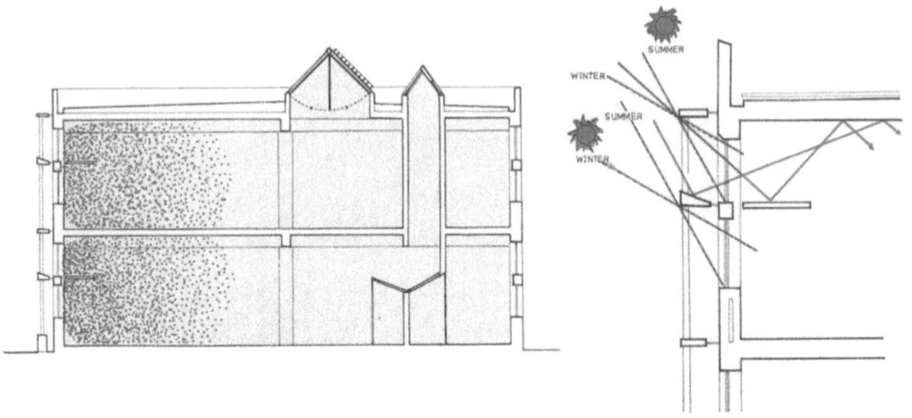

Figure 7. Mode of operation of light shelves. They provide daylight to a zone near the south facade.

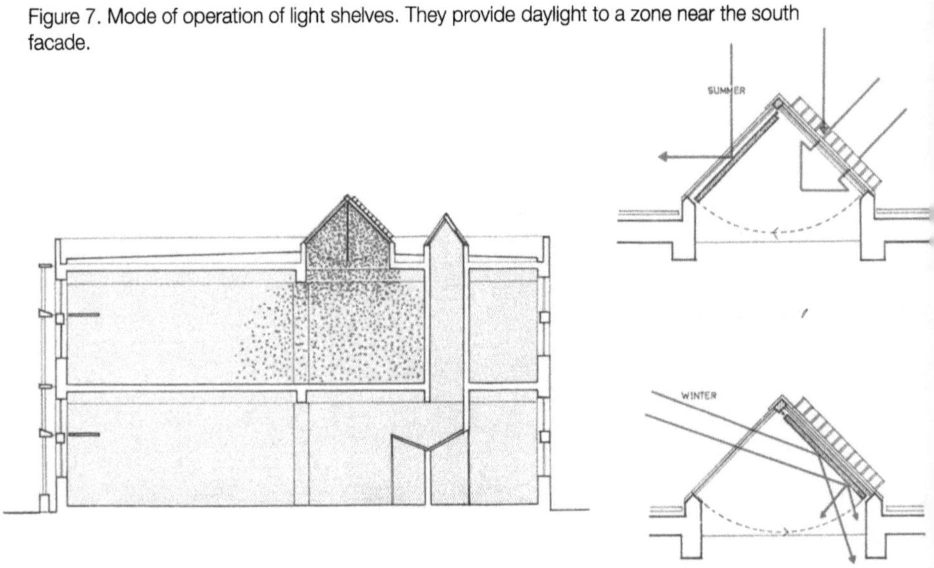

Figure 8. Mode of operation of the dual-mode rooflights. They provide daylight to the back of the upper-floor classrooms.

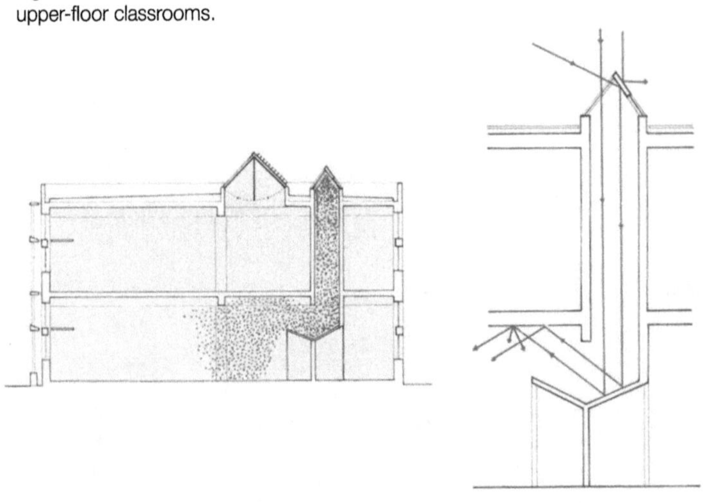

Figure 9. Mode of operation of the light ducts. They provide daylight to the back of the ground-floor classrooms.

Passive Solar Heating Systems

Direct Solar Gain

South-facing non-absorbing transparent glazing is provided on the south side with the aim of providing just enough direct gains to maintain comfortable internal conditions in the classrooms (see Figures 10 and 11).

Thermal Mass

The thermal energy is stored in the mass of the building so that large temperature swings are avoided and a steady optimum temperature is maintained. The thermal mass is provided by the classroom floors (which are made of concrete), the internal walls (which are heavy brick) and the external walls (which are well- insulated) - see Figure 4.

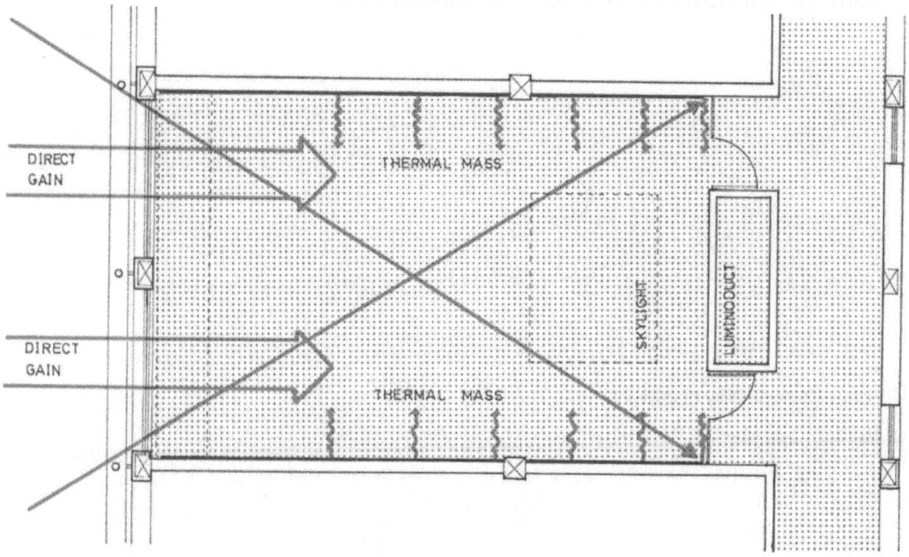

Figure 10. Mode of operation of passive solar heating systems - plan

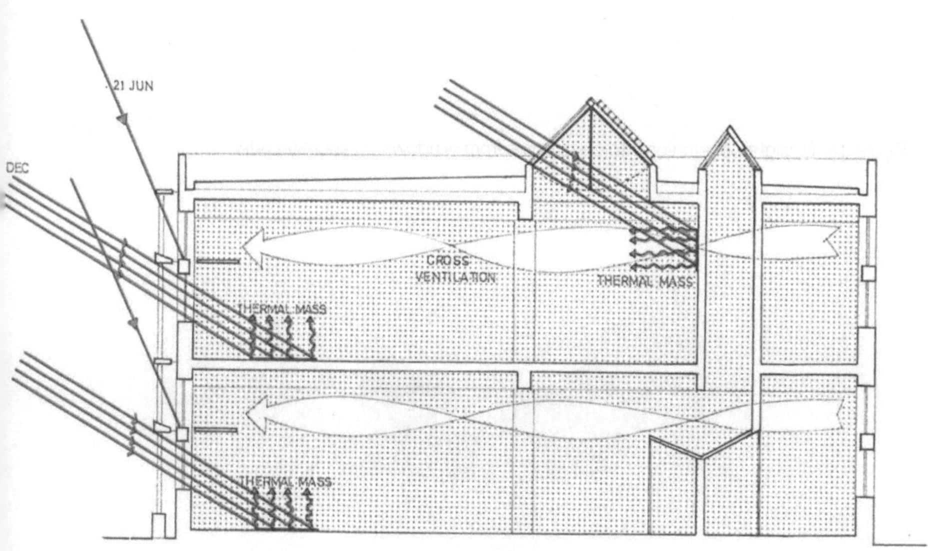

Figure 11. Mode of operation of passive solar heating systems - section

Systems for Preventing Overheating

Shading

To prevent entry of direct radiation into the classrooms in the warm periods of the year, horizontal external overhangs are provided above the glazing (see Figure 4) to prevent all but diffuse radiation entering the building.

Natural Ventilation

The upper parts of the classroom windows and doors are provided with opening panels to allow adequate cross ventilation to be achieved without draughts. Periodic air renewal can thus be effected quickly and efficiently.

Auxiliary Systems

The auxiliary heating system consists of a wood-fired boiler, a hot water distribution system and room radiators. Both it and the artificial lighting system are designed to ensure high efficiency energy conversion. A supervisory control and data acquisition (SCADA) system restricts their use to periods when they are strictly necessary to maintain comfort conditions. The data acquisition facility enables detailed evaluation of the building's behaviour during operation to be carried out.

ENERGY CALCULATIONS PERFORMED AND DESIGN TOOLS USED

Daylighting

Design Tools

The daylighting systems were designed with the help of a number of design tools. A physical model (scale 1:10) was used for experimental work under real sky conditions (see Figure 12). To measure the penetration of daylight from various apertures, the black cloth was used to close the other openings. With all the apertures closed, the illuminance was slightly above zero because the model was not completely opaque.

The LUZDIA computer model developed by the Laboratorio Nacional de Engenharia Civil, Lisbon, was used to calculate daylight factors. This simulates a rectangular room with vertical glazing. The effect of the light duct was simulated with the help of a scale model.

The daylighting evaluation work also made use of the concept of H indicators - the mean number of hours a year for which stated levels of illuminance are exceeded. For instance, H300 = 700 indicates that daylight illuminances will exceed 300 lux for 700 hours a year. The H indicators were calculated using a model described in a paper on dynamic characterization of daylit spaces presented by L.C. Carvalho to the European Commission conference on architecture held in Paris in December 1989.

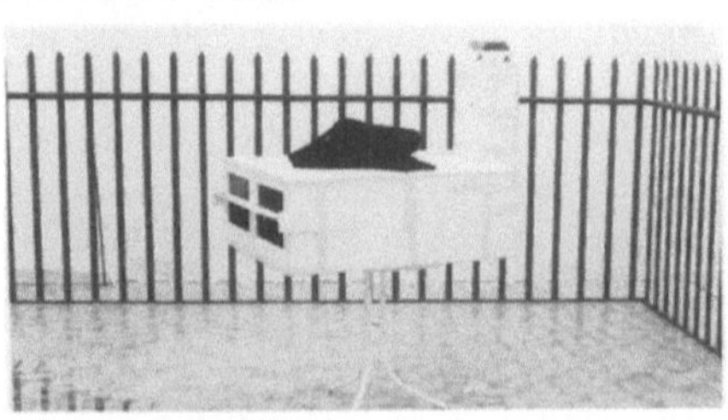

Figure 12. Scale model used for daylighting measurements under real sky conditions

Performance of Daylighting Systems

In the daylight assessment work, it was assumed that the school will be used 1850 hours a year, of which 1700 hours are during daylight, i.e. when the sun height is > 5° - see Figure 13.

Studies on the daylighting through the windows show that sunlight will penetrate deeply into the classrooms from the south facade. The chart in Figure 14 shows the illuminances (lux) under clear sky conditions on a workplane at a point on the symmetry axis 2.5 m from the back of the room. (Glare will be prevented by the curtains shown in Figure 4, which diffuse the direct sunlight.)

The effect of the light duct on introduction of light into the back of the ground floor classrooms was determined using the scale model (Figure 12). Figure 15 gives the results under clear sky conditions for the same workplane as Figure 14. The daylight factor at that point from the windows on the south facade is about 0.5%.

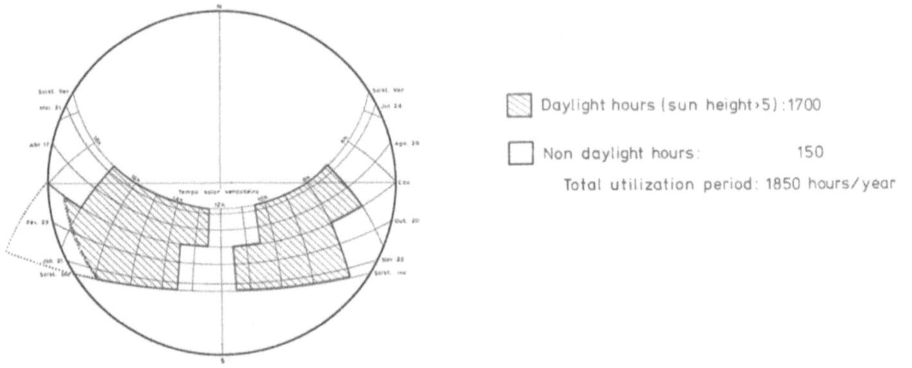

Figure 13. Availability of daylight

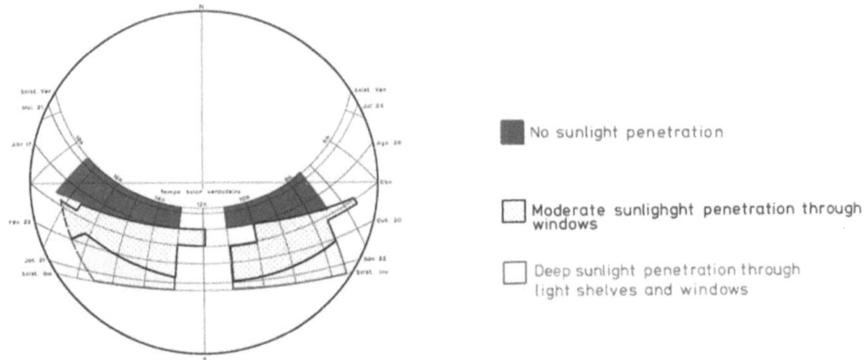

Figure 14. Daylight penetration into classrooms from windows on south facade

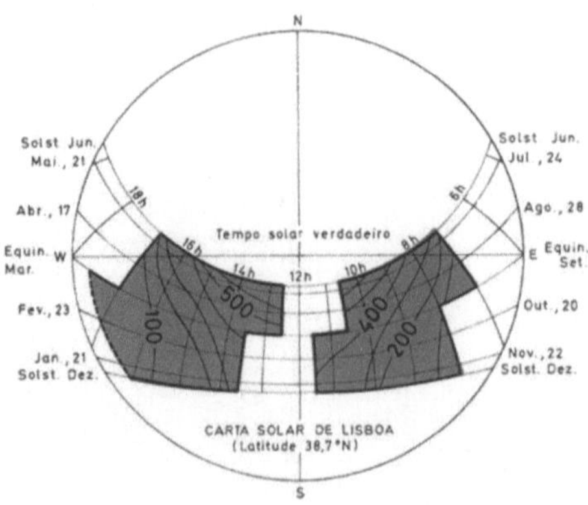

Figure 15. Daylight penetration from light ducts

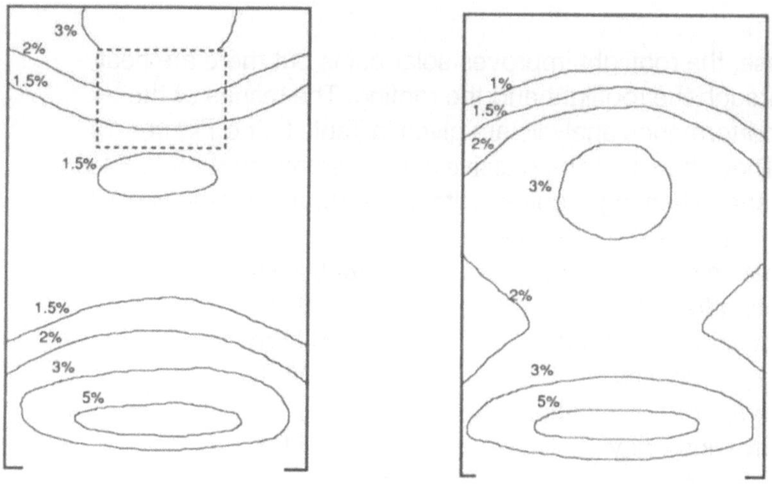

a) b)

Figure 16. Daylight factors for an upper-floor classroom when rooflight glazing is unshaded on (a) the south side and (b) the north side

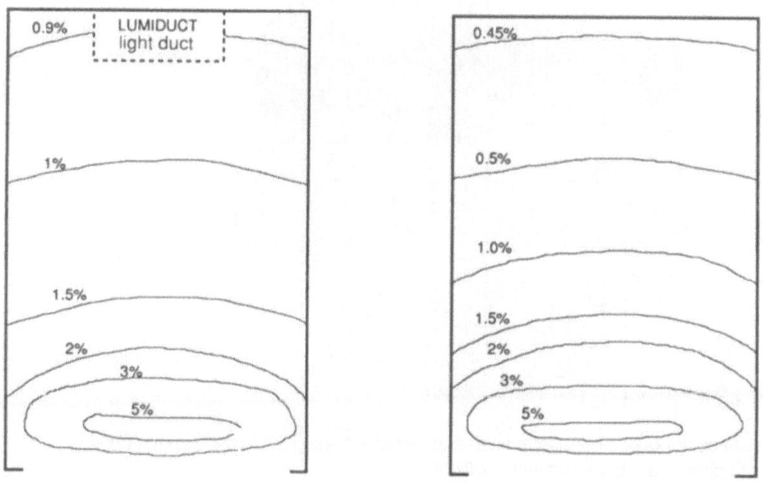

a) b)

Figure 17. Daylight factors for a ground-floor classroom (a) with and (b) without the light duct

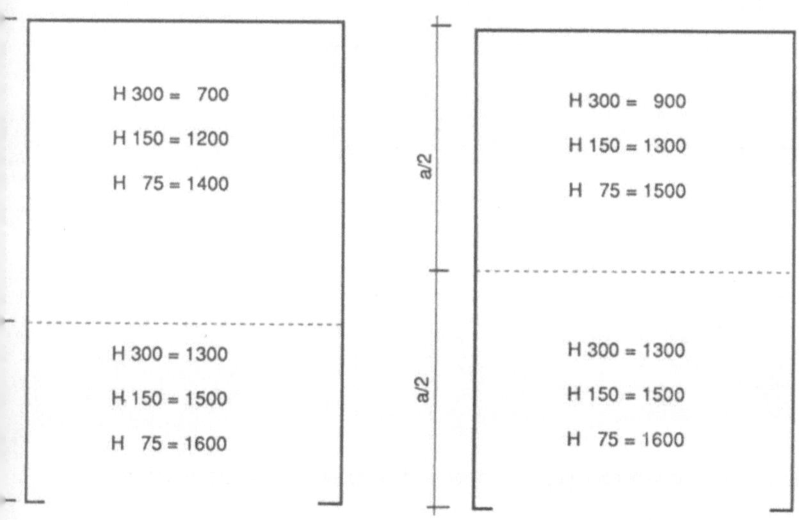

a) b)

Figure 18. H indicators for (a) a ground-floor and (b) an upper- floor classroom

The effect of the rooflights on introduction of daylight into the back of the upper-floor classrooms is shown in Figure 16. Diagram (a) presents the daylight factors for a room when the rooflight is in the winter position (i.e. the south side glazing is unshaded). Diagram (b) gives daylight factors when the north side glazing is unshaded.

Daylight factors for the ground-floor classrooms are shown in Figure 17. Diagram (a) gives the results for a room with the light duct in operation; diagram (b) gives daylight factors for the room without the light duct. It can be seen that the presence of the light duct reduces the difference in lighting between the back and window zones of the room.

The H indicators described in the Design Tools section opposite are given in Figure 18 for both ground- and upper-floor classrooms.

Thermal Performance

General

As indicated above, the building incorporates a number of passive solar features which improve thermal comfort. All the classrooms have 9.6 m^2 south-facing glazed windows to improve solar gains. The 3.45 m^2 glazed rooflights also contribute to solar gains. Overhangs two-thirds of the way up the south-facing windows ensure that direct solar radiation penetrates the classrooms in winter and only diffuse radiation enters in summer. The 220 mm dense brick internal wall and 200 mm concrete floor create a thermal mass which stores direct solar gains and prevents excessive temperature swings. Opening panels at the top of north and south windows enable cross ventilation to be achieved when necessary without draughts. The compact building form and high insulation levels minimize heat losses.

Design Tools

To design the passive solar heating systems, the building's thermal performance was analyzed using a simplified SLR-derived computer method developed especially for Portuguese conditions and published by the Portuguese General Directorate for Energy.

The analysis was conducted for three different cases: (1) an upper-floor classroom with a rooflight; (2) an upper-floor classroom without a rooflight; and (3) a ground-floor classroom adjacent to the entrance hall. The results are given opposite.

Upper-Floor Classroom with Rooflight

In this case, the rooflight improves solar gains but there are heat losses through the rooflight and the roofing. The results of the thermal performance analysis are given in Table 3 and Figures 19 and 20. They show that the passive solar systems contribute 64% of the total space heating requirements of the room over the year.

space heating demand/year	5800 kWh
solar heating fraction/year	64 %
auxiliary heating load/year	2100 kWh
total storage capacity	7650 Wh/K
internal temperature variation on a sunny winter day	6 K

Table 3. Thermal performance of upper-floor classroom with rooflight

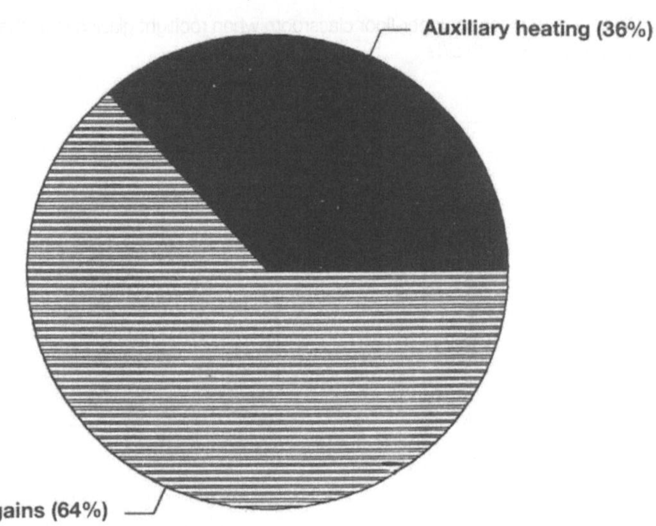

Figure 19. Annual contributions of passive solar and auxiliary heating systems to space heating requirements of upper-floor classroom with rooflight

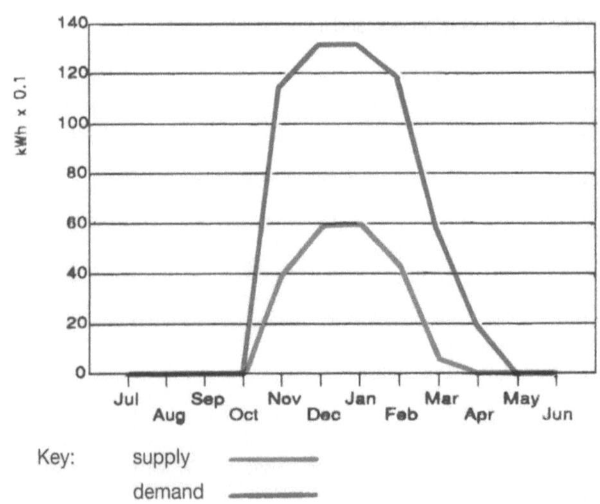

Figure 20. Net monthly space heating supply and demand for upper- floor classroom with rooflight

Upper-Floor Classroom without Rooflight

The results of the thermal performance analysis of an upper-floor classroom without a rooflight are given in Table 4 and Figures 21 and 23. In this case, there is solar gain through the south- facing windows only - but heat is lost only through the roof. Passive solar gains contribute 61% of the annual space heating requirements.

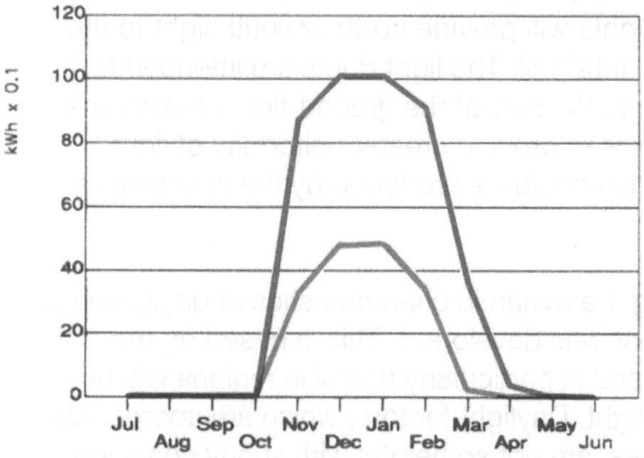

Key: supply / demand

Figure 21. Net monthly space heating supply and demand for upper- floor classroom without rooflight

space heating demand/year	4200 kWh
solar heating fraction/yr	61 %
auxiliary heating load/yr	1650 kWh
total storage capacity	7650 Wh/K
internal temperature variation on a sunny winter day	4.5 K

Table 4. Thermal performance of upper-floor classroom without rooflight

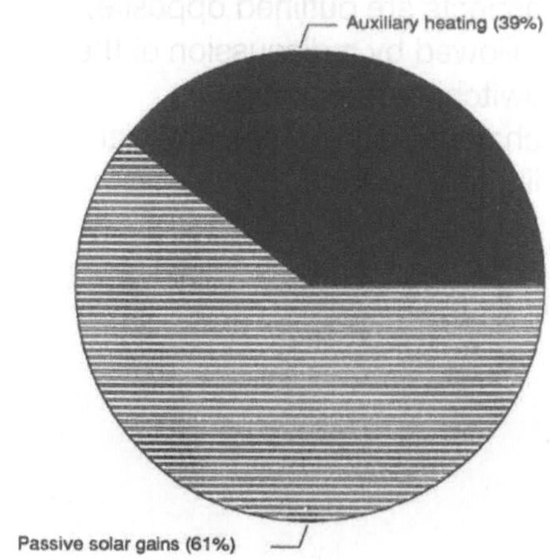

Figure 23. Annual contributions of passive solar and auxiliary heating systems to space heating requirements of upper-floor classroom without rooflight

Ground-Floor Classroom Adjacent to Entrance Hall

In this case, heat is lost through the entrance hall (which acts as a buffer zone) and the floor. The results of the thermal performance analysis are given in Table 5 and Figures 22 and 24. Solar gains contribute 60% of the annual space heating requirements.

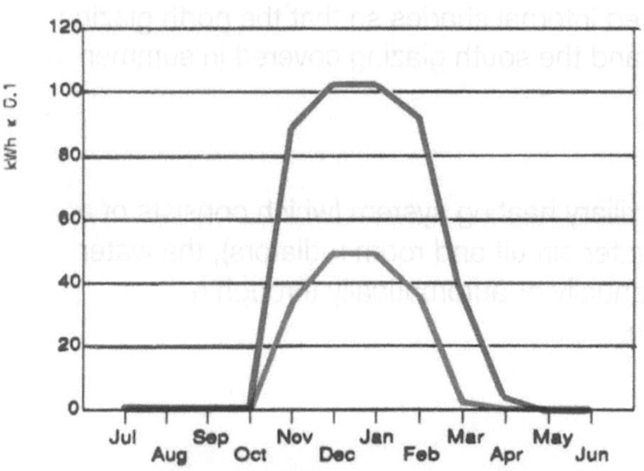

Key: supply / demand

Figure 22. Net monthly space heating supply and demand for ground-floor classroom adjacent to entrance hall

space heating demand/year	4300 kWh
solar heating fraction/yr	60 %
auxiliary heating load/yr	1700 kWh
total storage capacity	7650 Wh/K
internal temperature variation on a sunny winter day	4.5 K

Table 5. Thermal performance of ground-floor classroom adjacent to entrance hall

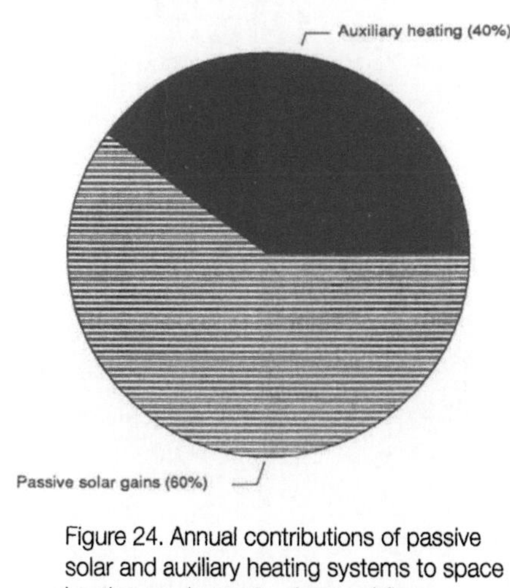

Figure 24. Annual contributions of passive solar and auxiliary heating systems to space heating requirements of ground-floor classroom adjacent to entrance hall

GENERAL DESIGN GUIDELINES/POINTS OF INTEREST RESULTING FROM THIS PROJECT

General

As a result of the performance evaluation studies described above, particular attention was paid to certain aspects of the design to ensure that the daylighting and heating systems are optimized. These aspects are outlined opposite, followed by a discussion of the switching/technology characteristics of the artificial lighting system.

Daylighting

To prevent glare without reducing daylighting, the south-facing apertures were fitted with diffusing devices.

Dual-mode rooflights and light ducts were incorporated at the north end of the building so that their performance in use can be studied. The idea is that the rooflights will provide north or south light to the classrooms at the occupants' will. The light ducts are intended to give daylight penetration to the rear of the ground-floor classrooms. The aim of these devices is to provide greater uniformity of light across the classrooms than could be produced by the windows on the south facade alone.

A system of representing the dynamic characteristics of daylight in a space throughout the year was developed. This is based on the concept of H indicators and is particularly useful in regions where there is much direct sunlight. Daylight factors, which are appropriate for use with overcast skies, are not so helpful with sunny conditions. The H indicator concept is easily understood by architects, artificial lighting designers and energy experts. It allows for easier dialogue between them and enables more comfortable and energy-efficient buildings to be designed.

Thermal Optimization

To ensure thermal optimization, the building was well insulated, particularly around the region of thermal bridges. Large south- facing windows were added to give high levels of direct solar gain. Overhangs were included on the south-facing windows to minimize summer-time overheating. Walls and floors were made of heavyweight material so solar energy can be stored. North- and south-facing windows and doors were given opening panels at the top to allow cross ventilation. The rooflights were fitted with highly reflective and well-insulated internal shades so that the north glazing can be covered in winter and the south glazing covered in summer.

Auxiliary Heating

To optimize use of the auxiliary heating system (which consists of a wood-fired boiler, a hot water circuit and room radiators), the water flow can be controlled manually or automatically through a distributed control system.

Data Acquisition

A computer-based system with distributed data acquisition, control and processing capabilities has been included in the building. Local units placed in different parts of the building will be linked together and to a central processing and control unit to allow co-ordinated actions. The central unit will be linked to the University of Coimbra so the performance of the various passive solar systems can be evaluated in use.

Control of Artificial Lighting Systems

The supervisory control and data acquisition system will be used, among other things, for automatic control of light fixtures to maintain adequate lighting levels and at the same time minimize electricity consumption.

The fixtures are arranged in a number of zones which are each controlled to give co-ordinated use of daylight and artificial light. In addition, electronic control gear (ECG - solid state ballast) is used throughout to improve conversion efficiency.

To test the electricity consumption of various possible artificial lighting systems, the following four cases were simulated: light circuits with conventional ballasts and a traditional (inadequate) switching circuit; the same but using ECG; an efficient switching system and conventional ballasts; and, finally, an efficient switching system with ECG. The results are given in Figure 25, where efficient switching is identified as "smart" and traditional switching as "careless". The analysis assumed a typical fixture with two fluorescent tubes (2 x 58W). The pie chart in Figure 26 compares the energy consumption of the two extreme situations. Savings of 54% were achieved.

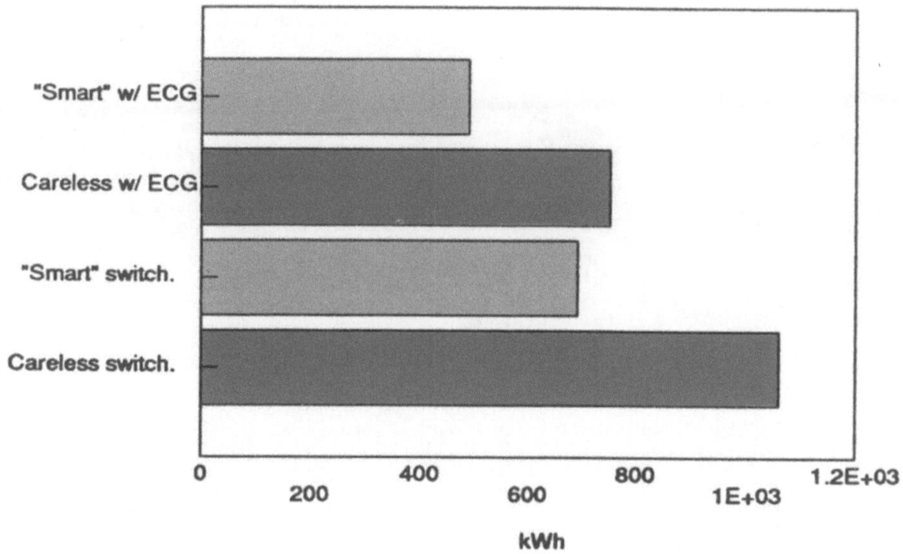

Figure 25. Comparison of annual electricity consumption per classroom of four lighting systems with different switching/technology characteristics

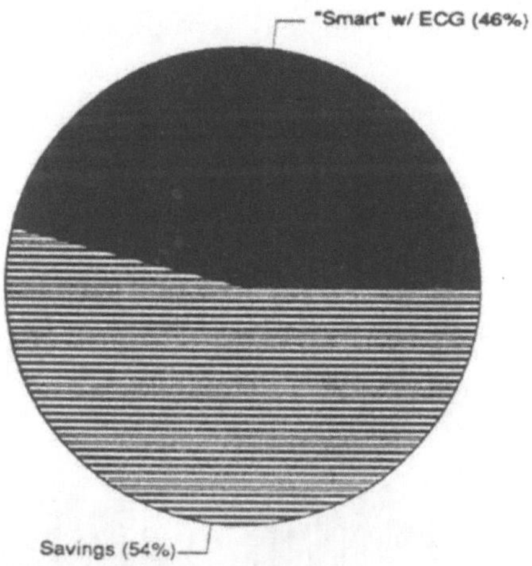

Figure 26. Savings in electricity consumption achieved by use of electronic ballasts

Final Comments

Participation in the Building 2000 programme enabled passive solar techniques to be incorporated successfully in this project. The experts made important contributions during the design phase; they played more than a strict consultancy role. The joint efforts of the University of Coimbra Department of Electrical Engineering and the Portuguese Ministry of Education, together with the enthusiasm of the Agueda town council for an innovative building design, enabled the work to be completed successfully. The Portuguese Ministry of Education department responsible for school building design is currently contemplating using many of the passive solar features incorporated in the Valongo do Vouga school in their future buildings.

BUILDING 2000 Building 2000 brochures are published by Directorate General XII of the Commission of the European Communities to show how design studies can help architects and other building designers use passive solar principles to the best effect to produce attractive energy-efficient buildings. Each brochure describes studies carried out with the support of the Commission during the design phase of one of thirty-six non-domestic buildings in the EC Member States. The studies were on such topics as daylighting, heating, cooling, ventilation, comfort, control systems and urban design. They were carried out with the help of acknowledged European experts in these fields and drew heavily on lessons learned and techniques developed through the Commision's research and development programme on solar energy applications to buildings.

Commission of the European Communities/Directorate-General for Science, Research and Development

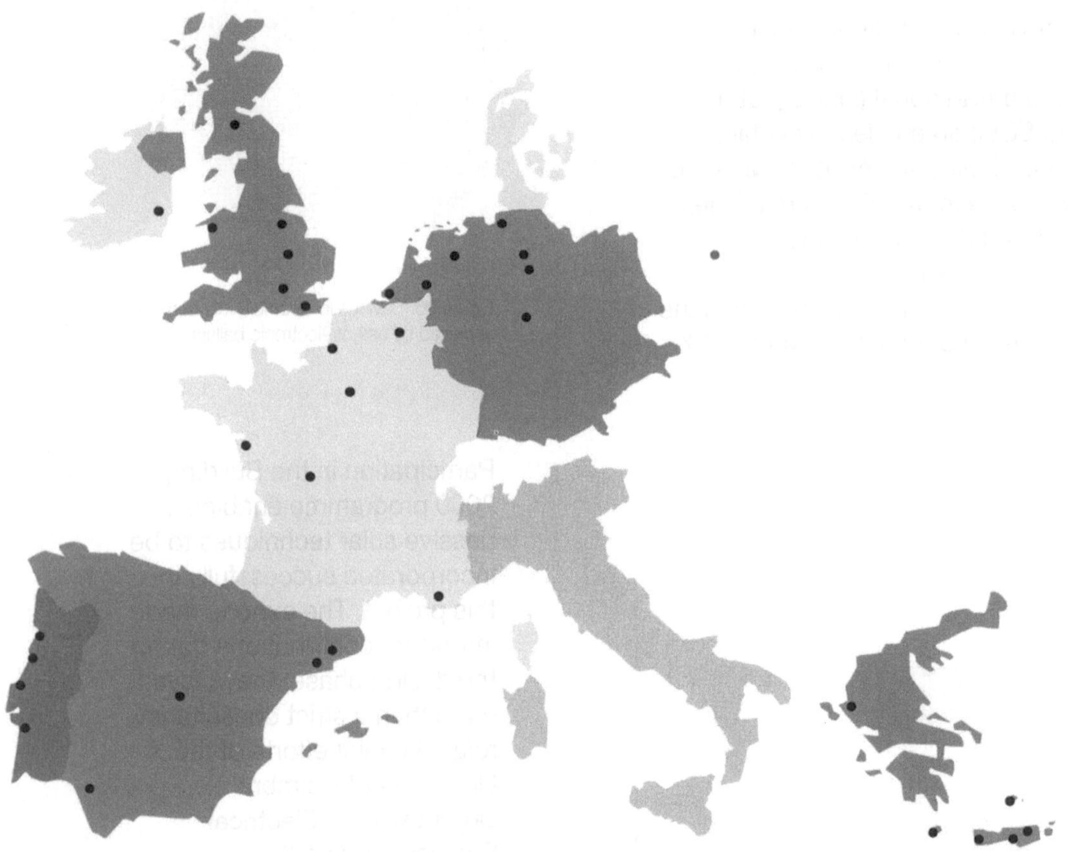

List of Design Team Participants and Advisers

Client
Town Council of Agueda
Ministry of Education

Architects
Jorge Gouveia, João Mateus and Luís Cunha
Direcção Geral dos Equipamentos Educativos
Praça de Alvalade 12
1200 Lisboa

Thermal Comfort Consultant
Francisco Moita, Architect

Daylighting Consultant
Lícinio Carvalho, Engineer

Preliminary Architectural Studies
Lança Cordeiro, Architect
J Farelo Pinto, Architect

Monitoring, Artificial Lighting
A Gomes Martins, Professor
António Almeida, Engineer
Aníbal de Almeida, Professor
Department of Electrical Engineering
University of Coimbra

This set of Building 2000 brochures illustrates how architects and other building designers can successfully apply passive solar principles to produce energy-efficient buildings.

BUILDING 2000 Participants

Project Director
Theo C. Steemers

Coordinator
Cees den Ouden

Technical Steering Committee
Dean Hawkes
Nick Baker
Alex Lohr
Jean P. Lepoivre

Regional Liaison Agents
(D) Jörn Behnsen
(E) Vicente Sifre
(F) Michel Raoust
(GB) Alan Hildon
(GR) Matheus Santamouris
(P) Eduardo Maldonado

Further information or copies of the brochures can be obtained from prof. ir. Cees den Ouden, EGM Engineering BV, P.O. Box 1042, 3300 BA Dordrecht, The Netherlands.

BUILDING 2000

Commission of the European Communities

- Use of passive solar and other energy-saving features in refurbishment of system-built school.
- Results can be applied to similar buildings in schools network.
- Auxiliary heating savings of up to 18% achieved by correct choice of window glazing and window size.
- Artificial lighting energy reduced by 20-70% by installation of rooflights.
- Visual comfort improved by careful selection of colours and reflectances of surfaces.
- Summer-time peak temperatures reduced by selection of glazing layout.

DOWNSVIEW COUNTY PRIMARY SCHOOL (INFANTS)

SWANLEY/KENT/UNITED KINGDOM

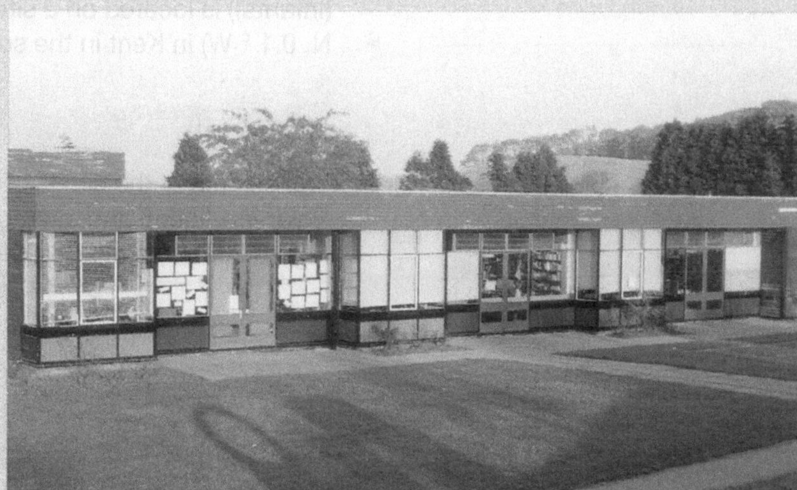

Building 2000 is a series of design studies illustrating passive solar architecture in buildings in the European Community.

ISSUE

S 6

DEC 1990

PROJECT DESCRIPTION

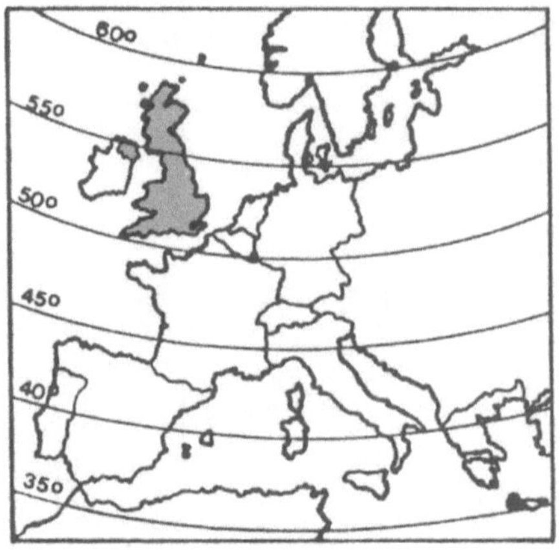

Figure 1. Location of Swanley

Building Type

From 1958 to 1973 there was an extensive school building programme in the UK which made use of various construction systems. In southern England alone, 4,400 schools were built using system methods with a design life of 60 years. The standards for these schools were set at a time when fuel prices were a third of their current values in real terms and environmental comfort expectations were low. The first schools built in the programme, therefore, had very high levels of glazing and poor insulation. They were cold in winter and uncomfortably hot in summer. Many are due for major refurbishment now, after 25-30 years, and the aim of this project was to explore the possibility of incorporating passive solar features at the time of refurbishment in order to improve thermal and visual comfort levels and cut fuel bills. One representative school was examined in detail. Experience from over 400 other schools was included in the study.

Location

The representative school (Downsview County Primary School (Infants)) is located on a site 24 m above sea level at Swanley (51.5 ° N, 0.1 ° W) in Kent in the south east of England.

Site Microclimate

The climate at the site is of the temperate maritime, lowland type. Solar irradiation and average external temperature data are given in Figure 2. There are 2080 degree days (base 15.6 °C) a year.

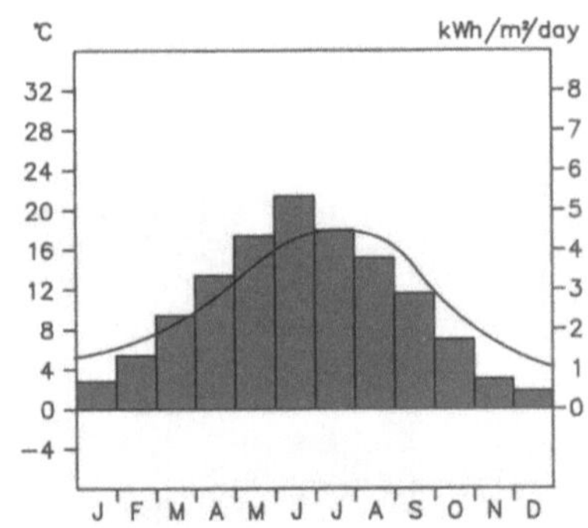

Key: solar irradiation on horizontal plane
average external temperature

Figure 2. Solar irradiation and external temperature data

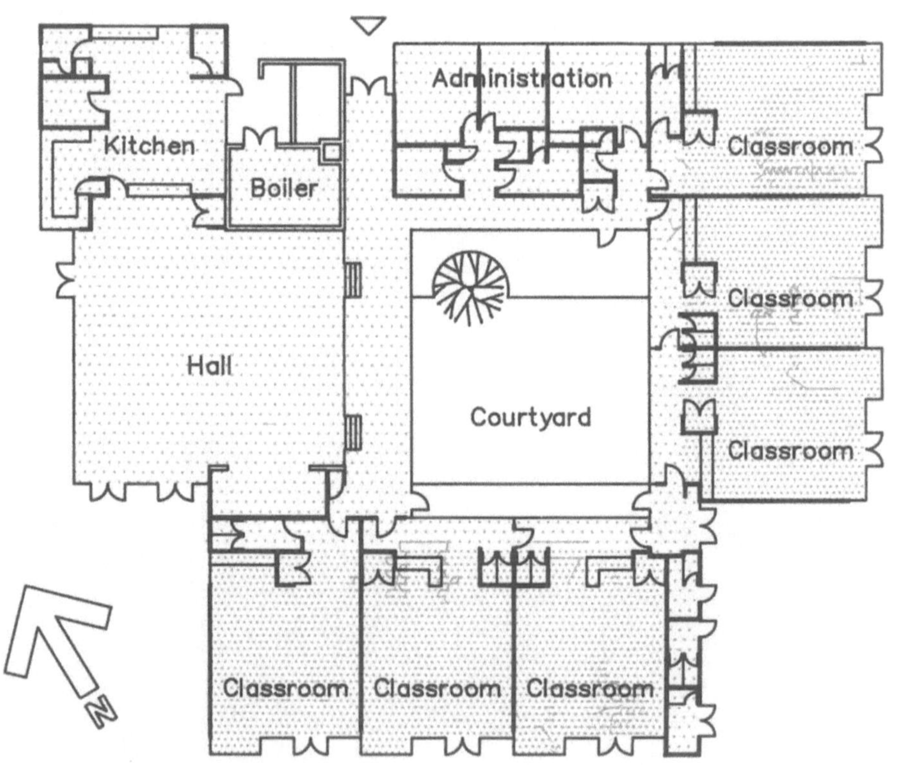

Figure 3. Plan of Swanley school

Design and Construction Details

The Swanley building is typical of those system-built schools in southern England where a structural steel frame supports the internal fabric and external curtain walls. Full details of these buildings are given in Table 1.

Premature failure has occurred in many of the buildings due to unsatisfactory detailing, poor materials and inadequate funds for maintenance. There is, therefore, a need to renew major components. Flat roofs, for instance, require recovering with high performance felts. Softwoods used in window walls have rotted, giving the opportunity for complete replacement of facades.

Building form	Mostly single-storey, 6% floor area two-storey
Width of window wall	7.0 metres (range 6.0-8.0)
Depth of classroom	9.0 metres (range 7.0-9.5)
Ceiling height	2.8 metres (range 2.5-3.2)
Volume	176 m^3
Partition walls	1000 kg/m^3 concrete block
Floor	200 mm concrete uninsulated
Floor covering	70% linoleum tiles/30% carpet
False ceiling	13 mm plasterboard with 200 mm gap to roof
Roof	Metal deck, fibreboard and roofing felt
Window wall	Single glazed, 70% fenestration
Spandrel panels	1.6 mm steel panels, height 0.8 metres, painted
Air infiltration	Crack length 28 m
Wall reflectance	0.4 (varies according to display material)
Ceiling reflectance	0.6 (matt white finish)
Floor reflectance	0.2 (dark grey tiles/dark brown carpet)
Occupancy pattern	0830-1600 hrs, excluding school holidays
Lighting energy	20 W/m^2, 9 to 15 kWh/m^2 year
Heating energy	160 kWh/m^2 year (boiler input)

Table 1. Details of system-built classrooms in southern England

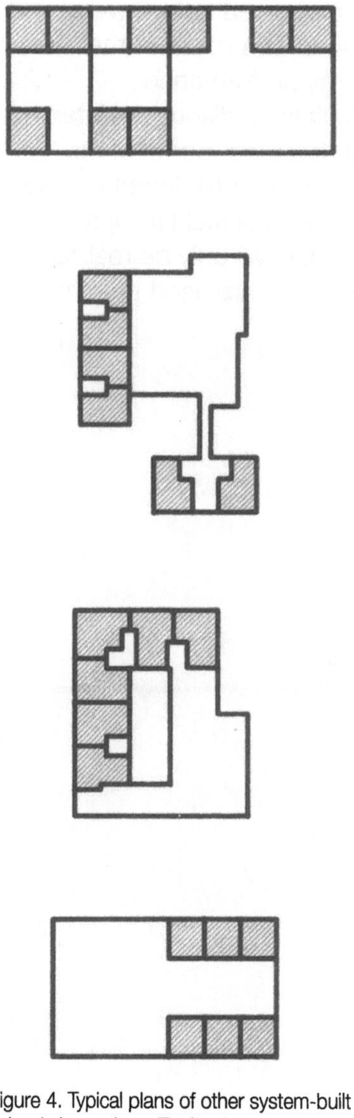

Figure 4. Typical plans of other system-built schools in southern England

The plan of the Swanley building is given in Figure 3. The shaded areas show the classrooms studied in depth. Typical plans for other system-built schools are shown in Figure 4.

DESCRIPTION OF PASSIVE SOLAR FEATURES/COMPONENTS

Introduction

A number of options are available for introducing passive solar techniques at the time of renewing major components in order to reduce energy costs and improve comfort. These options, which are described below, were defined as a result of the energy performance calculations discussed later in this brochure. Some of the options are interdependent so that their full fuel savings potential will only be realized when they are used together.

Windows

The sizing and position of windows needs to be optimized with respect to the provision of natural light, solar gains and heat losses.

Solar Gains and Heat Losses

The modelling studies described below showed that in the Swanley classroom with its existing level of glazing, the solar gains through the windows amount to 27% of the total heat requirements during the heating season but are balanced out by heat losses through the windows. Substitution of part of the glazing by opaque insulation will reduce both solar gains and heat losses.

Changing the existing single glazing to double glazing will halve heat losses and slightly reduce solar gains so that, overall, 12% of the annual heating fuel is saved.
Further improvements can be made by using low-emissivity glass, with its 30% lower heat losses. With such glazing, discomfort from cold-spill from large areas of glazing is only noticeable for 10 days a year, compared with 45 days for ordinary double glazing.

Any work to reduce glazing area should maintain and, if possible, increase the window head heights to ensure daylight penetration to the back of the room. Glazing at a height below 0.7 m does not provide much light and can usefully be replaced by materials of better thermal performance.

Natural Lighting

Many of the system-built schools in the south of England were constructed to a specification which called for a minimum 2% daylight factor, i.e. an illumination at desk height of at least 2% the external daylight. This was often achieved by using large areas of glazing in the window wall and small high level windows at the rear of the classroom. This led to wide differences in illumination level across the room. In practice, only 40% to 90% of floor areas were found to have daylight factors of 2% or above. The situation was made worse by the low reflectance display materials often put on the walls which covered up the light-coloured surfaces specified by the building designer. Because the human eye is far more tolerant of low lighting levels when the illumination is relatively uniform, a variation of more than 3:1 in average illumination levels between the front and rear of the room should be avoided. The desire to switch on artificial lighting depends on the light level in the darkest part of the room. Experience suggests that when the decision to switch lights on has been made all light switches are used, rather than only those in the area required, and there is then a great reluctance to turn lights off when daylight levels improve. Daylighting design must, therefore, aim to provide adequate and uniform lighting levels.

Rooflights

Heat Losses

Heat losses through rooflights become significant when they constitute about 30% of the total area of glazing. At night, under clear sky conditions, the heat losses from horizontal rooflights rise by about a half because of the increased radiative losses. To counter this, any low-cost option to install higher-performance glazing should be taken. This will have the added advantage of reducing condensation on the glass.

Natural Lighting

The energy performance study showed that introduction of four 1 m^2 rooflights (see below) would lead to a 70% reduction of lighting electricity use while having minimal effect on auxiliary heating requirements.

Infiltration

The design infiltration rate for the Swanley school was 8 litres a second for each person during occupied hours. This approximates to 5 air changes per hour (AC/h). Measurements showed that, in practice, 2 AC/h was typical and it is this figure which was used in the performance calculations.

Unwanted air leakage during unoccupied hours during the heating season can be reduced by draught-stripping windows and external doors. Because louver windows are difficult to make air-tight any such installations should, where possible, be replaced by fully draught-stripped opening windows or, where appropriate, by fixed glazing. Draught-stripping can reduce air infiltration from 2 AC/h to 0.5 AC/h, thus helping to retain heat overnight and reduce preheat requirements next day.

In summer, additional ventilation to provide 2-3 AC/h during unoccupied hours can cool the fabric and room contents overnight and thus improve the internal temperature conditions next day. This can, however, be difficult to provide without causing security problems.

A total of one square metre of ventilation aperture from opening rooflights will decrease summer-time temperatures for the 30% or so summer days when there is not enough wind to provide adequate cross ventilation.

Fans for ventilating internal WCs can be run 24 hours a day in summer which should provide 2AC/h in the associated space. The increased fan run time can be linked to the heating system control system to provide summer-time ventilation only.

External Paving

The use of light-coloured paved surfaces outside classrooms with a reflectance of between 0.4 and 0.6 can significantly improve the amount of light reaching the back of classrooms, especially if there are no rooflights or windows at the back of the room. (Higher reflectances can cause glare on sunny days.) Where possible, the paving should extend to at least the height of the window head.

Internal Surfaces

The perceived need to use artificial lighting depends on light quantity and quality. These are determined by the area and positioning of windows and rooflights and by the reflection of light from internal surfaces. Floor surfaces (especially those near windows) should have reflectances of 0.3- 0.4; this presents a compromise between susceptibility to showing scuff marks and adequate light reflectance. Walls benefit from light colours and teachers should be advised to refrain from using dark-coloured papers and pin-up material. Light-coloured inward projecting window sills can provide an area of brightness intermediate between high-intensity daylight and lower interior lighting levels. Window reveals should also be coloured to provide a similar intermediate brightness area. The area below the sill should also be kept light so that the eye can easily perceive detail in this area. If dark colours are used, glare from the window encourages occupants to use artificial light. Light-coloured glazing bars should always be used to avoid the glare which results when dark colours are seen against the sky.

Building Fabric

The thermal performance calculations showed that in the original Swanley classroom 15% of the heat losses during the heating season were lost through the ground floor and 33% through the walls and roof. Thus even small reductions in heat losses through the building fabric can have a significant effect on auxiliary heating bills. Improvements in thermal insulation not only saves energy but also increases thermal comfort by raising the surface temperature of the walls, etc.

Ground Floor Slab

Opportunities to improve the U-values of the ground floor itself are rarely available after the original construction stage. Addition of carpets, however, will cut heat losses through the ground floor slab from 15% to 9% of the total. The reduction in heat losses will be slightly greater with a radiator or bottom-exit fan convector auxiliary heating system.

Walls

The U-values of the original classroom walls are in the range 1.0-1.6 $W/m^2 K$. Total heat losses through the walls are 15%-25% of the total, depending on layout. Filling the cavity can improve U-values to 0.6 $W/m^2 K$ or below and should be carried out at the same time as improvements to windows or roofs, if it has not already been done.

Roof

School buildings erected before 1965 usually had flat roofs with U-values in the range 1.5-2.0 $W/m^2 K$. Refurbishment to 1989 (0.6 $W/m^2 K$) or 1990 (0.45 $W/m^2 K$) standards provides adequate insulation and helps to reduce peak summer-time temperatures.

ENERGY PERFORMANCE CALCULATIONS AND DESIGN TOOLS USED

Introduction

The above options were established with the help of a number of energy performance studies of a typical classroom carried out using the SERI-RES computer model. Some input parameters for the original classroom are given in Table 2. Its thermal performance when oriented south is given in Table 3.

area	68 m^2
U-values	
walls	1.0 W/m^2 K
floor	1.1 W/m^2 K
glazing	5.4 W/m^2 K
infiltration rate in heating season	
school hours	2 AC/h
night-time	1 AC/h

Table 2. Some input parameters for the original classroom

	kWh	kWh/m^2	% of total
Auxiliary heating (pa)	6030	88.8	
Lighting demand (pa)	970	14.3	
Gains (Oct to Apr)			
Auxiliary heating	5910	87.1	52
Lighting demand	820	12.1	7
Solar gains	2980	43.9	27
Casual gains	1600	23.5	14
Total	11310	166.6	100
Losses (Oct to Apr)			
Windows	3080	45.4	27
Walls and roof	3650	53.7	33
Ground	1700	25.1	15
Air infiltration	2510	36.9	22
Interzone	370	5.5	3
Total	11310	166.6	100

Table 3. Thermal performance of original south-facing 68 m^2 classroom

Effect of Changing Orientation

The effect on thermal performance of changing the orientation of the room from south to north, east and west is shown in Table 4. There is an 11% increase in auxiliary heating requirements for north- and east-facing orientations. Heat losses through the window and ground floor are higher for north, east and west orientations.

Classroom window orientation	Auxiliary heating	Lighting energy	Solar gains	Window losses	Ground losses
South-facing	base	base	base	base	base
North-facing	+11.4	+1.2	-62.0	-4.4	-11.0
East-facing	+11.2	+0.3	-46.1	-3.2	-7.7
West-facing	+6.1	+1.0	-36.9	-2.4	-6.1

Table 4. Effect of orientation on thermal performance of original classroom

Effect of Changing Insulation and Infiltration

The effects on auxiliary heating requirements of the south-facing room of making some realistic changes to insulation and infiltration levels are given in Table 5. These include increasing the insulation of the external wall from 1.0 W/m^2 K to 0.53 W/m^2 K or 0.37 W/m^2 K, changing the single- glazing (U-value 5.4 W/m^2 K) to double-glazing (U-value 2.9 W/m^2 K) and reducing night-time infiltration from 1 to 0.5 AC/h while leaving the air change rate during occupied hours at 2 AH/h. It was not felt to be realistic to upgrade the insulation levels in the roof and floors.

	Auxiliary heating kWh pa/m^2	Saving kWh pa/m^2
Base case (original south-facing classroom)	88.8	
Wall U-value 0.53 W/m^2 K	83.8	5.0
Wall U-value 0.37 W/m^2 K	82.3	6.5
Double glazing	74.3	14.5
Night infiltration of 0.5 AC/h	81.8	7.0
All of the above with wall U-value 0.37 W/m^2 K	63.7	25.1

Table 5. Effect on auxiliary heating of south-facing classroom of upgrading insulation and infiltration rates

	Minimum daylight factor (%)	Auxiliar heating kWh pa/m^2	Change kWh pa/m^2	Lighting demand kWh pa/m^2	Change kWh pa/m^2
Base case (sill height = 0.7m)	0.91	88.8	-	14.34	-
20% core	1.35	88.3	-0.5	16.1	+1.7
2-step switching	0.91	85.4	-3.4	20.8	+6.5
HF with dimming	0.91	91.5	+2.7	11.6	-2.8

Table 6. Effect of artificial lighting control systems on lighting and heating energy use

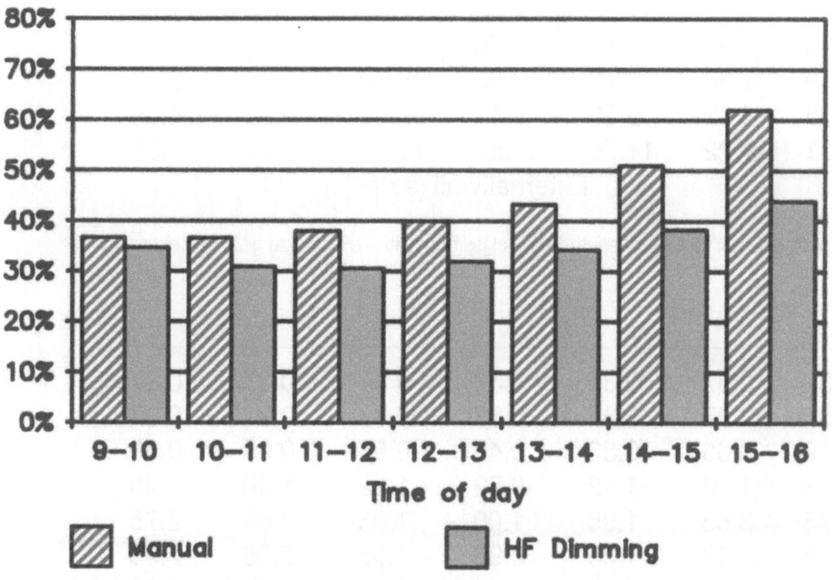

Figure 5. Hour-by-hour artificial lighting use as fraction of maximum lighting demand with manual control system and with dimming control system with HF luminaires

Effect of Changing Percent Glazing and Introducing Rooflights

The effect on auxiliary heating and lighting electricity requirements of changing the percent of glazing in the south-facing external wall from 70% in stages down to 20% is shown in Figure 7 along with the effect of introducing (1) two 1 m^2 rooflights at the rear of the classroom and (2) four 1 m^2 rooflights as shown in Figure 6, i.e. two at the rear and two in the middle of the room. It was found that a 70% reduction in lighting electricity use is possible with the better daylighting provided by the four rooflights. Reducing the percent glazing from 70% to 20% reduces summer-time overheating and increases auxiliary heating requirements by 10% because of the lower solar gains.

Artificial Lighting Control Systems

Three cases were examined in order to assess the effect of artificial lighting control systems on lighting and auxiliary heating energy use. In the first, a 20% core of the room was lit continuously and lighting of the perimeter controlled manually. In the second case, there was a simple two-step photocell switching system. In the final case, there was a dimming control system with high frequency (HF) fluorescent luminaires which had an input of 13 W/m^2 (compared with 20 W/m^2 for ordinary fluorescent lights). The results are given in Table 6. Compared with the original manual operation, the photocell switching system is not expected to give savings in lighting electricity use. This is because, in a manual system in schools, lights are not usually switched on until internal light levels fall below 50-100 lux whereas a photocell system would be calibrated to maintain the design illumination level of 300- 500 lux.

The variation in artificial lighting use hour-by-hour for the classroom with manual lighting control and with HF fluorescents is shown in Figure 5. For the manual case, the average demand rises through the day because the model predicts that, if lights are switched on they tend to stay on for the rest of the occupied period. With the dimming system, there is less variation and there is a minimum around midday because of the high level of natural lighting at that time.

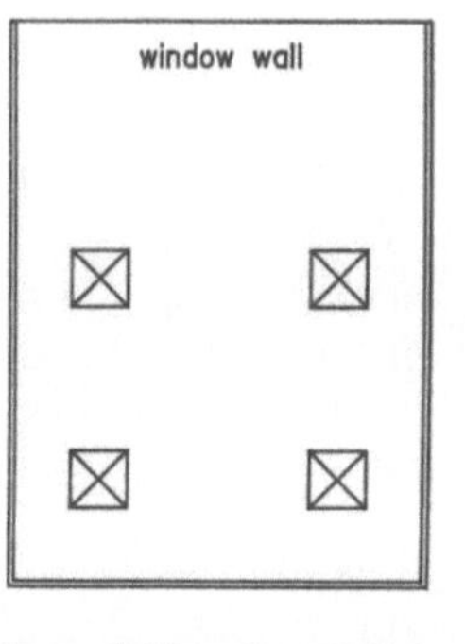

Figure 6. Plan of south-facing classroom showing position of rooflights

Daylight Factors

Tables 7-9 give the daylight factors resulting separately from (a) 70% double glazing on the window wall, (b) two 1 m rooflights in the middle of the classroom and (c) two 1 m rooflights at the back of the classroom. The daylight factors for a different percent of window glazing or different rooflight sizes can be calculated from the tables by scaling the different glazing sizes.

0.64	0.64	0.65	0.65	0.65	0.64	0.63
0.68	0.69	0.71	0.71	0.71	0.69	0.68
0.75	0.77	0.80	0.80	0.80	0.77	0.75
0.86	0.91	0.95	0.95	0.95	0.91	0.86
1.06	1.15	1.21	1.23	1.21	1.15	1.06
1.43	1.61	1.72	1.76	1.72	1.61	1.43
2.19	2.58	2.79	2.86	2.79	2.58	2.19
4.05	4.92	5.30	5.41	5.30	4.92	4.05
9.31	11.22	11.75	11.81	11.75	11.22	9,31

External wall (south)

Table 7. Daylight factors for south-facing classroom with 70% double glazing in external wall

0.37	0.37	0.37	0.37	0.37	0.37	0.37
0.37	0.37	0.37	0.37	0.37	0.37	0.37
0.59	0.65	0.53	0.46	0.53	0.65	0.59
1.43	1.80	1.12	0.72	1.12	1.80	1.43
2.75	3.65	1.96	1.00	1.95	3.65	2.75
2.25	2.96	1.65	0.90	1.65	2.96	2.25
0.91	1.07	0.76	0.57	0.76	1.07	0.91
0.45	0.46	0.43	0.41	0.43	0.46	0.45
0.37	0.37	0.37	0.37	0.37	0.37	0.37

External wall (south)

Table 8. Daylight factors for south-facing classroom with two 1 m^2 rooflights in middle of classroom

2.51	3.32	1.81	0.96	1.81	3.32	2.49
2.70	3.55	1.92	0.99	1.92	3.54	2.70
1.23	1.52	0.98	0.67	0.98	1.52	1.23
0.53	0.57	0.49	0.44	0.49	0.57	0.53
0.37	0.37	0.37	0.37	0.37	0.37	0.37
0.37	0.37	0.37	0.37	0.37	0.37	0.37
0.37	0.37	0.37	0.37	0.37	0.37	0.37
0.37	0.37	0.37	0.37	0.37	0.37	0.37

External wall (south)

Table 9. Daylight factors for south-facing classroom with two 1 m^2 rooflights at rear of classroom

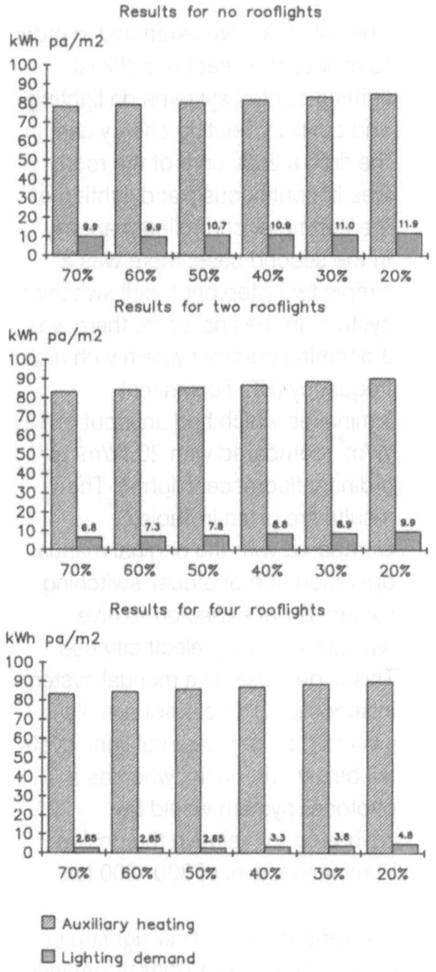

Figure 7. Effect on auxiliary heating and lighting requirements of changing percent glazing in external wall and introducing rooflights

Building	Unimproved	Fully improved				
Boiler power (kW) (per classroom)	9.0	4.8	5.5	6.3	7.5	9.0
Boiler hours run						
96-100% load	299	318	195	117	58	15
91-95	71	67	49	38	18	7
86-90	19	14	10	6	3	2
81-85	3	6	1	1	0	0
76-80	7	5	7	5	1	2
71-75	5	8	4	5	3	4
66-70	4	8	8	8	8	4
61-65	9	8	8	7	6	4
56-60	8	1	0	0	0	0
51-55	55	15	8	8	10	7
46-50	71	10	11	11	14	17
41-45	107	12	16	13	33	25
36-40	118	15	12	23	42	42
31-35	98	0	4	40	37	34
26-30	96	8	55	64	92	71
21-25	106	22	67	97	97	91
16-20	98	50	93	91	92	96
11-15	64	94	107	87	94	99
6-10	43	107	96	96	85	89
0-5	42	110	98	100	103	105
Hours run/year	1323	878	849	817	796	714
Total kWh	5919	2091	1870	1744	1712	1515
Boiler standing losses %	7	3	3	3	3	7
Standing loss kWh pa	1111	167	187	204	239	600
Energy input into each classroom:						
Preheat kWh	3449	1724	1402	1144	868	549
Occupied kWh	2306	213	402	581	844	966
Frost kWh	156	154	66	19	0	0
Solar kWh	1272	938	938	938	938	938
Casual	2464	2412	2412	2412	2412	2412
Boiler stdg. loss %	8.4	8.0	9.8	11.3	13.1	15.2

Table 10. Boiler utilization tables

Boiler Controls

The total boiler power required in a refurbished school compared to the original building calculated using the Excalibur computer program is shown in Table 10. Interpolation will show the contribution required from individual boilers in multi-boiler installations. When condensing boilers are used, the hours run at low output can be used.

After a school has been refurbished it will have a lower demand for auxiliary heating than the original design. The existing boiler therefore can be considerably oversized. If no changes are made to the boilers and boiler controls, standing losses from over-sized boilers can increase annual consumption by 7% although the shorter preheat times will reduce annual fuel consumption. With higher boiler margins, the building fabric will continue to absorb heat during occupied hours with increased kWh consumed during this time. To minimize standing losses, boiler controls should be installed to ensure that only those boilers required to satisfy the load are hot. Heating boilers need only be hot 600-850 hours per heating season in an improved building, compared with 1100-1400 hours before refurbishment.

Heating Controls

As the total energy requirements decrease due to lower fabric and ventilation losses, heat gains from lighting, occupants, and solar heat reduce heat required from boilers. The heating controls must sense these heat gains and reduce heating power. As far as possible, heating controls should include thermostats, sensors or thermostatic valves in each room. If existing fan convectors are used, the pipe and convection losses will still supply heat to the room at approximately 20% of full output. In an improved classroom, this will cause the room to overheat for 45 to 70 days per heating season and increase fuel consumption by 14 to 23%.

GENERAL DESIGN GUIDELINES/POINTS OF INTEREST RESULTING FROM THIS PROJECT

Conclusions

The following conclusions combine the study results with good energy efficiency practice. Each possible action should be checked for conformity with applicable regulations and the energy savings results related to the extent of existing improvements. Heating energy use in a fully improved school can be reduced by 30 to 50% while lighting energy use can be reduced by 70%.

Window Wall Replacement

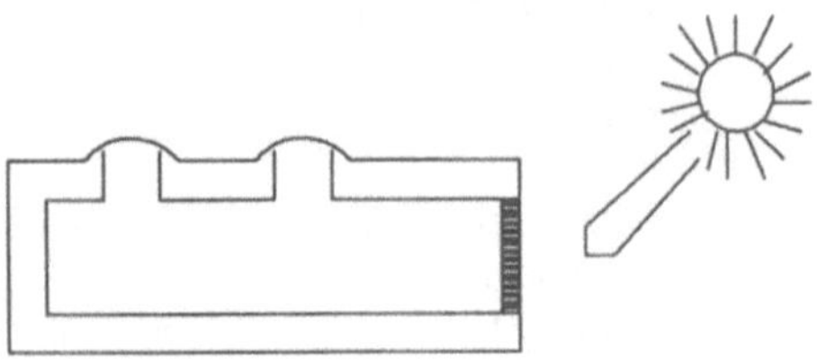

If no rooflights and little or no rear lights:

Use double glazing;
Put window head as high as possible;
Set sill height to desk height - approx. 0.7 m;
Insulate panels below sills;
Use light colours to decorate below window sill;
Keep glazing in range of 40% to 70%.
External light shelves can increase light at back of the room and provide reduction in summer overheating. Some are not robust.

If rooflights are installed - as above except:

Keep glazing in range of 25 to 40%.

Rooflights

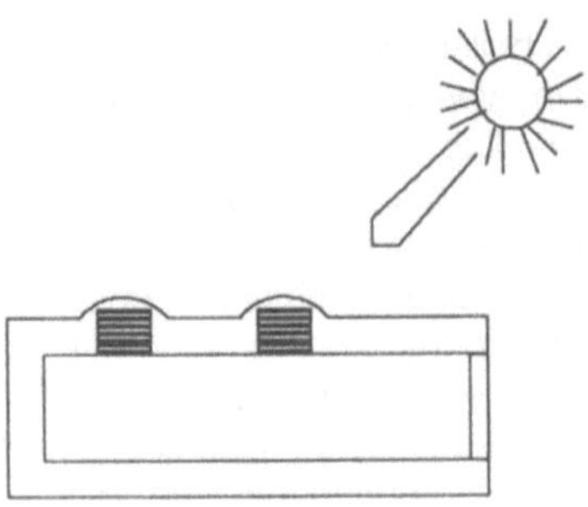

4 rooflights lead to 70% reduction in lighting energy use;
2 middle of room rooflights with existing rear lights are equivalent to 4 rooflights for lighting energy savings;
2 rear of room rooflights alone do not significantly reduce lightingenergy;
If costs allow, use low emissivity or triple glazing to reduce the effectof cold sky radiation and reduce condensation on the glass;
Keep distance between rooflight glazing and ceiling to a practical minimum.

Roof

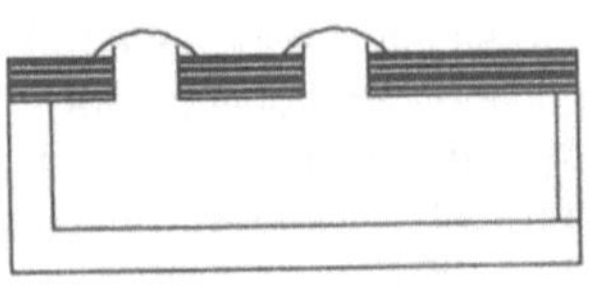

Insulation to a U-value better than 0.6 has little effect on annual energy use, though local regulation may require higher standards;
Install 4 rooflights if no lighting to rear of room, or two if there are windows at the rear of the room.

Walls

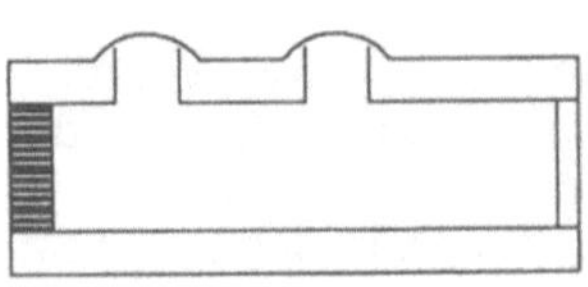

Insulation to a U-value better than 0.6 has little effect on annual energy use, though local regulations may require higher standards.

Air Infiltration

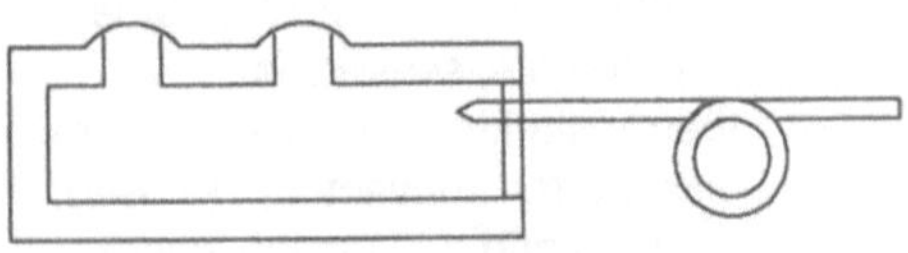

Draught-stripping should be fitted to all windows and all external doors;
Louver windows are very difficult to seal effectively and should be replaced with fixed or opening windows;
Draught lobbies are not effective unless at least one of the doors stays closed at all times. To be effective the minimum separation between doors must be less than 4 m;
Self closing shutters should be fitted to all extract fans.

Atria

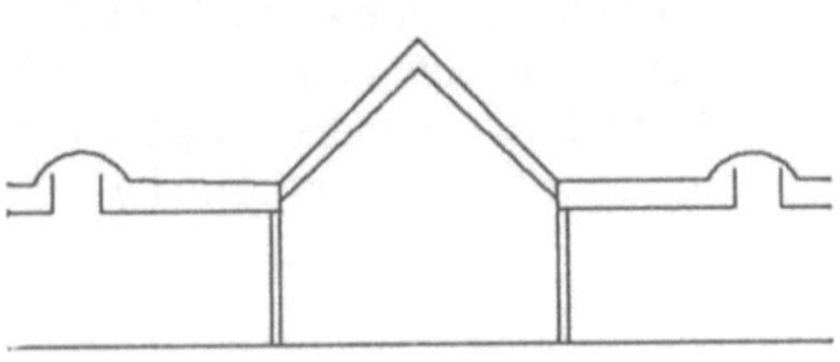

Initial studies indicated that the retrofit of an atrium to a courtyard was not viable in energy terms, and was not pursued.

Artificial Lighting

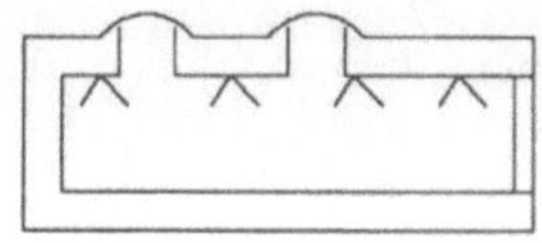

Replacement of original fluorescent luminaires by modern high efficiency units can save 40% of lighting energy use;
Rewiring of lighting installations should enable switching of lamps individually or in rows parallel to the window;
If there is limited daylighting in the room, lights will tend to be on all day and minimal savings will be achieved by lighting controls.

Decorations

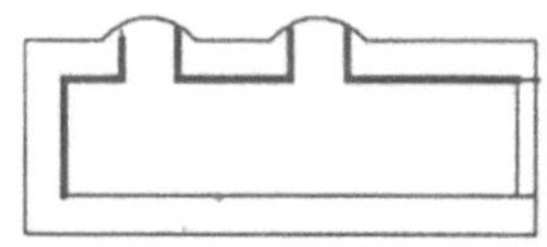

Use light colours for walls and ceilings;
Keep areas below window sill in light colours;
Keep window reveals and sills in light colours, not in white.

Use lighter coloured tiles and carpets close to windows.

Floors

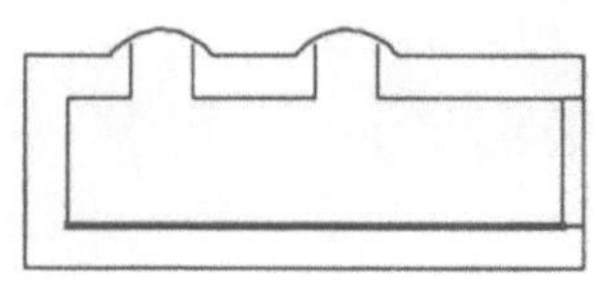

Use light coloured paving outside windows;
Do not allow growing trees to obstruct daylight.

External

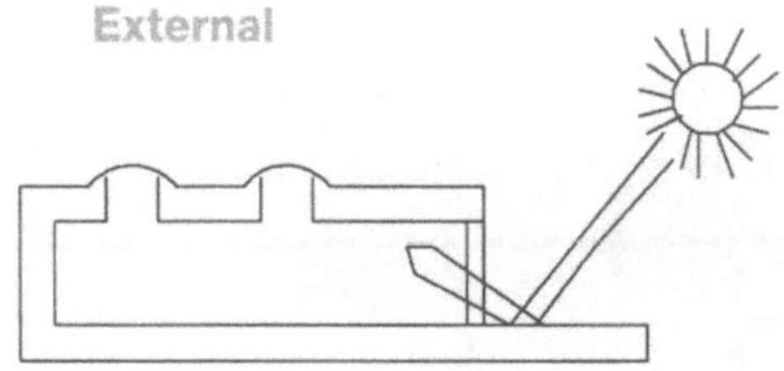

Boiler standing losses must be controlled by ensuring that boilers are only hot whilst heat is required and cooling or cold at all other times;
Boiler standing losses can be 30% of annual heating fuel use;
Temperature controls should be installed in every room to ensure that benefit is obtained from internal and solar gains.
Standing losses from fan convectors and pipework must be controlled. Losses may increase annual consumption by 14% to 23% in improved rooms.

Heating Controls

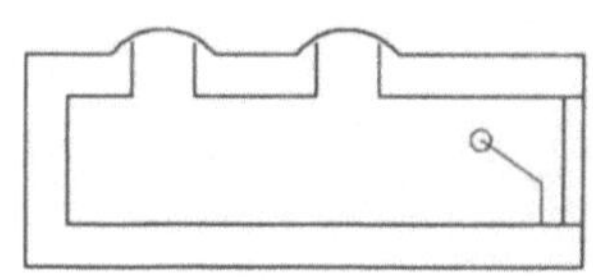

The study results described in this brochure indicate that significant savings relating to heating and lighting energy can be achieved at zero or minimal additional capital cost whenever refurbishment works are undertaken. To ensure that these opportunities are properly considered, building owners and operators are advised to include the following form of words in their brief to designers and maintenance organizations:

Passive Solar Design

1) Design optimization to maximize benefits from passive solar energy shall be undertaken for these works. The client is to be advised of implications on future works and, subject to approval being given, due provision made where necessary.

2) The designer is to provide summaries of calculations on heating fuel use, level and quality of daylighting, use of energy for artificial lighting and summer-time environmental conditions, to show that full consideration has been given to utilization of passive solar energy.

BUILDING 2000 Building 2000 brochures are published by Directorate General XII of the Commission of the European Communities to show how design studies can help architects and other building designers use passive solar principles to the best effect to produce attractive energy-efficient buildings. Each brochure describes studies carried out with the support of the Commission during the design phase of one of thirty-six non-domestic buildings in the EC Member States. The studies were on such topics as daylighting, heating, cooling, ventilation, comfort, control systems and urban design. They were carried out with the help of acknowledged European experts in these fields and drew heavily on lessons learned and techniques developed through the Commision's research and development programme on solar energy applications to buildings.

Commission of the European Communities/Directorate-General for Science, Research and Development

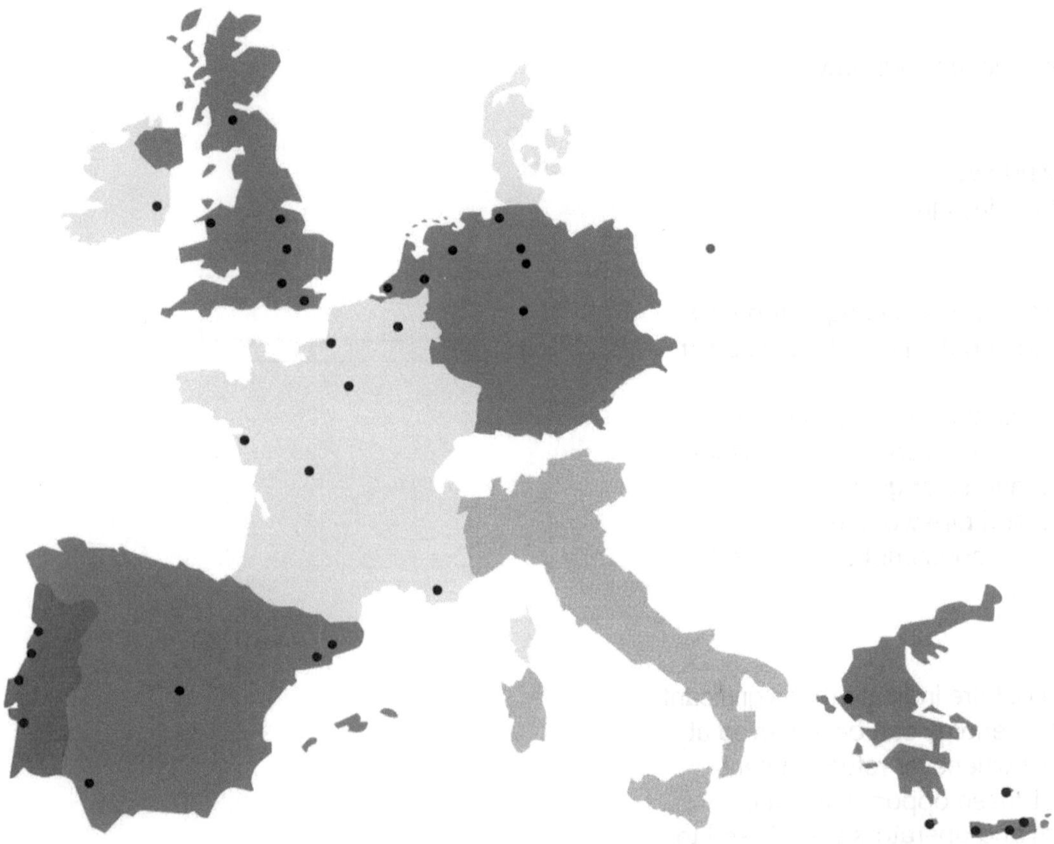

List of Design Team Participants anc Advisers

Client
Kent County Council
Maidstone
Kent
United Kingdom

Energy Performance Analysis
SERI-RES analysis of heating and lighting use:
David Bartholomew, David Newton ar
David Radcliffe
SEMA Group
London
United Kingdom

Excalibur analysis of heating energy use and boiler loading:
Colin Ashford
Kent County Council

Architectural Advice
Peter Sharman
Building Design
Kent County Council

Lighting Advice
Joe Lynes
Hull College of Architecture

Project Technical Control
Colin Ashford
Building Design
Kent County Council

This set of Building 2000 brochures illustrates how architects and other building designers can successfully apply passive solar principles to produce energy-efficient buildings.

BUILDING 2000 Participants

Project Director
Theo C. Steemers

Coordinator
Cees den Ouden

Technical Steering Committee
Dean Hawkes
Nick Baker
Alex Lohr
Jean P. Lepoivre

Regional Liaison Agents
(D) Jörn Behnsen
(E) Vicente Sifre
(F) Michel Raoust
(GB) Alan Hildon
(GR) Matheus Santamouris
(P) Eduardo Maldonado

Further information or copies of the brochures can be obtained from prof. ir. Cees den Ouden, EGM Engineering BV, P.O. Box 1042, 3300 BA Dordrecht, The Netherlands.

Commission of the European Communities

BUILDING 2000

LABORATORIES
AND
UNIVERSITIES

L1
High Tech, Laboratory, Leuven, Belgium
L2
Business Innovation Centre, Porto, Portugal
L3
Architecture Faculty Building, Lisbon, Portugal
L4
School of Medicine, Ioannina, Greece
L5
Geosciences Research Laboratory, Ampus, France

Building 2000 is a series of design studies illustrating passive solar architecture in buildings in the European Community.

VOLUME 1 / **KLUWER ACADEMIC PUBLISHERS**

BUILDING 2000

Commission of the European Communities

- Multi-use laboratory suitable for high tech projects
- Low energy design with passive solar features to appeal to young innovative firms
- Solar walls, internal atria and conservatory provide buffer spaces which in winter preheat incoming air for mechanical ventilation and reduce heat losses. They also allow good daylight penetration.
- Thermal mass of building and convection ventilation in buffer spaces provide free cooling in summer

HIGH TECH LABORATORY

LEUVEN/BELGIUM

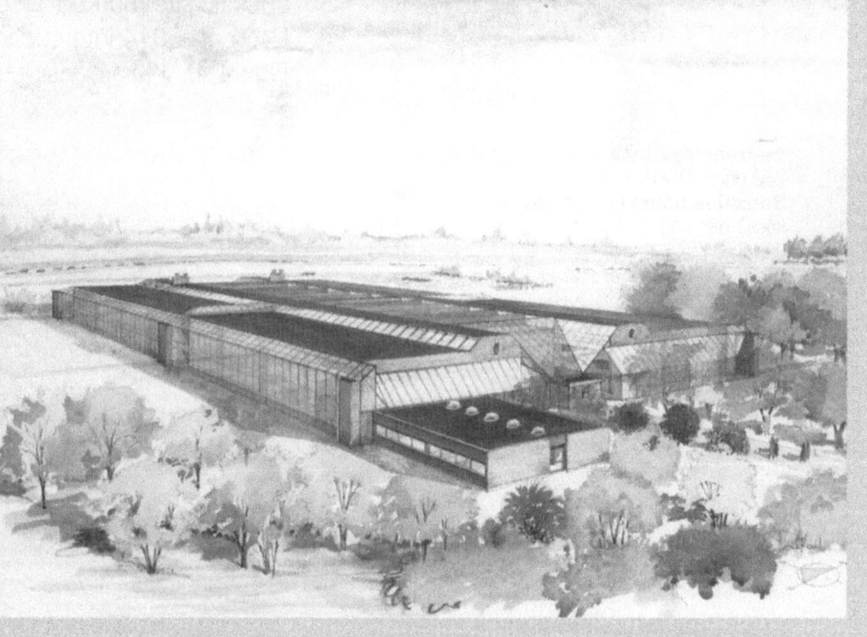

Building 2000 is a series of design studies illustrating passive solar architecture in buildings in the European Community.

ISSUE
L 1
FEB 1991

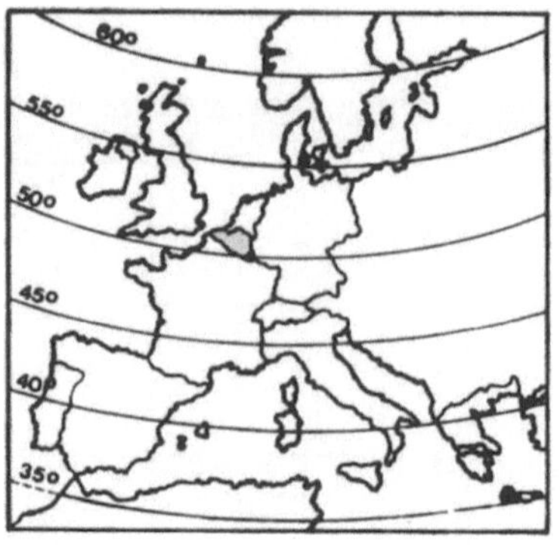

Figure 1. Location

Building Type

This is a building suitable for the development of research-based or innovative prototype projects in high tech fields such as microelectronics, robotics or biotechnology. The aim was to create a flexible laboratory building which could be used by a number of small or medium-sized companies of the type set up jointly by industry and the universities.

Location

The building is for the Catholic University of Leuven which is located in the middle of Belgium, about 20 km east of Brussels (Figure 1). The specific site marked for the building is between the approach to the E40 and E314 intersection and the university science campus (see Figure 4). It is a considerable distance from other buildings.

Degree days (base 15° C) over heating period of 269 days 2080	
Sunshine hours heating season	
(Oct-Apr)	617
Jan-Dec	1555
Mean daily temperatures	
Jul max.	21.6° C
Jan min.	-0.3° C

Table 1. Some key climate data

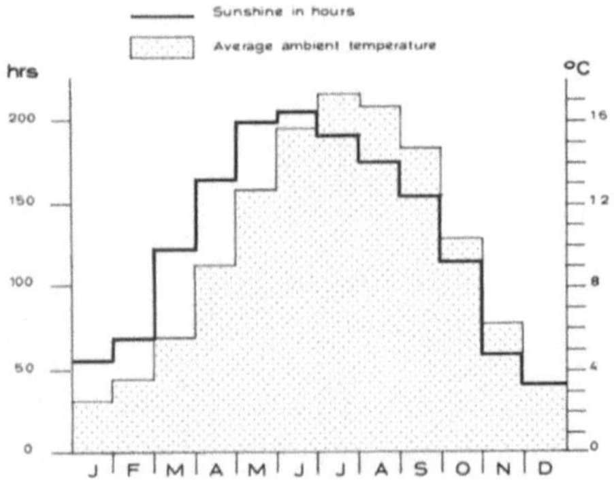

Figure 2. Sunshine hours and average ambient temperatures

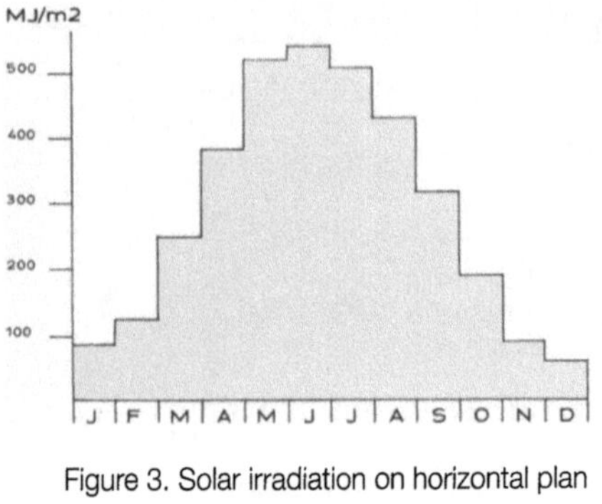

Figure 3. Solar irradiation on horizontal plan

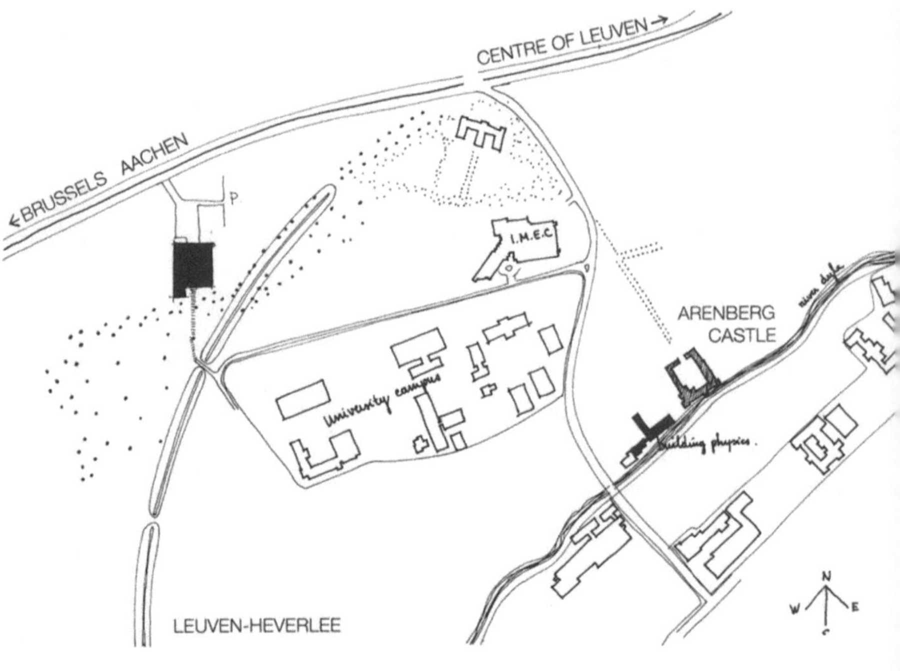

Figure 4. Site plan

Site Microclimate

The site is flat and fairly open, apart from some tall trees to the south which provide shelter against the prevailing wind and summer overheating. Some climate data are given in Table 1 and Figures 2 and 3.

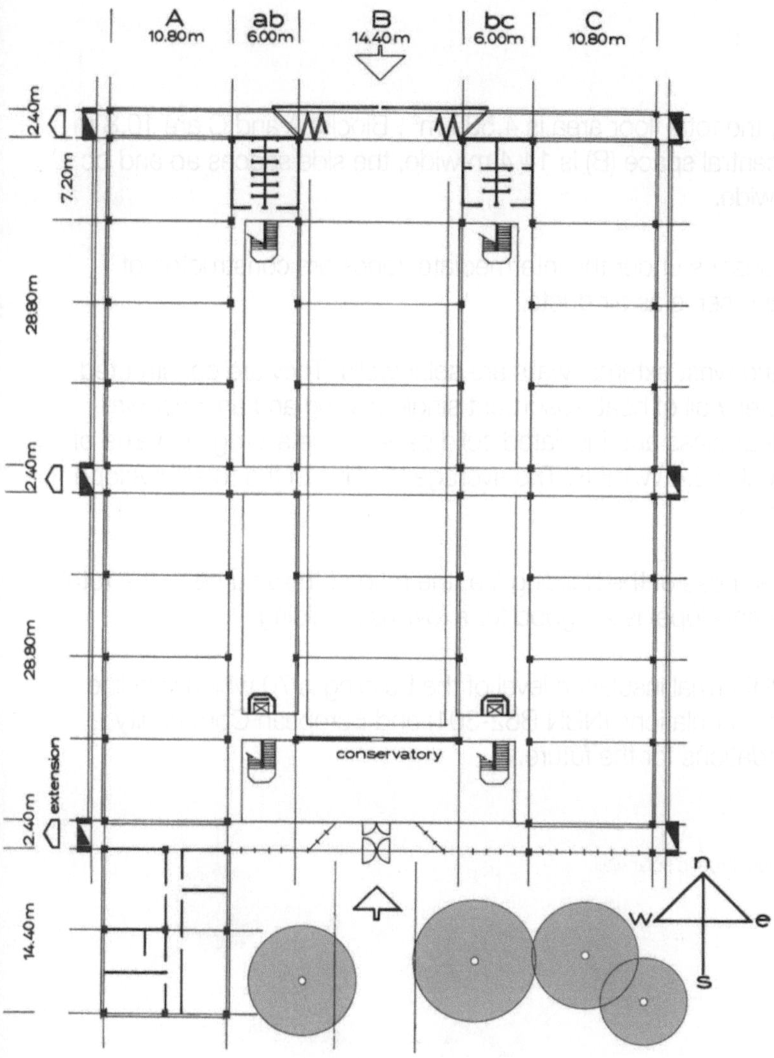

Figure 5. Plan of Leuven building

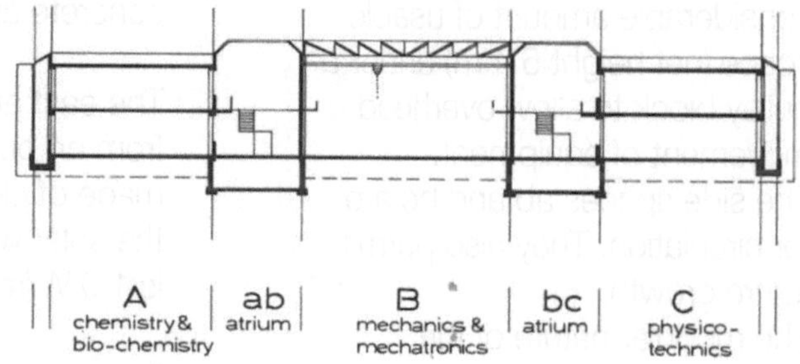

Figure 6. Section of Leuven building

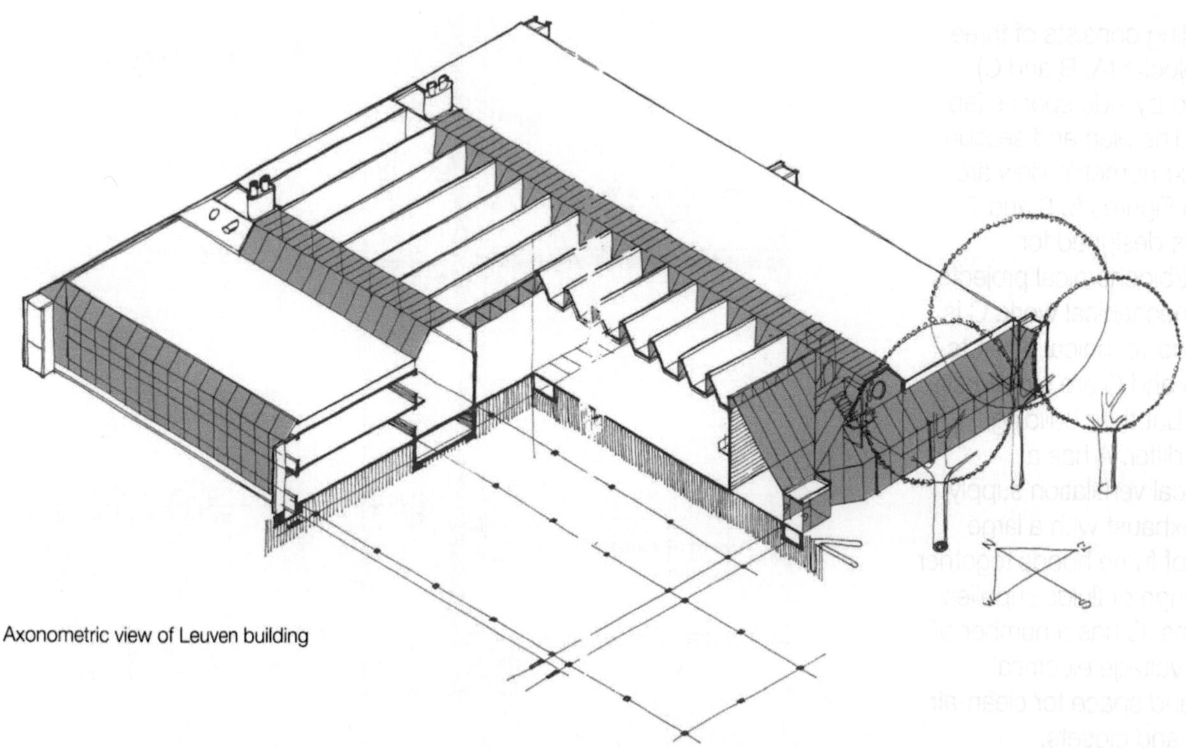

Figure 7. Axonometric view of Leuven building

Design and Construction Details

The central block (B) has a large rear entrance and a considerable amount of usable space (net height 5.4 m) under a pulley block to allow overhead movement of equipment.
The side spaces ab and bc are for circulation. They also permit future growth.
The modular nature of the building makes for easy construction and flexibility in use. Three large or eight small projects can be carried out simultaneously.

There is office space at the front; washrooms/cloakrooms are at the rear. These are topped by technical areas.
The building has a reinforced concrete frame with tinted glass cladding.

In front and behind the four standard bays are 2.4 m zones for circulation and service runs - ducts, pipes and cable trays.

The building consists of three parallel blocks (A, B and C) separated by side spaces (ab and bc). The plan and section and an axonometric view are shown in Figures 5, 6 and 7.
Block A is designed for chemical/biochemical projects. B is for mechanical work. C is for physico-technical projects. Blocks A and C are structurally identical but their building services differ. A has a mechanical ventilation supply and an exhaust with a large number of fume hoods together with a range of fluids supplies and drains. C has a number of different voltage electrical circuits and space for clean-air cubicles and closets.

At Leuven, the total floor area is 4,500 m^2 . Blocks A and C are 10.8 m wide; the central space (B) is 14.4 m wide; the side spaces ab and bc are 6.0 m wide.

The crawl spaces under the intermediate zones are constructed of concrete and serve as air ducts.

The east and west external walls are solar walls. They are constructed from an outer wall of heat-absorbent single glazing and an inner wall made of clear glass and insulated solid panels. The average U-value of the solar wall is 2.4 W/m^2 K. The average U-value of the total envelope is 1.0 W/m^2 K.

The compactness of the building (i.e. the ratio of the volume to the total area of the envelope) is 2 - good for a low-rise building.

The overall thermal insulation level of the building is 70 which is in line with Belgian regulations (NBN B62-301) and European Community recommendations for the future.

Figure 8. Section through solar wall

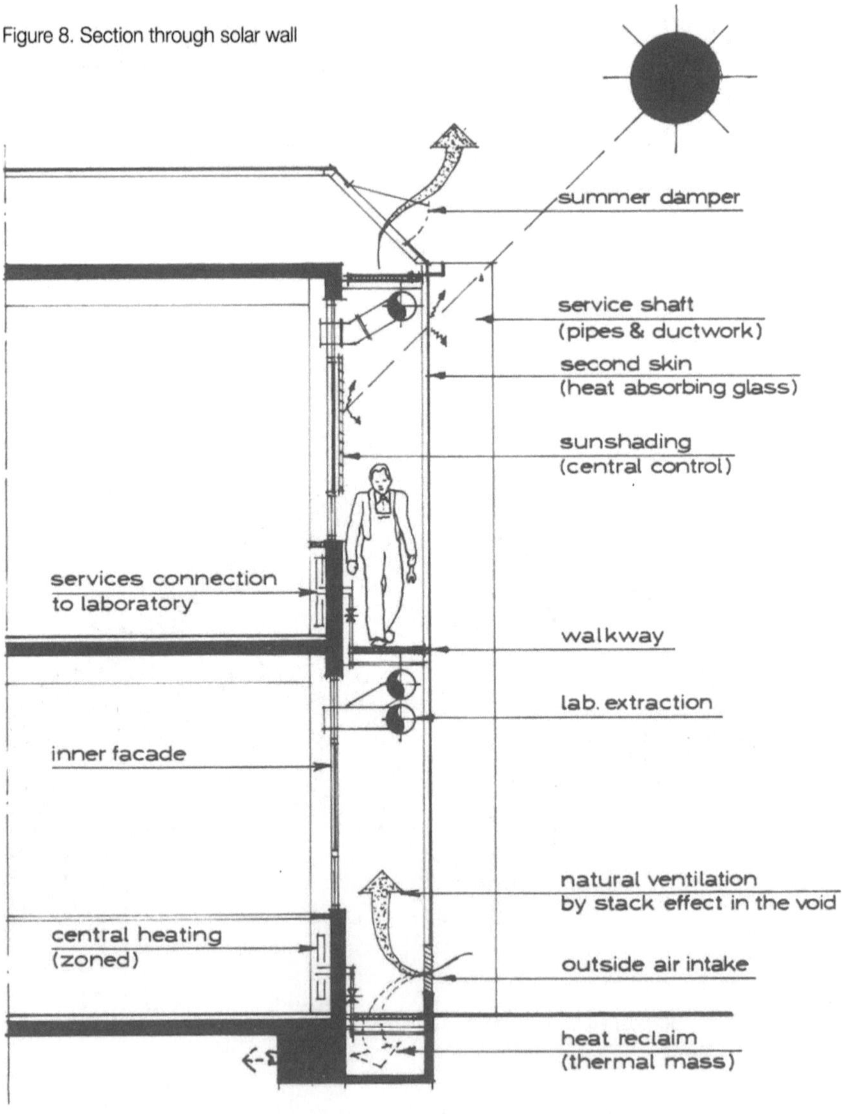

DESCRIPTION OF PASSIVE SOLAR FEATURES/COMPONENTS

Passive Solar Features and their Operation in Winter and Summer

The solar walls, conservatory and atria shown in Figures 5, 6 and 8 are part of a passive solar system aimed at creating a comfortable environment in the building and reducing the amount of conventional energy needed to run it.

The system has two different operating modes, according to season.

Mode of Operation in Winter

In winter, the solar walls, the conservatory on the south side and the internal atria serve as buffer spaces between the outside environment and the building interior. Solar gain in the glazed areas is distributed evenly over the buffer spaces by fans. Fresh air is introduced into the bottom of the solar walls, prewarmed by passage through the concrete mass of the air ducts in the crawl space and then warmed further in the solar walls by solar radiation and conduction heat losses from the laboratories. The warmed air is taken into the air handling units to reduce the load on the heating plant.

With the louvres at the top of the solar wall shut, the double wall acts as additional insulation.

The solar wall and the glazed areas in the roof maximize daylight entry into the building. Automatic switching of the artificial lighting in the atria, conservatory spaces and external zones of the laboratory floors is achieved by means of photocells.

Mode of Operation in Summer

In summer, the conservatory is shaded by the existing high trees. Catwalks, ducts and service runs serve as a sunscreen in the solar walls. Natural stratification of warm air in the buffer zones prevents the laboratories from overheating. Motorized vents open at the top of the solar walls and atria so that unwanted heat can be vented to the outside by the stack effect.

The laboratories are ventilated mechanically. Air intake is through louvres in the north walls of the plant rooms. No mechanical cooling is required.

There is natural ventilation in a N-S direction through the atrium zones. Air intake is at the bottom of the conservatory. The air movement is reinforced by the prevailing wind from the SW.

At night, outside cold air is drawn in by fans through the cavity of the concrete floors and the solar walls.

Mechanical Features and Design Temperatures

The air handling units are on the north side of the building, in the ab and bc zones at the first floor level. The laboratory floors are mechanically ventilated with air change rates in the range 5-10 changes per hour. Central heating is via perimeter radiators. The building is zoned for better control.

The design temperatures during the heating season are 20° C ± 1° C in the laboratories and 18° C ± 2° C in the hall and circulation areas.
The summer internal design temperature is 23° C ± 3° C.

The air supply ducts run in zones ab and bc and extraction ducts run in the cavity of the solar walls. Reclaim of heat in the handling units is indirect so there is no contamination of the supply air.

The average lighting load in the laboratories is 400 lux.
The plant rooms for the laboratories' building services are on the north side in the basement in the ab and bc zones.

Pipes for water, sewage, compressed air, gases and electric cables run in the ab and bc zones and in the twin wall cavity and feed the laboratory equipment from the perimeter. This allows for growth and change in the laboratory without disturbance of the floor underneath.

Maintenance can be carried out in the twin wall cavity and in the ab and bc zones without interfering with laboratory activities.

ENERGY CALCULATIONS PERFORMED AND DESIGN TOOLS USED

Energy Performance of Leuven Laboratory

The predicted annual thermal performance of the Leuven laboratory is shown in Figure 9. Key points regarding heating and lighting energy consumption are as follows.

Heating

The total annual consumption of fuel for heating the 4,500 m² building is 405 kWh/m². Over half of this is required for mechanical ventilation.

Useful solar gains amount to about 70 kWh/m² a year and compensate for most of the efficiency losses of the heating system.

Lighting

Electricity consumption for artificial lighting is estimated to be 17 kWh/m2 a year. This is low because, although the building is quite compact, each block (A, B and C) has good natural lighting from the solar walls and the rooflights in zones ab and bc.

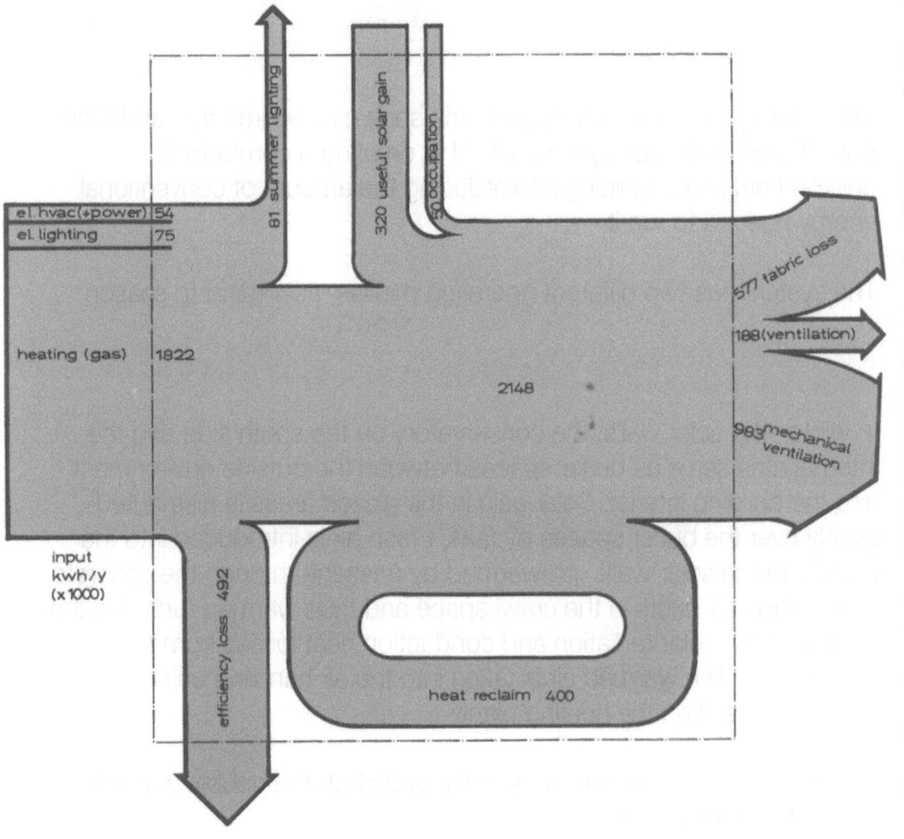

Figure 9. Sankey diagram of energy flows in Leuven laboratory

Case Study on Conversion Project

To gain experience of the thermal performance and use of solar walls in laboratory buildings before embarking on the design of the Leuven project, studies were carried out on a recently- completed project involving conversion of a five-storey 5,000 m² building into new research laboratories and offices with a total floor area greater than that of the original facilities.

The original building is shown in Figure 10. It was put up in the 1960s before the energy crisis and had single glazing and precast concrete external walls of poor construction. Vertical risers for air ducts and pipes were at 3 m centres on the inside perimeter walls, just in front of the windows. In winter the building had been uncomfortable because of excessive ventilation and heat loss by conduction through the external walls. In summer, it was subject to overheating.

The converted building is shown in Figure 11. The building has been given a new glazed skin, placed some 900 mm outside the original facade. The extract ducts, feeder pipes and cables for laboratory services have been put in the cavity between the twin walls (see Figure 12). The top of the cavity is surmounted by a new roof structure which houses the extract fans. Integrated into the roof structure are motorized louvres operated by temperature sensors in the cavity. There are catwalks for maintenance access at each floor in the cavity.

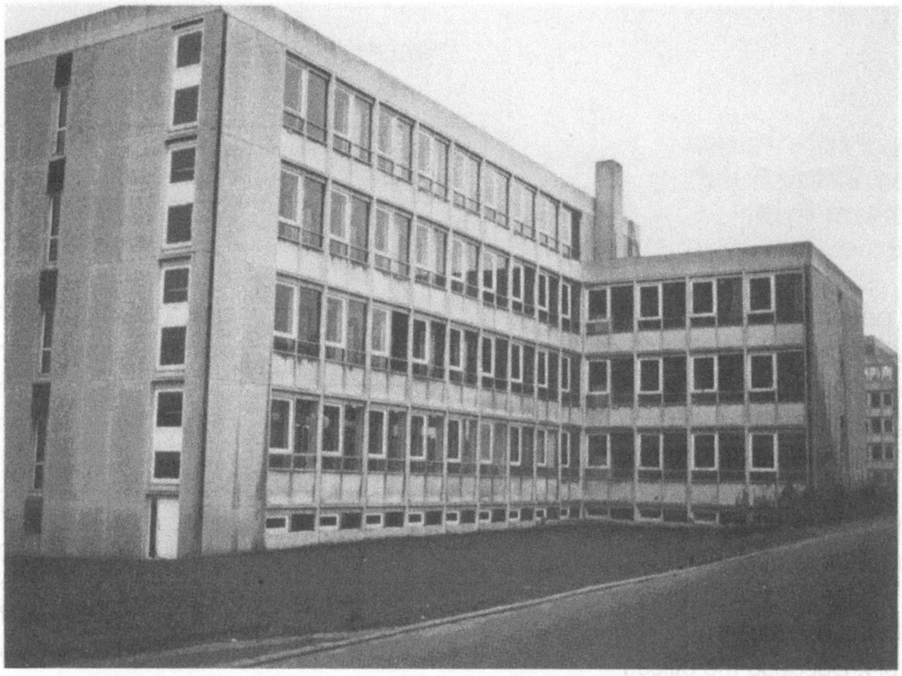

Figure 10. Case study building before conversion

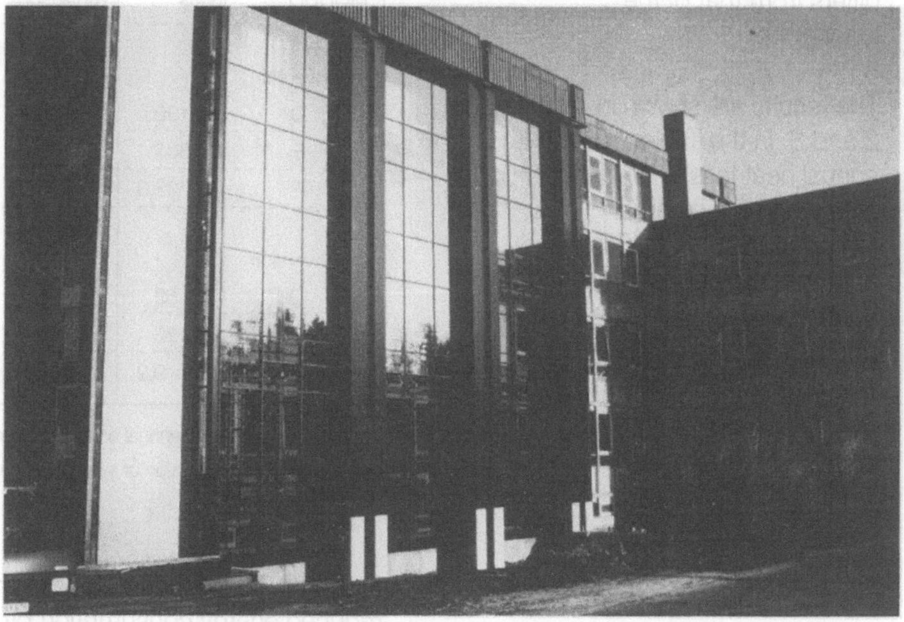

Figure 11. Case study building after conversion

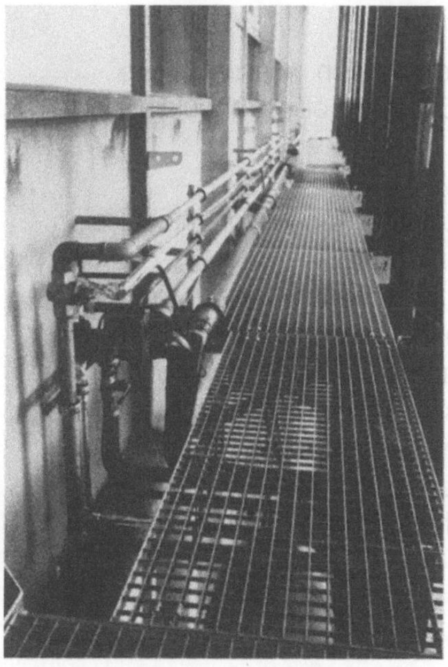

Figure 12. View inside cavity of solar wall

The aim of the double skin is to achieve:

- a lower U-value of the facade - 2.4 W/m^2 K compared with the existing 5 W/m^2 K;
- reduced ventilation losses and less draught from winds;
- protection of the existing facade against rain and environmental degradation;
- avoidance of cold bridges and condensation problems;
- solar gain in winter by closing the roof louvres;
- avoidance of overheating in summer by opening the roof louvres. A solar chimney is created by reflected and trapped heat. The heat is vented before it reaches the laboratory spaces.
- less glare and more equal distribution of daylight in the laboratories;
- an increase in net usable floor area by placing vertical ducts and risers outside the laboratory floor.

Energy Performance of Case Study Building Before and After Conversion

Energy performance calculations were carried out for a module of the case study building with floor area 22 m² and external wall area 12 m² before and after conversion. Two cases were taken: (a) when the module was designed for use as office space and (b) when it was used as a laboratory. Because the offices are not mechanically ventilated, their external wall construction differs from that of the laboratories.

The results are shown in Tables 2 and 3. In the Tables, Q is the annual heat loss in W/m² floor area when the external temperature is -8° C. E is the annual energy consumption for heating and ventilating in kWh per m² floor area. F gives the annual fuel consumption in l/m².

OFFICE MODULE	Fabric loss	Mechanical ventilation	Solar gain	Total	Saving	Total in %
Before reconversion						
Q	130	–		130		100
E	303	–		303		100
F	22,6	–		22,6		100
After reconversion						
Q	52	–		52	78	40
E	122	–	–26	96	207	32
F	9,2	–	– 1,9	7,3	15,3	32

Table 2. Comparison of annual energy consumption of office module before and after conversion. (For explanation of symbols and units, see main text opposite.)

LAB MODULE	Fabric loss	Mechanical ventilation	Solar gain	Total	Saving	Total in %
Before reconversion						
Q	130	300		430		100
E	303	698		1.001		100
F	22,6	52,1		74,7		100
After reconversion						
Q	52	281		333	97	77
E	122	654	–61	715	286	71
F	9,2	49	– 4,6	53,6	21,3	71

Table 3. Comparison of annual energy consumption of laboratory module before and after conversion. (For explanation of symbols and units, see main text opposite.)

From the Tables it can be seen that the conversion measures have reduced energy consumption by over 65% in the office module and nearly 30% in the laboratory module.

Solar gains amount to over 20% of the total heating load in the office module and nearly 8% in the laboratory module.

If the incoming fresh air is taken through the solar wall cavity into the air handling units in winter, the average temperature inside the cavity is some 4° C lower than if the cavity is closed. This results in a greater heat loss through the internal skin but the extra heat together with the solar gains is used to prewarm the air supply for the mechanical ventilation. Thus the efficiency in using the solar energy increases without loss of comfort.

GENERAL DESIGN GUIDELINES/POINTS OF INTEREST RESULTING FROM THIS PROJECT

Solar Walls

Experience shows that solar walls are very suitable for laboratory buildings. They have a range of advantages which are not available in a traditional type of construction. They can be used for both new buildings and rehabilitation projects.

As has been seen already, solar walls bring energy savings. In a refurbishment project such as that described in the case study, they can provide annual energy savings of 330 kWh/m^2 wall plus 50-100 kWh/m^2 from solar gain, depending on the quantity of preheated air used. This represents a yearly savings of around 40 l fuel per m^2 wall. As with most passive solar features, however, the saving is achieved only at some additional capital cost. In the case study the extra cost of the second skin amounted to 12,500 BF/m^2 (291 ECU/m^2). At today's prices, it would take 20 years to repay this additional investment from the resulting energy savings. This cannot be justified on economic grounds. However, solar walls can bring additional benefits which, under particular circumstances, can make their incorporation in a building worthwhile. Examples of these advantages are listed opposite. A monetary value can be put on many of them so that the payback time for the additional capital cost of the wall can be reduced to 5-10 years.

Summary of Advantages of Solar Walls

Solar walls can offer:

- energy savings;
- good comfort control in winter; no cold radiation from large glazed areas;
- good comfort control in summer by removal of solar heat before it can reach the interior, extraction of heat by stack effect and avoidance of mechanical cooling;
- good technical detail in the external wall with no problem regarding cold bridges and wind infiltration;
- good daylighting without harming the U-value of the wall;
- good sound insulation against traffic noise;
- good architectural possibilities in, for example, the upgrading of existing buildings.

Advantages particularly relevant to laboratories include:

- extra space for service runs and ducts without cluttering up the interior;
- the ability to incorporate additional pipe couplings, etc., very easily through the external walls as laboratory activities grow or change;
- easy maintenance of service runs, valves, etc., via catwalks in the cavity.

Summary of Advantages of the Atrium

An atrium can bring the following advantages:

- energy savings because of good floor to wall ratio, the possibility of natural ventilation without mechanical cooling and useful solar gains;
- good comfort control in winter and summer because of the buffer zone;
- good daylighting;
- a low floor-to-floor height because there are fewer mechanical services in ceilings or floors;
- cheaper exterior wall construction in the courtyard because there is no need for protection from rain or wind, although these advantages can be lost by the need for fire precautions;
- very good land-use ratio;
- good architectural possibilities in, say, the upgrading of existing courtyard buildings;
- extra multi-purpose space and circulation routes.

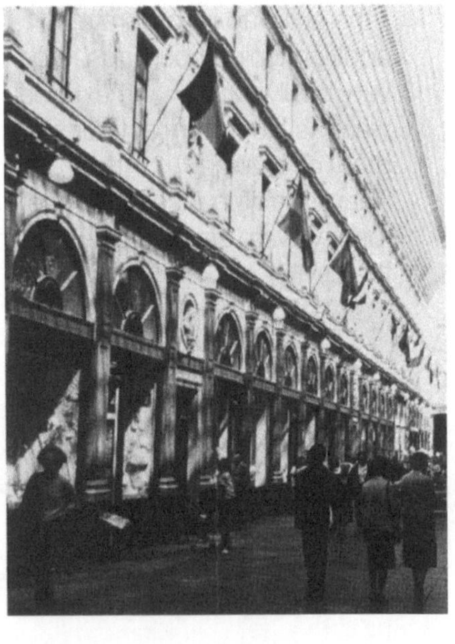

Figure 13. The 19th century St. Hubertus shopping gallery in Brussels.
An example of a first generation atrium building.

The Atrium

The recent revival of interest in the atrium, like the solar wall, has resulted from the need to take a fresh look at energy use since the energy crisis.

As with the solar wall, the acceptability of the present-day atrium rests on a number of arguments and a summary of the advantages which the presence of such a feature can bring to a building is given opposite.

Building Form and Energy Use

In deciding whether or not incorporation of an atrium is useful in a particular case, it is important to take into account the relationship of building form and energy use. For instance, although development of a design with a good compactness ratio can bring low energy use in a small building like a house or a one- or two-storey pavilion-type building, it does not necessarily have energy significance in bigger or more complex projects.

Figure 14 illustrates a number of different building forms. Table 4 gives the predicted heat losses of some examples of each of these. It also indicates whether or not a particular building form is capable of making use of natural lighting or natural ventilation.

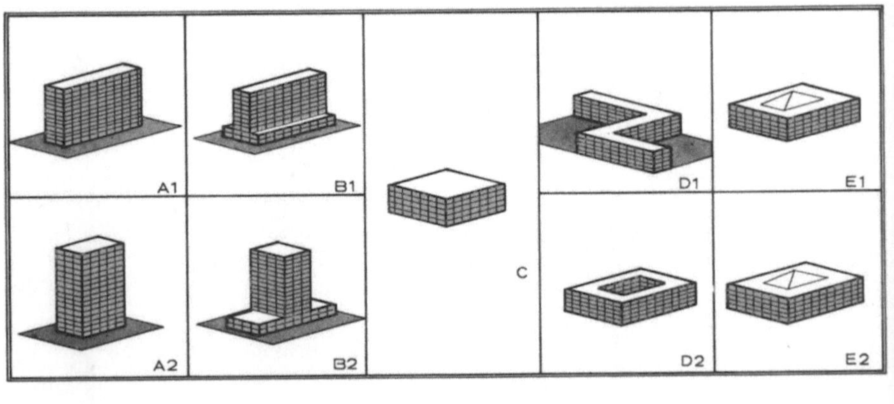

Figure 14. Some examples of different building forms

	A1	A2	B1	B2	C	D1	D2	E
totaal floor surface in modules (7,2 x 7,2 m)	240	240	240	240	240	240	240	240 +15
number of storeys	13	16	11	16	5	5	5	5
floor to floor distance can be kept low						x	x	x
total facade surface in modules (7,2 x 7,2 m)	146	149	132	145	80	130	120	80
total surface of the envelope in modules	183	179	205	216	176	226	216	191
heat losses by conduction in % referred to type C	136	138	136	147	100	142	134	112
natural lighting natural ventilation is possible	x					x x	x x	x x

Table 4. Building form and energy use

It is clear that the deep plan building type C has the best floor-to-wall ratio. The energy needed to compensate for the heat lost by conduction from the external envelope is 30-40% more for the high rise (types A and B) or narrow plan (type D) forms.

However, the deep plan building has a number of well known disadvantages relating to annual energy consumption. The large amount of electrical energy needed for lighting and ventilating the enclosed spaces can be higher than the energy saved from lower heat losses. Furthermore, the internal heat gains from the artificial lighting, people and equipment can necessitate cooling and air conditioning, with their accompanying energy consequences. In addition, the large package of building services/ducts which has to be brought into such a building results in an increase in floor-to-floor height which, in turn, brings additional external wall area and increased heat loss.

The advantage of the courtyard type of building is that it takes no more land than the deep plan type. By covering the courtyard with glass it is possible to combine the advantages of the narrow plan and deep plan type. Further, glazing the roof causes the courtyard walls to change from external to internal walls, producing a very compact building. In addition, the atrium provides a means of introducing natural light and natural ventilation into the rooms next to the courtyard. The atrium itself becomes an extra multi-purpose space in the building.

BUILDING 2000 Building 2000 brochures are published by Directorate General XII of the Commission of the European Communities to show how design studies can help architects and other building designers use passive solar principles to the best effect to produce attractive energy efficient buildings. Each brochure describes studies carried out with the support of the Commission during the design phase of one thirty-six non-domestic buildings in the EC Member States. The studies were on such topics as daylighting, heating, cooling, ventilation, comfort, control systems and urban design. They were carried out with the help of acknowledged European experts in these fields and drew heavily on lessons learned and techniques developed through the Commission's research and development programme on solar energy applications to buildings.

Commission of the European Communities/Directorate-General for Science, Research and Development

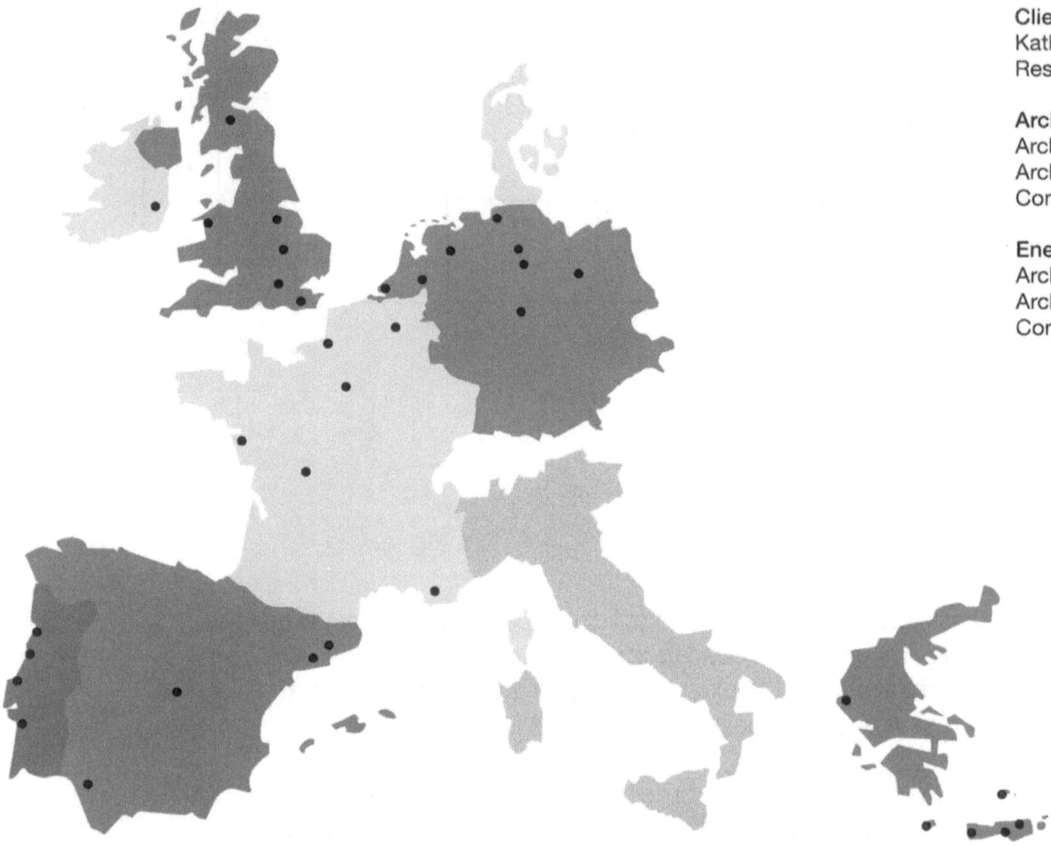

List of Design Team Participants and Advisers

Client
Katholieke Universiteit Leuven
Research and Development

Architect
Archiduk bvba
Architects, Engineers and Energy Consultants

Energy Consultants
Archiduk bvba
Architects, Engineers and Energy Consultants

This set of **Building 2000** brochures illustrates how architects and other building designers can successfully apply passive solar principles to produce energy-efficient buildings.

BUILDING 2000 Participants

Project Director
Theo C. Steemers

Coordinator
Cees den Ouden

Technical Steering Committee
Dean Hawkes
Nick Baker
Alex Lohr
Jean P. Lepoivre

Regional Liaison Agents
(D) Jörn Behnsen
(E) Vicente Sifre
(F) Michel Raoust
(GB) Alan Hildon
(GR) Matheus Santamouris
(P) Eduardo Maldonado

Further information or copies of the brochures can be obtained from prof. ir. Cees den Ouden, EGM Engineering BV, P.O. Box 1042, 3300 BA Dordrecht, The Netherlands.

BUILDING 2000

Commission of the European Communities

- Building containing offices, manufacturing areas and classrooms with different environmental needs
- Heavy mass construction reduces indoor temperature swings, improving utilization of solar gains in winter and reducing or eliminating need for cooling in summer
- Carefully-designed glazing system optimizes daylighting throughout year while allowing solar gains to enter in winter and providing shading in summer
- Solar and internal gains contribute 65% of heating requirements
- Building is divided into three zones two of which do not require mechanical cooling

BUSINESS INNOVATION CENTRE

PORTO/PORTUGAL

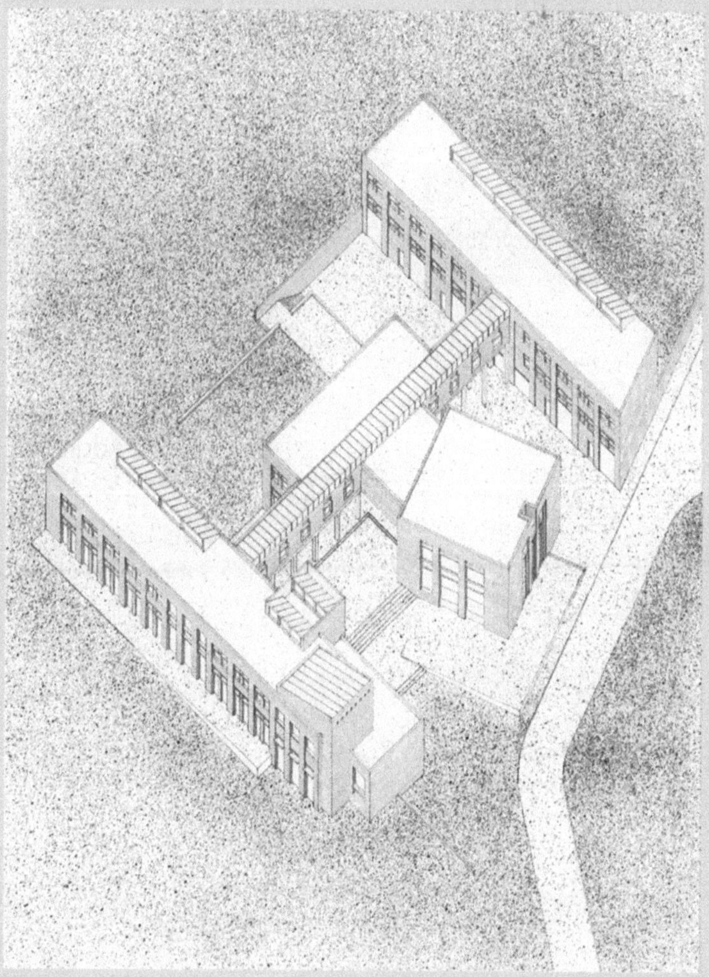

Building 2000 is a series of design studies illustrating passive solar architecture in buildings in the European Community.

ISSUE

L 2

FEB 1991

PROJECT DESCRIPTION

Location

The building is located between the economics and engineering buildings on the University of Porto campus in Porto (latitude 41° 1' N), a coastal city in northern Portugal (see Figure 1).

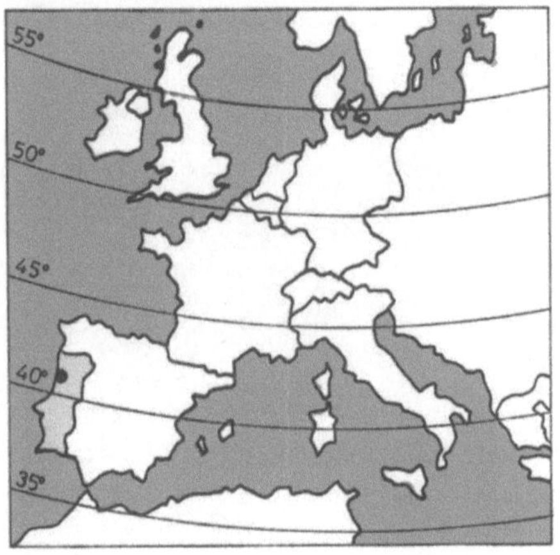

Figure 1. Location of Porto

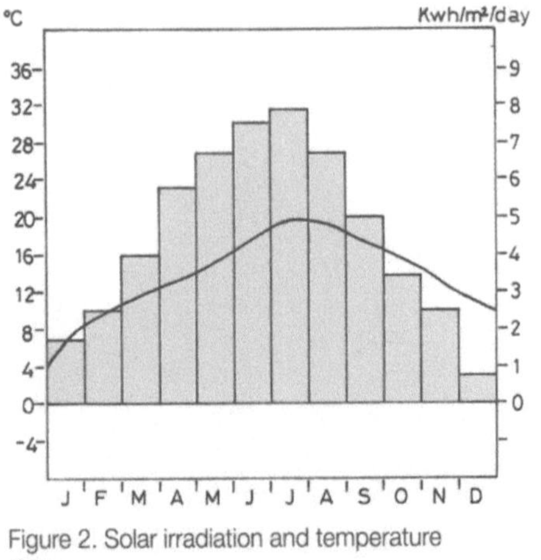

Figure 2. Solar irradiation and temperature data

Key: solar irradiation on horizontal plane
— average external temperature

Degree days per year (base 18 °C)	1615
Global irradiation on horizontal plane	
in kWh/m^2/year	1642
(in MJ/m^2/year	911)
Sunshine hours per year	2667

Table 1. Some key climate data

Building Type

This building consists of a mix of offices, manufacturing areas and classrooms. It contains a business area for companies developing innovative products and services, the Community Documentation Centre and the University Extension Foundation. The business development area comprises office accommodation, communal support facilities, "clean" manufacturing workshops and classrooms where company managers can undergo management, planning and accountancy, etc., training. The documentation centre has an extensive library and facilities for accessing electronic databanks. The University Extension Foundation provides technology transfer and continuing education services.

Site Microclimate

The climate is mild throughout the year. In winter, it is sufficiently cold for some form of heating to be required but not so cold that a large heating system is necessary. In summer, there can be some extremely warm days with temperatures in the upper thirties (°C). However, the average temperature in July (the warmest month) is only 20 °C, well below the point where mechanical cooling is required for comfort. Daily temperature swings are about 12 °C so that even very hot days will have cool nights. Humidity is low and the weather is sunny. Winds are moderate, predominantly from the north-west. Additional climate data are given in Figure 2 and Table 1.

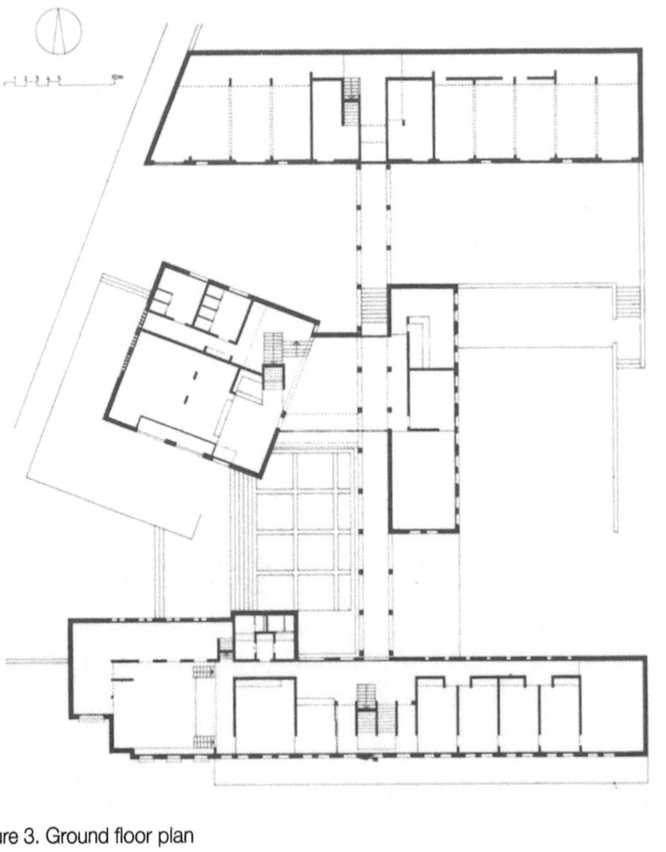

Figure 3. Ground floor plan

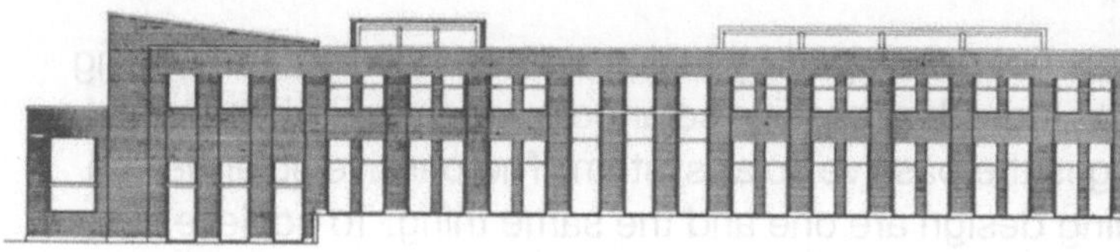

Figure 4. South facade

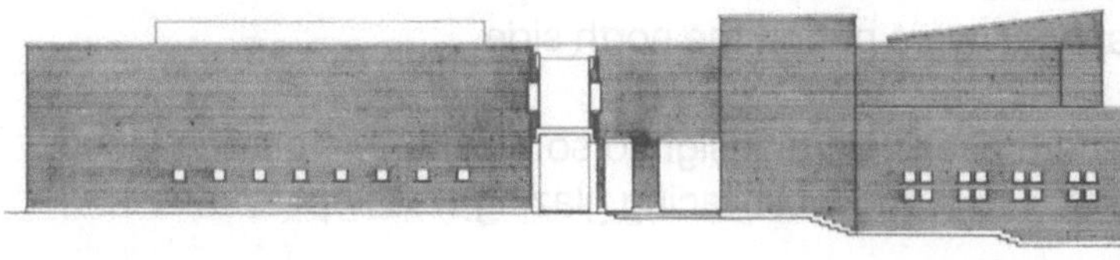

Figure 5. North facade

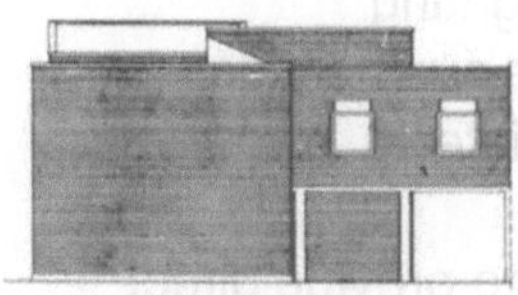

Figure 6. East facade

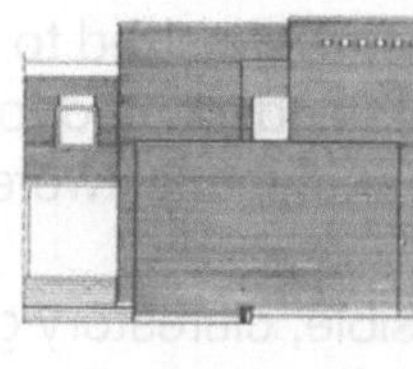

Figure 7. West facade

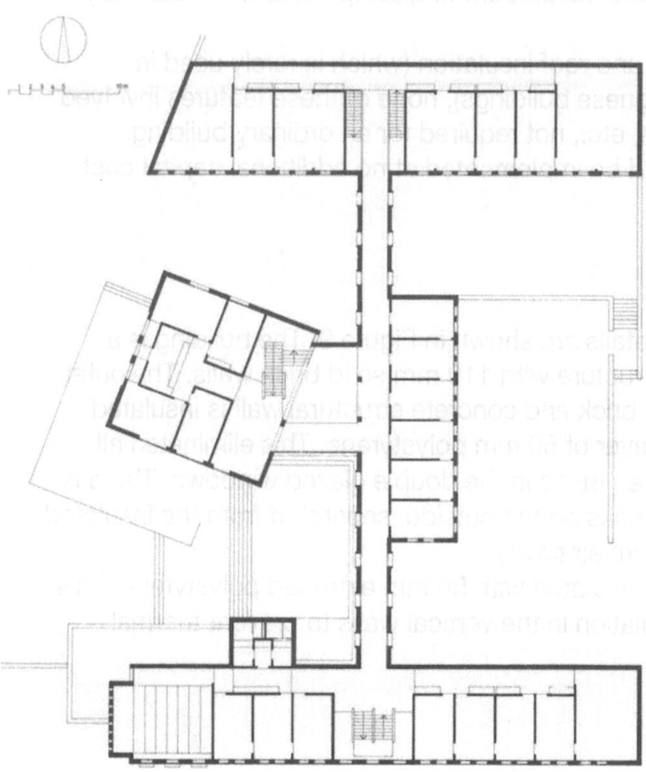

Figure 8. First floor plan

Design Strategy

The building was designed to capitalize on the mildness of the climate and provide a comfortable indoor environment with low conventional fuel consumption throughout the year. Air conditioning was only to be provided for those areas where a precisely-controlled environment was essential. To achieve this, a climate-sensitive design was necessary. The building (which contains 2700 m^2 occupied space and is on two storeys) was therefore divided into three zones (in all of which the main rooms are essentially south-facing) with all the spaces requiring air conditioning concentrated into the central zone. The ground floor and first floor plans are shown in Figures 3 and 8 and the south, north, east and west facades in Figures 4-7.

DESCRIPTION OF PASSIVE SOLAR FEATURES/COMPONENTS

Details of Passive Solar Design

The design was not developed by taking a basic building and adding on a number of individual passive solar components. Rather, the whole building is the passive solar system. The passive solar design and the building design are one and the same thing. To achieve this, a series of steps were taken.

First, the building was shaped to maximize the amount of exposure to the south and the interior spaces were kept shallow. Corridors and other buffer spaces were put on the north side.

Then, the interior spaces were designed so most of the glazing faced south and north. East- and west-facing glazing was kept to a minimum.

A well-insulated massive construction was selected for the envelope. Glazing areas were sized to meet heating and daylighting requirements and located to minimize summer-time overheating. Overhangs and shades were added where appropriate.

Where possible, clerestory glazing was introduced to provide direct gain and daylighting to those parts of the building without a south exposure.

Massive indoor partitions were added to give internal temperature control to areas where the amount of glazing made this necessary.

Apart from the wall and roof insulation (which is rarely used in conventional Portuguese buildings), none of these features involved the use of materials, etc., not required for an ordinary building. Therefore they could be implemented at no additional capital cost.

Figure 9. Construction details

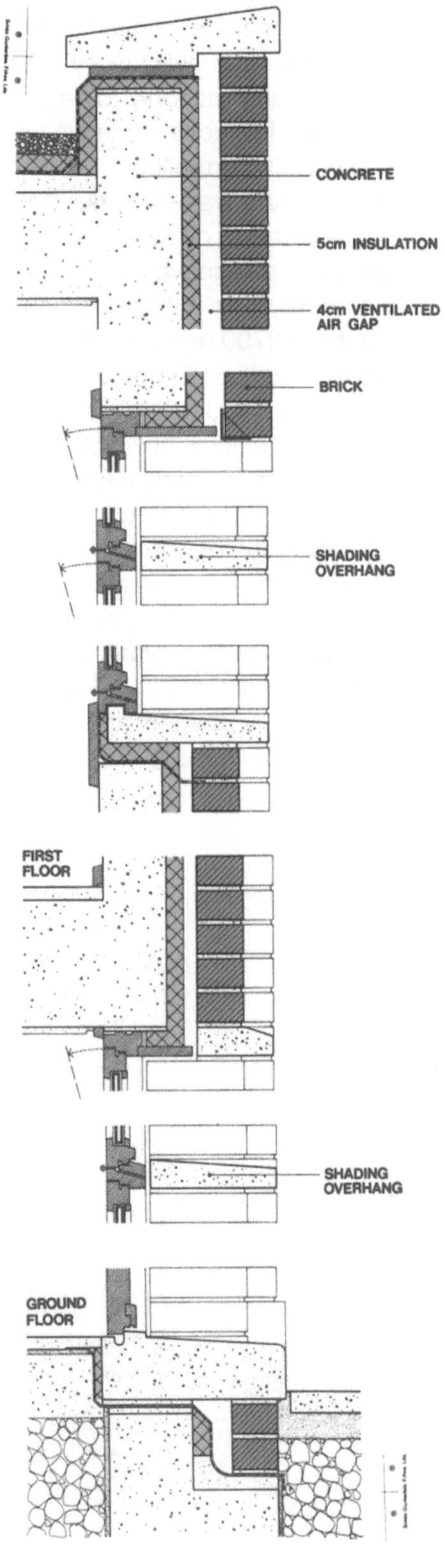

Construction Details

The construction details are shown in Figure 9. The building is a precast concrete structure with 110 mm solid brick walls. The outer surface of the solid brick and concrete structural wall is insulated with a continuous layer of 50 mm polystyrene. This eliminated all thermal bridges, even those in the double glazed windows. There is a second layer of bricks on the outside, separated from the insulated inner wall by a 40 mm air cavity.
The roof is flat and insulated with 50 mm extruded polystyrene. This is linked to the insulation in the vertical walls to prevent thermal bridging.

Daylighting

To achieve visual comfort in the indoor spaces and reduce electricity consumption for artificial lighting, daylighting design received careful consideration from the outset.

This was one reason why most south-facing rooms are shallow with a depth not greater than 6 - 8 m. In addition, most spaces are daylit from two sides.

The daylighting of typical ground- and first-floor rooms in the south zone is illustrated in Figure 10. Light enters through carefully-designed south-facing windows and through apertures from the first-floor north-facing corridor. In the first-floor room, this aperture is a vertical window near the ceiling. In a ground-floor room, the opening is a horizontal row of transparent bricks in the corridor floor. The first-floor corridor is lit by vertical clerestory glazing. The ground-floor corridor is lit by windows on the north facade.

The north zone of the building mostly consists of manufacturing areas lit by south-facing windows. The corridor on the north side of this zone is lit by north-facing windows and a vertical clerestory in the roof.

In the central zone, no room is more than 4 m deep so adequate daylighting is achieved by glazing on one side only.

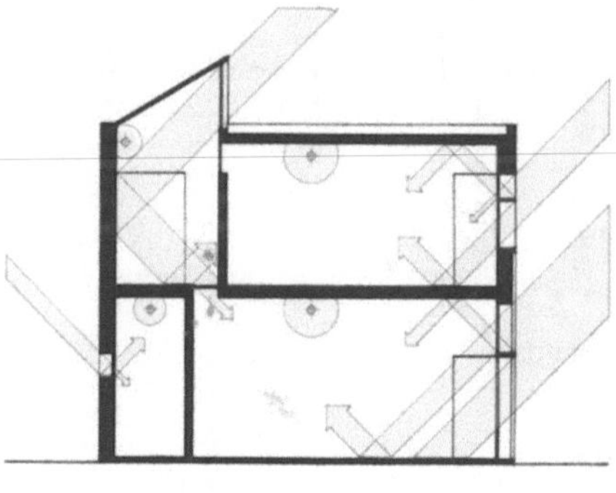

Figure 10. Daylighting of typical first- and ground-floor rooms in the south zone

Artificial Lighting

To complement the daylighting, two levels of artificial lighting are provided: general lighting of 100 lux, automatically controlled by light- sensitive sensors; and task lighting for individual work stations. Efficient lighting fixtures reduce internal gains.

Auxiliary Heating, Cooling and Ventilating Systems

The building contains two types of auxiliary systems for environmental control.

The central zone has a conventional air conditioning system with zonal control. A chiller provides cooling in summer and a gas boiler provides heating in winter.
The north and south zones are provided with heating and mechanical ventilation only. Heating is achieved through a network of hot water radiators, the hot water being produced by the gas boiler used for the central zone.

The ventilation air is treated in a central air handling unit to a temperature slightly below that of the indoor set point in winter and to 20 °C in summer. It is then distributed through ducts throughout the building. The treated air enters each room below the heating radiator at floor level so that it is mixed within the occupied space (i.e. to a height of about 2 m) and displaces the existing air to the corridor.

Control of Mechanical Ventilation Air

The temperatures of the incoming ventilation air have been selected in order to conserve energy.

In winter, to encourage solar radiation to make the maximum possible contribution to space heating requirements, the inlet temperature of the ventilation air is reduced when solar gains are available. It would, however, be uncomfortable for occupants if cold fresh air was brought directly into the heated space. Therefore, a compromise temperature of 16 - 18 °C (i.e. just below the indoor set point) is used for the inlet ventilation air.

In summer, the use of slightly cooled inlet air (of, say, 20 °C) contributes to occupant comfort, reduces cooling loads during warm daytime hours and eliminates the need for additional mechanical space cooling.

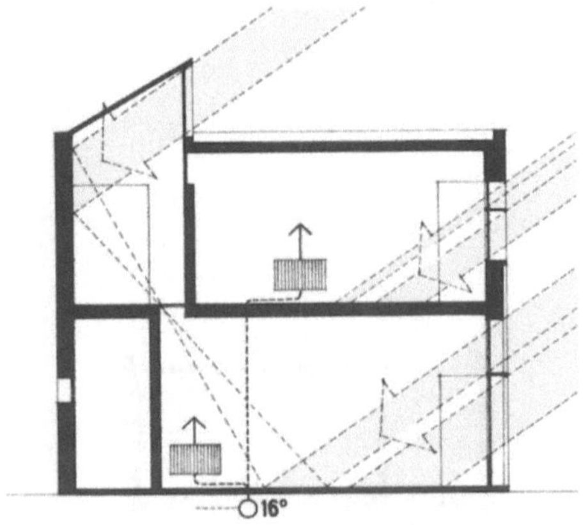

Winter day

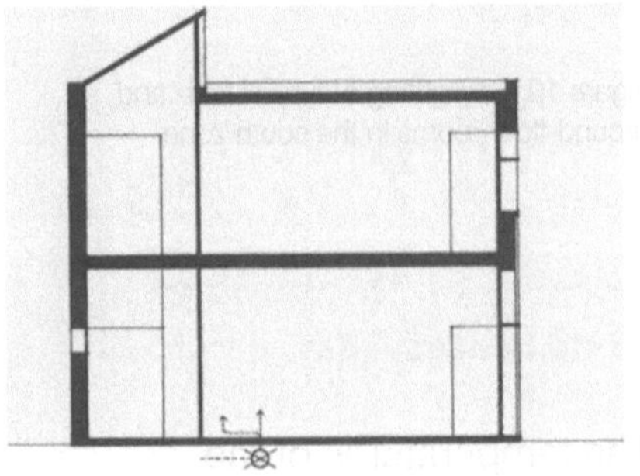

Winter night

Figure 11. Operating mode in winter

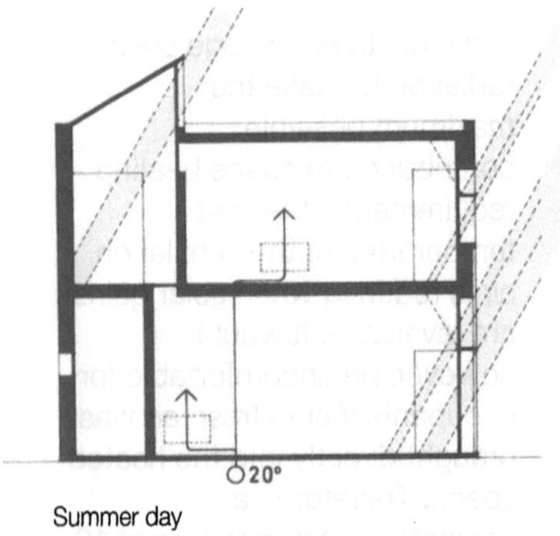

Summer day

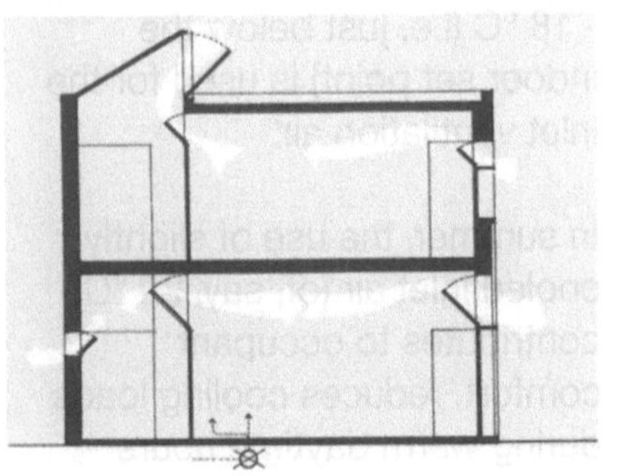

Summer night

Figure 12. Operating mode in summer

Operating Mode in Winter

The mode of operation of the building in winter is illustrated in Figure 11.

During the day, solar radiation enters the building and warms it up, some heat being stored for later release at night. Heating is supplied if necessary only during hours of occupancy so that there is a slow drop in temperature during the night. As the windows are not provided with night insulation, there is no difference between the day and night modes except for the operation of the heating system.

The building does not require a special control strategy. The high level of insulation on the building envelope ensures that the average internal temperature is always at least a few degrees above the mean outdoor temperature. When there are solar gains, the mean indoor temperature is 4 - 8 °C above the mean external temperature. Achievement of steady indoor temperatures is helped by the building's large inertia. The value of this strategy has already been proved in another local building (the CTO described in the Commission of the European Communities' Project Monitor brochure number 9) where, even during the coldest months, temperatures close to the comfort range can be reached naturally indoors.

Operating Mode in Summer

The summer-time mode of operation is shown in Figure 12. Apart from the mechanical ventilation system for the whole building and the air conditioning system in the central zone, the building will run itself.

Careful placement and well-designed overhangs ensure that all windows are shaded in the warmer summer months and cooling loads are low. In addition, east- and west-facing windows have vertical movable shades for improved solar control. These measures, together with the high building inertia, lead to small internal temperature variations.

The heat accumulated during the day is dissipated at night through natural ventilation. Special operable windows have been designed which allow a sufficient area to be opened to ventilate the building without impairing building security. After working hours, external (including clerestory) windows are opened together with the openings between south-facing rooms and north-facing corridors. Cool night air then cools down the heavy indoor mass, ready for thenext day.

ENERGY PERFORMANCE CALCULATIONS AND DESIGN TOOLS USED

Design Tools

Design of the passive solar aspects of the building was carried out with the help of a number of computer-based design tools.

The initial work was carried out using the annual SLR (Solar Load Ratio) method described in the US Department of Energy Passive Solar Design Handbook published in 1981. This uses very simple correlations to size window areas to give acceptable solar fractions. It also gauges the thickness of indoor partitioning which will give indoor temperature stability. Although this method consistently underestimates winter performance of a massive building with a high level of inertia (such as this) it does nevertheless point designers in the right direction and gives them a degree of confidence that the final design will be a successful one.

For the detailed analysis of the final design, a simulation was carried out using the ESP package developed in Scotland. This enables multizone dynamic thermal simulations of a building to be made under real operating conditions. Unlike the simpler SLR method it allows particular ventilation, lighting and temperature control strategies to be specified and permits detailed consideration to be made of window shading devices, input occupancy patterns, and so on. Thus a much more precise thermal characterization of the building can be made.

The daylighting studies were carried out using the advanced Genelux ray- tracing simulation tool.

Results of the SLR Analysis

The preliminary SLR study gave an idea of the contribution which solar gains would make to the heating load of the north, centre and south zones of the building. The results are given in Table 2.

Zone	Solar fraction
north	0.62
centre	0.35
south	0.45

Table 2. Contribution of solar gains to total annual heating requirements determined by SLR method

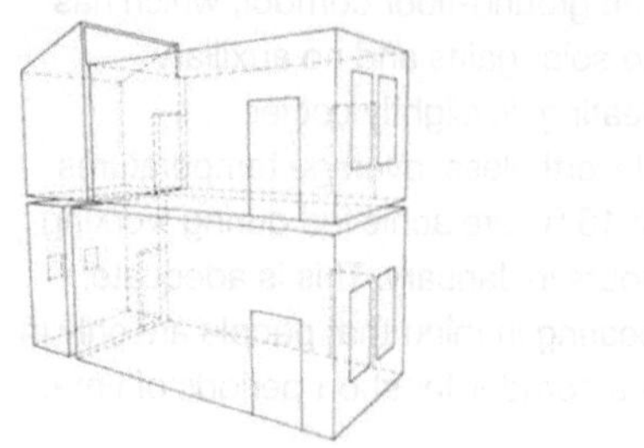

Figure 14. The office/corridor section used for the detailed study using ESP

The Detailed Study Using ESP

Introduction

Because ESP uses real weather data both winter and summer performance studies were possible. A typical four-zone section of the south part of the building, rather than the whole building, was selected for the study. This is illustrated in Figure 14. It includes a ground-floor and an upper-floor office and the relevant parts of the north corridors. By extrapolating the results of this simulation an idea could be obtained of the performance of the whole building.

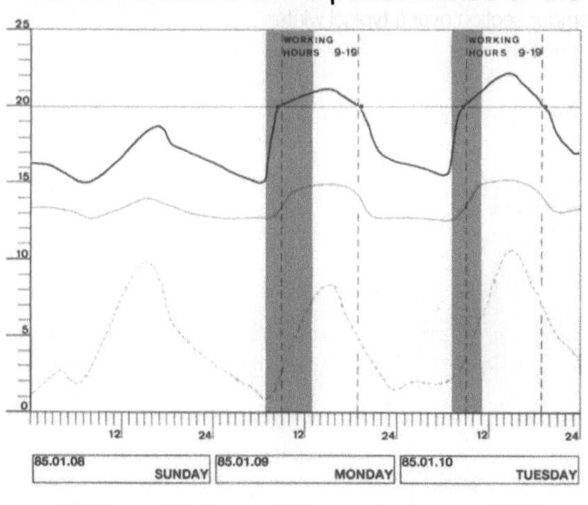

Figure 13. Temperatures achieved in office/corridor section on three January days

Winter Performance

In winter, heating is only needed during working hours. At night, the heating system is shut down and the indoor temperature allowed to coast. As the building has a large inertia, the temperature drop is fairly small. To make the building comfortable for occupants when they arrive at 9 am, the auxiliary heating is turned on two hours earlier. It can be turned off an hour before occupancy ends without impairing comfort.

The winter-time ESP simulation of the office/corridor section took account of the above heating strategy. The complete operating conditions are given in Table 3. The internal temperatures achieved are shown in Figures 13 and 15.

The winter-time ESP simulation of the office/corridor section gave the following temperature patterns.

In the offices, temperatures are often above the 20 °C set point because of solar gain during the day. Maximum temperatures, however, are well within the comfort range. (See Figure 15.)

In the first-floor corridor, the temperature pattern is similar to that in the offices even though there is no auxiliary heating (see Figure 15). This is because of the solar gains through the clerestory windows. A comfortable average temperature of 19 °C is obtained during January, the coldest winter month.

The ground-floor corridor, which has no solar gains and no auxiliary heating, is slightly cooler. Nevertheless, average temperatures of 16 °C are achieved during working hours in January. This is adequate, bearing in mind that people are only in the corridor for short periods of time.

The thermal performance of the section as determined by an ESP simulation over a typical winter (1985/86) is shown in Table 4. A breakdown of the energy use of the complete building is given in Figure 16. The total consumption for heating over this period is 6 kWh/m .

Set point	20 °C
Heating period	7 am to 7 pm
Infiltration rates in air changes per hour (AC/h)	
offices	0.7
upper corridor	0.4
lower corridor	0.5
Internal gains from 9 am to 8 pm	600 W
Mechanical ventilation rates from 9 am to 8 pm	40 m² /h
at 16 °C supplied to the offices and exhausted to the corridors	

Table 3. Operating conditions for ESP simulation over typical winter

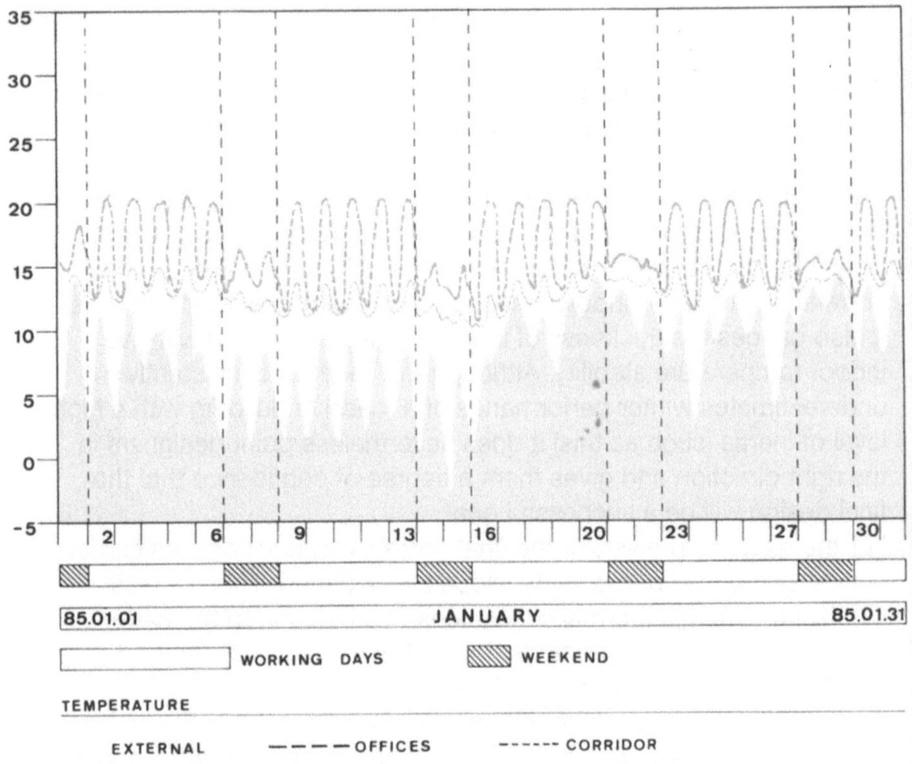

Figure 15. Temperatures achieved in office/corridor section in January

Period	Mean external temp (°C)	Gv (kWh/m²)	Degree days	Total load (kWh)	Aux energy (kWh)	Solar fraction and internal gains
Nov	12.0	7.95	237	159	53	0.66
Dec	11.4	9.40	275	201	66	0.66
Jan	8.7	11.65	378	362	134	0.63
Feb	12.4	9.57	216	96	37	0.61
Mar	11.1	11.67	279	140	48	0.66
Heating season	11.2	10.07	1385	958	338	0.65

Table 4. Thermal performance of office/corridor section over a typical winter

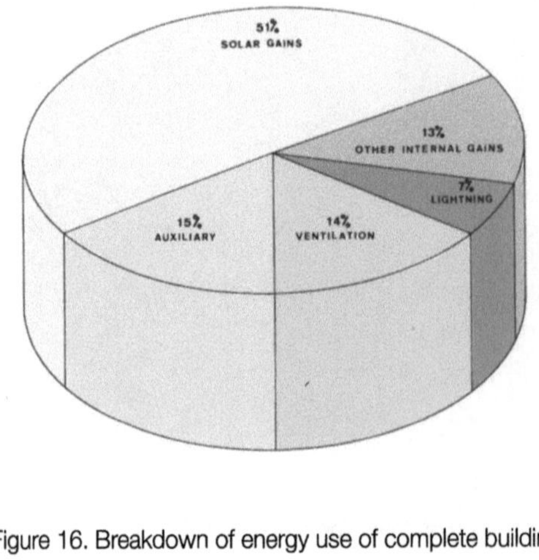

Figure 16. Breakdown of energy use of complete building

Infiltration rates in air changes per hour (AC/h)	
offices	0.7
upper corridor	0.4
lower corridor	0.5
Internal gains from 9 am to 8 pm	600 W
Ventilation rates	
from 9-20 hours	40 m /hour at 20 °C supplied to the offices and exhausted to the corridors
at night	4 ACH by natural ventilation

Table 5. Operating conditions for ESP simulation over typical summer

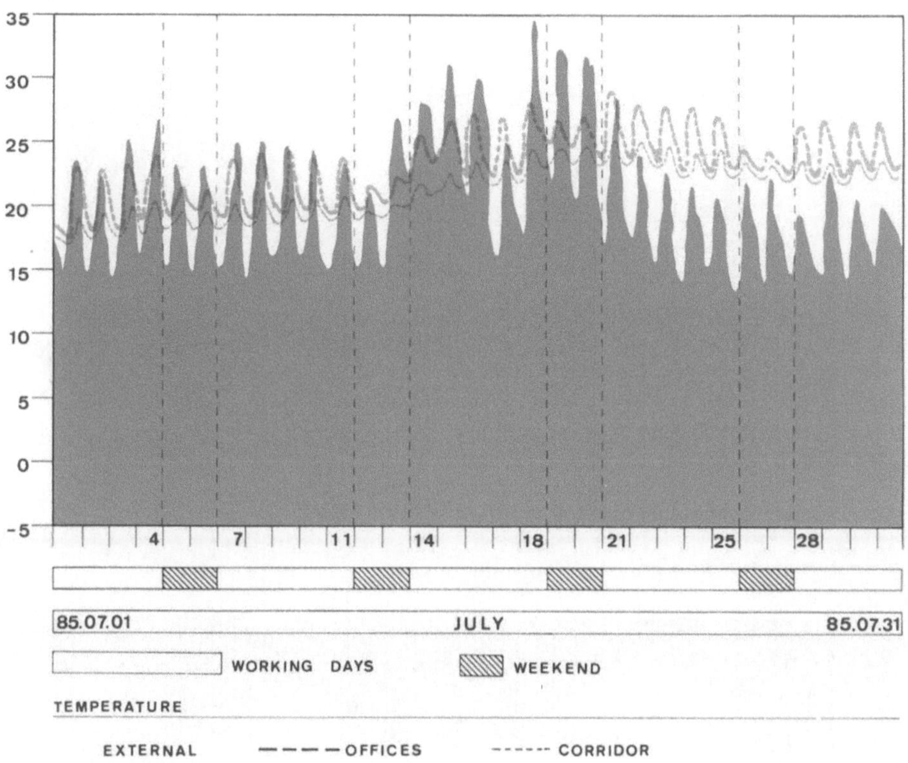

Figure 17. Temperatures achieved in office/corridor section in July

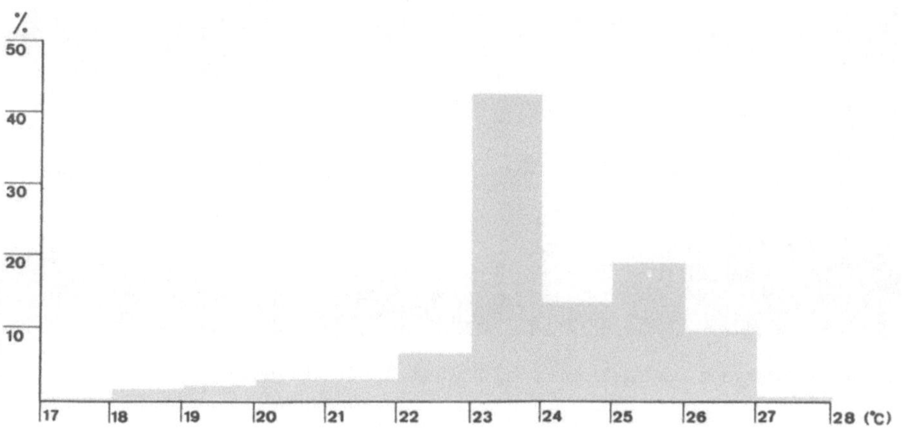

Figure 18. Frequency distribution of indoor temperatures in summer

Summer Performance

The operating conditions used for the ESP simulation of the summer-time performance of the south zone office/corridor section are given in Table 5. The simulation showed that, even without mechanical cooling, indoor temperatures remain at comfortable levels (see Figure 17). Even under extreme conditions such as those experienced in July 1986 (the warmest month on record in Porto) peak indoor temperatures would have reached 28 °C only in the unlikely event of all the lights being left on continuously all day.

The frequency distribution of internal temperatures in the section is shown in Figure 18. Because of the high level of inertia in the building, daily indoor temperature swings are about 5 °C. Night-time cross ventilation cooling is quite effective in bringing down the internal temperature to about 20 °C by early morning. Thanks to internal effective shading of direct solar radiation, the remaining gains are not sufficient to overheat the office spaces severely. The number of hours when the temperature exceeds 27 °C - the maximum normally acceptable for an air conditioned space - is relatively small.

The temperature in the first floor corridor is similar to that in the offices. The lower corridor, however, which only has a small area of glazing, remains in the 20- 22 °C range most of the time.

Studies on the central zone of the building showed that air conditioning keeps the indoor temperature within the 25-27 °C range, with an estimated energy consumption of 15 kWh/m^2 per year.

Results of Daylighting Study

The study of daylight distribution using the Genelux simulation tool involved refining an initial design by means of a number of iterations until a final design was reached which had optimized geometry and surface properties. The study was only possible through the expert assistance provided by the Building 2000 experts.

Figure 19 shows the daylight distribution in part of a typical offices/corridor section of the building under overcast sky conditions when the outdoor horizontal illuminance is 10000 lux. North is to the left and south to the right of the diagram. It shows that the glazing in the upper corridor floor, together with the windows on the south facade, allows enough daylight to be introduced into the lower room to provide good levels of natural lighting there.

The average daylight factor on the floor of the lower room with clear double glazing, no window frames, no furniture and 70% reflectors is 6%. If dirt, window frames, sunlight protection and furniture are taken into account this figure drops to about 2%. This is quite acceptable for Porto.

The photograph of the same spaces (Figure 20) shows that good daylight factors are obtained in most parts of the rooms.

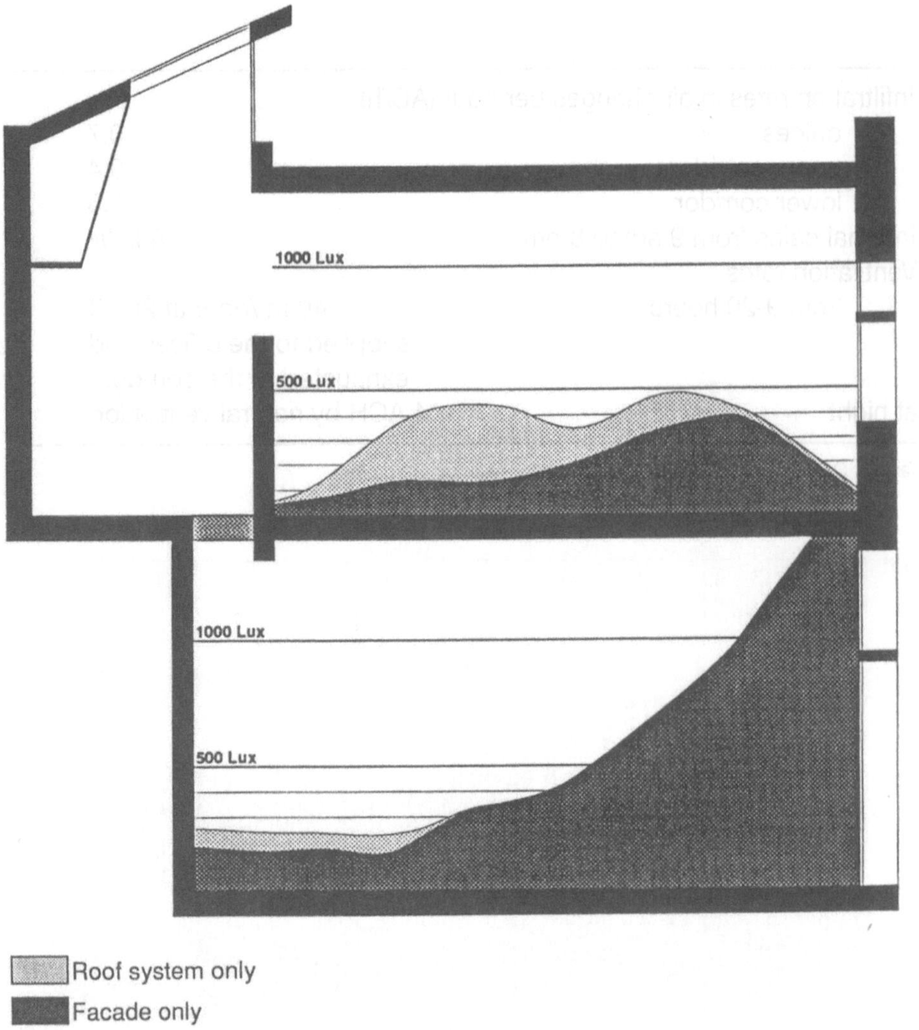

Figure 19. Daylight distribution in typical office/corridor section of building

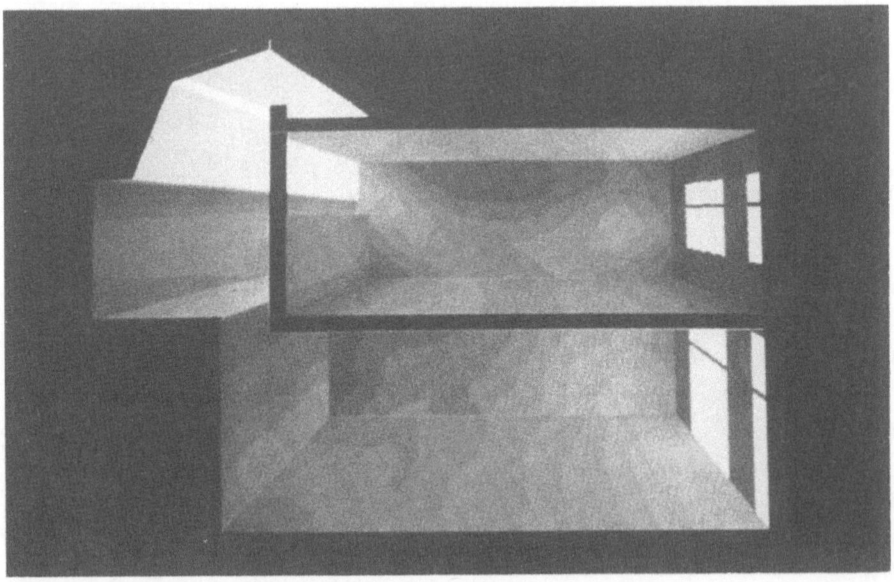

Figure 20. Photograph showing natural lighting of typical office/corridor section

GENERAL DESIGN GUIDELINES/POINTS OF INTEREST RESULTING FROM THIS PROJECT

General Lessons Learned from this Project

The design of this building involved straightforward application of textbook rules. Well-known principles of climate-sensitive, energy-efficient design were applied which took into account technical issues such as thermal performance, comfort, lighting, etc., and integrated them into the formal architectural design process. The building combines heavy mass construction with direct gain in winter and shading and night-time cross ventilation cooling in summer, plus efficient daylighting to reduce artificial lighting throughout the year.

In addition to this general point, three specific ideas deserve a special mention: passive solar integration; the importance of teamwork; and the actual energy performance of the building. These are discussed below.

Passive Solar Integration

To achieve the best possible solutions to the problem of providing comfort conditions in passive solar buildings, the building use requirements must be defined as well as possible from the outset. In warm southern climates, passive technologies must be evaluated from the early stages in the overall context of the building design taking into account winter heating, summer cooling and daylighting. Auxiliary heating and air conditioning should be evaluated in terms of the actual needs of the building.

Teamwork

A successful design can only be produced if a competent team containing specialists in all the important fields - architecture, structural, daylighting and thermal engineering - is established and works together right from the start. The creation of a successful energy-efficient building begins with the correct design options and it is essential that there is proper interaction between the different members of the design team.

In this particular building two important decisions made early on had a decisive influence on thermal performance. If they had been left to later in the design process major changes in design philosophy would have to have been made.

The first of these decisions began with a request from the thermal engineer that this be a heavy-mass building in all respects (external and internal walls, floors and ceilings). This led the architect to propose a structural option which needed detailed discussion and influenced all other aspects of the design.

The second began with the realization that, although most spaces would not need air conditioning in summer, some would. This led to the placing of the latter together in a central zone so that the cost of the air conditioning system could be reduced and its efficiency improved.

Building Energy Performance

In a temperate climate such as that found in most parts of Portugal it is possible to design buildings which require very little heating and no air conditioning and still provide occupants with comfortable conditions throughout the year. This objective should be achieved by means of intelligent building design rather than the adding of more or less sophisticated passive or active components to a less well thought out design.

Although design experience is indispensable in deciding the major options, design optimization through detailed simulation work is the only real way of proving that the design goals are achieved. In the present case, daylighting and thermal design could not havebeen optimized without the simulation studies reported earlier in this brochure. However, computer simulations only provided quantitative answers and helped in decisions related to sizes/materials. The technical solutions always have to be devised by the designers themselves.

BUILDING 2000 Building 2000 brochures are published by Directorate General XII of the Commission of the European Communities to show how design studies can help architects and other building designers use passive solar principles to the best effect to produce attractive energy-efficient buildings. Each brochure describes studies carried out with the support of the Commission during the design phase of one of thirty-six non-domestic buildings in the EC Member States. The studies were on such topics as daylighting, heating, cooling, ventilation, comfort, control systems and urban design. They were carried out with the help of acknowledged European experts in these fields and drew heavily on lessons learned and techniques developed through the Commision's research and development programme on solar energy applications to buildings.

Commission of the European Communities/Directorate-General for Science, Research and Development

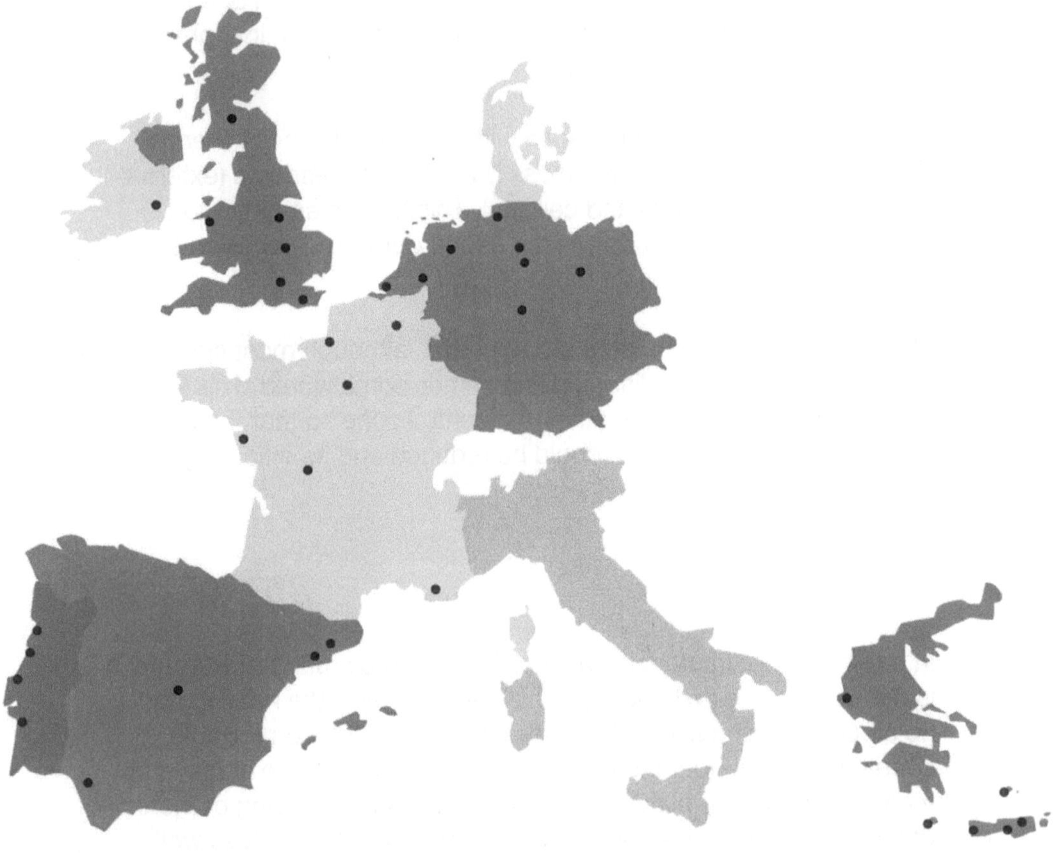

List of Design Team Participants and Advisers

Client
Fundação Gomes Teixeira/Business Innovation Centre
Universidade do Porto
R. Campo Alegre 877
4100 Porto
Portugal

Architects
Camilo Cortesão
Mercês Vieira
R. 15 de Novembro 61
4100 Porto
Portugal

Technical Support

Civil Engineering
TECNOPOR
Av. Boavista 44-5 E
4100 Porto
Portugal

Mechanical Engineering and Energy
INEGI
R. dos Bragas
4099 Porto Codex
Portugal

Electrical Engineering
FASE
R. Damião de Gois 389-A
4000 Porto
Portugal

Daylighting (Building 2000)
Dr Marc Fontoynont
36 Rue Cuvier
F-69006 Lyon
France

This set of **Building 2000** brochures illustrates how architects and other building designers can successfully apply passive solar principles to produce energy-efficient buildings.

BUILDING 2000 Participants

Project Director
Theo C. Steemers

Coordinator
Cees den Ouden

Technical Steering Committee
Dean Hawkes
Nick Baker
Alex Lohr
Jean P. Lepoivre

Regional Liaison Agents
(D) Jörn Behnsen
(E) Vicente Sifre
(F) Michel Raoust
(GB) Alan Hildon
(GR) Matheus Santamouris
(P) Eduardo Maldonado

Further information or copies of the brochures can be obtained from prof. ir. Cees den Ouden, EGM Engineering BV, P.O. Box 1042, 3300 BA Dordrecht, The Netherlands.

BUILDING 2000

Commission of the European Communities

- School of Architecture building for Technical University of Lisbon new campus
- Good daylighting in all rooms
- Cross Ventilation and shading for summer cooling
- Solar gains to provide heating in winter
- Automatic control of artificial lighting

ARCHITECTURE FACULTY BUILDING

LISBON/PORTUGAL

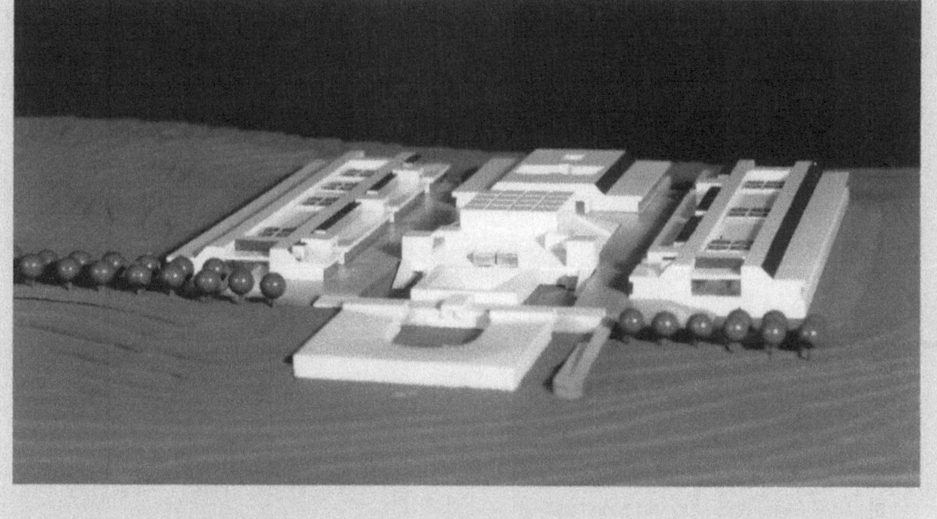

Building 2000 is a series of design studies illustrating passive solar architecture in buildings in the European Community.

ISSUE L 3

FEB 1991

PROJECT DESCRIPTION

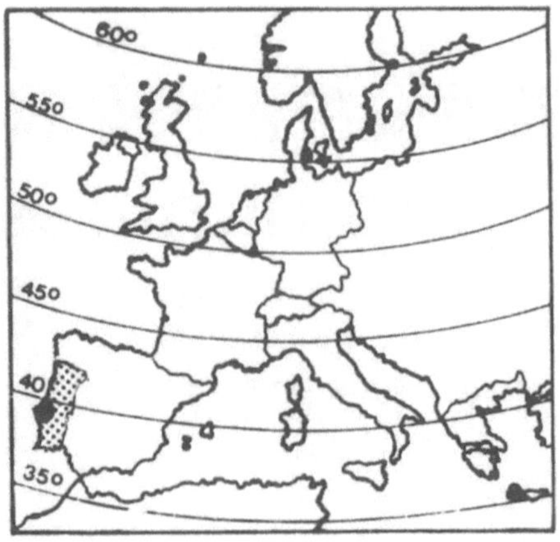

Figure 1. Location of Lisbon

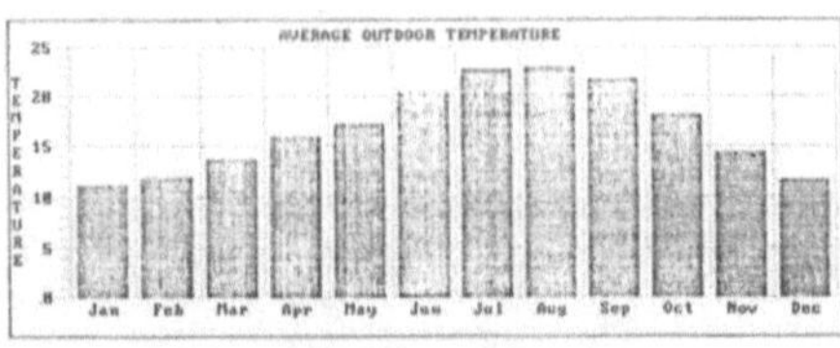

Figure 2. Mean external temperatures

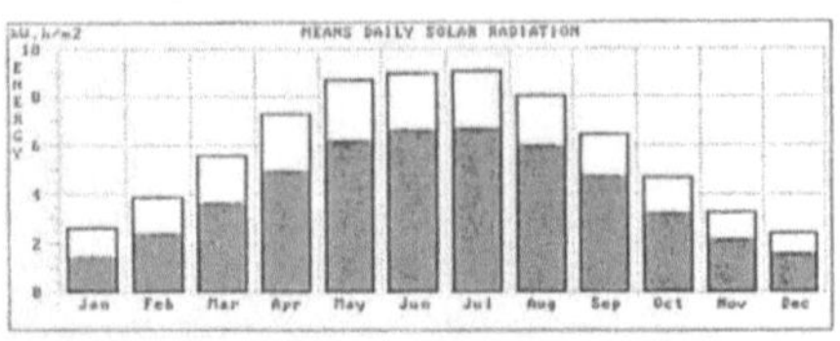

Key: diffuse
direct

Figure 3. Mean daily solar irradiation data

Sunshine hours per year	2788
Ambient temperatures	
Jan maximum	21 °C
Jan minimum	-0.5 °C
Jan mean minimum	10.25 °C
Aug maximum	39 °C
Aug minimum	10 °C
Aug mean maximum	24.8 °C

Table 1. Some key climate data

Building Type

The new campus for the Technical University of Lisbon being constructed in the next few years will include a new School of Architecture for the Faculty of Architecture. The complete faculty will consist of a number of buildings. It is the main building containing the studios and classrooms which is the subject of this brochure.

Location

The new campus is 110 m above sea level at Alto da Ajuda on the outskirts of Lisbon (latitude 38 ° 7' N). This is in a quiet and open part of a green area, near the river Tejo. The site covers 25,000 m^2 of land and slopes to the south. It is unobstructed to the south and is protected by large trees on the northern and eastern edges.

Site Microclimate

The climate is typically Mediterranean with moderate temperatures. Some key climate data are given in Figures 2 and 3 and Table 1. There are large daily temperature swings in summer. Winds are from the north-west. Breezes tend to be from the south-west and north-east.

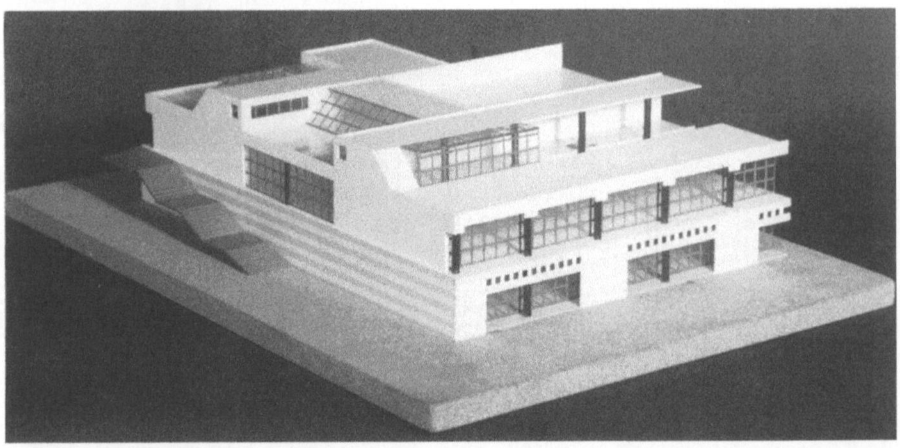

Figure 4. View of model of studio building from the south-east

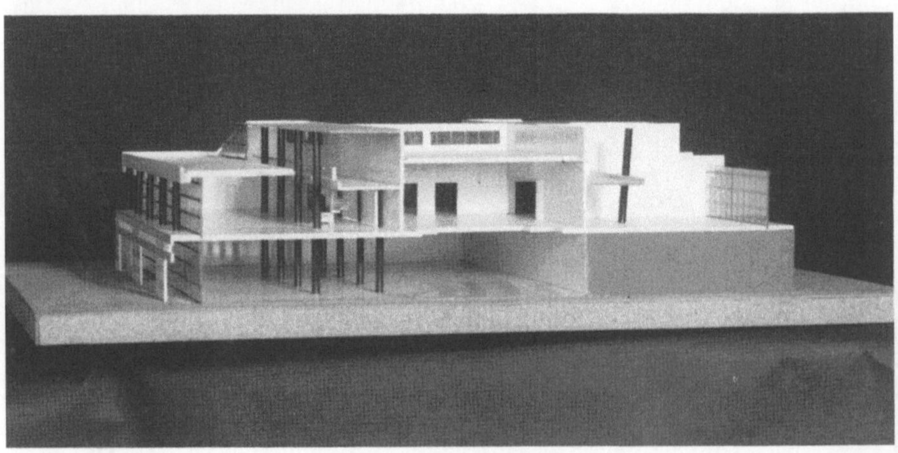

Figure 5. Section

Design and Construction Details

The studio building consists of a covered patio bordered by the entrance block on the east and two-storey studio/classroom blocks to the north, south and west. The north block is for first- and second-year students, the south block for third- and fourth-year students and the east block for fifth-year students (see Figure 7).

The building has a precast concrete frame with 330 mm cavity walls made of hollow bricks finished with sand/cement render. The ground floor consists of 150 mm concrete slabs on a 200 mm gravel bed and damp proof course. It is finished with ceramic tiles. The first-floor balcony/gallery floor is made of precast concrete slab finished with the same ceramic tiles. The roofs have precast concrete frames with insulation and weather-proof membranes. Flat roofs are finished with mosaic concrete tiles (400 mm x 400 mm) on a 60 mm air gap. Internally, the ceilings are finished with panels. Windows consist of double layers of brown-tinted glazing and are weather- stripped with aluminium.

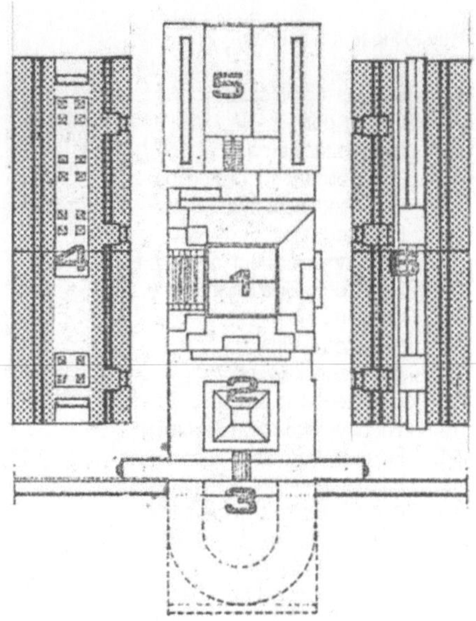

Key:

1. Patio
2. Entrance building
3. Entrance
4. South block of studios/classrooms
5. West block of studios/classrooms
6. North block of studios/classrooms

Figure 7. Plan of studio building

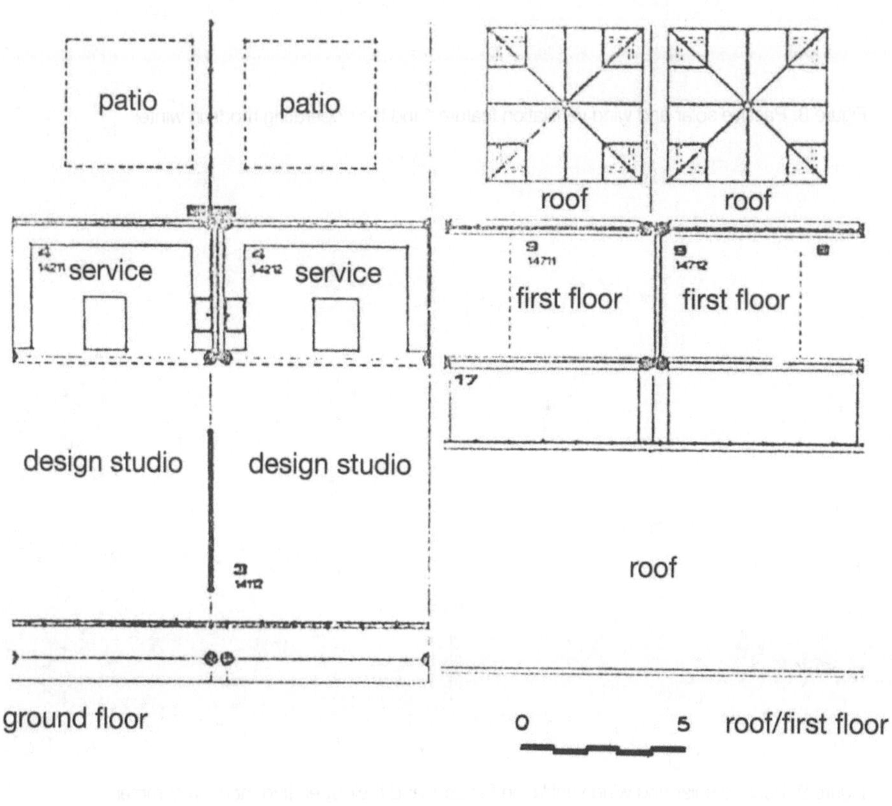

Figure 6. Dimensions of various parts of the building

DESCRIPTION OF PASSIVE SOLAR FEATURES/COMPONENTS

Key to Figures 8 and 9:

A. Studios facing south
B. Central patio
C. Studios facing north
1. Roller blind/night protection
2. Rock bed (heat store)
3. Insulated duct
4. Perpendicular duct (heat store)
5. Insulated hollow pillar
6. Wind turbine
7. Insulated ducts
8. Adjustable louvers
9. Insulated false ceiling
10. Window/ventilation (summer)
11. Fresh air from north
12. False ceiling
13. Ducts. Air panel with shutters
14. Thermal and ventilated ceiling
15. Reflected louvers
16. Shading in summer
17. Thermal mass (ceramic tiles)

The aim was to produce an energy-efficient building with good daylighting where thermal conditions close to comfort level can be achieved throughout the year with minimum use of auxiliary heating in winter or mechanical ventilation in summer.

The study of the heating, cooling and daylighting aspects of the design described later in the brochure was carried out after the preliminary architectural work had been completed. The results of the studies, therefore, were incorporated into the original design. The passive solar and wind ventilation features and their mode of operation to produce heating in summer and cooling in winter are illustrated on a north-south section of the studio building in Figures 8 and 9.

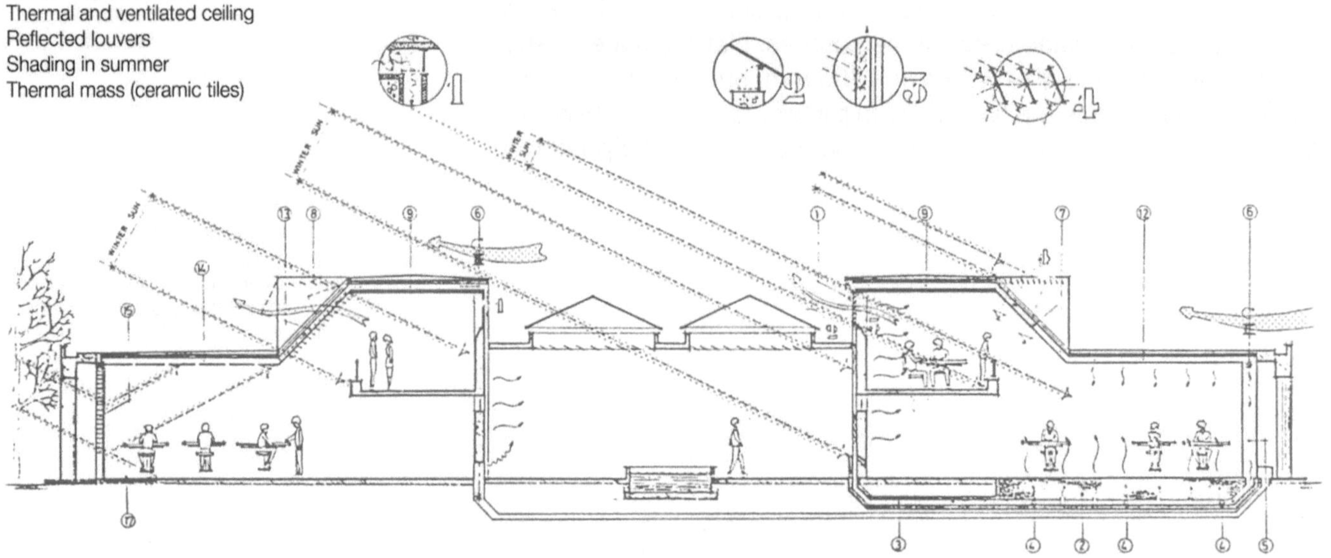

Figure 8. Passive solar and wind ventilation features and their operating mode in winter

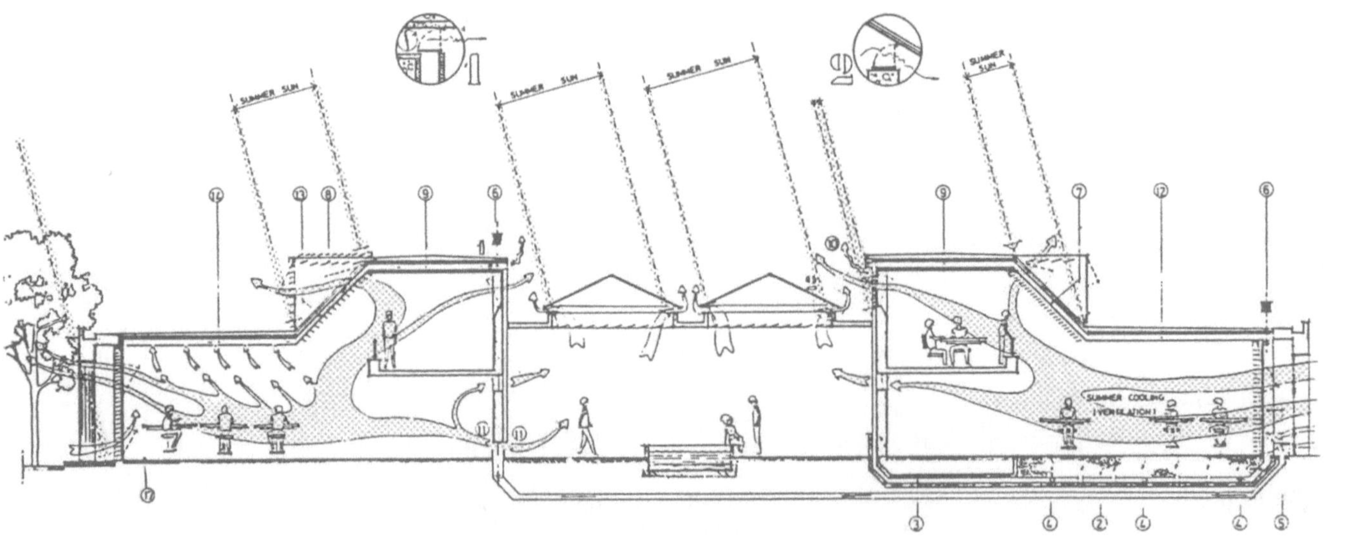

Figure 9. Passive solar and wind ventilation features and their operating mode in summer

ENERGY CALCULATIONS PERFORMED AND DESIGN TOOLS USED

Design Tools

The passive solar, wind and daylighting aspects of the design were studied with the help of a number of design tools. The thermal performance was evaluated using LT and Methode 2000. The daylighting analysis was carried out by determining daylight factors (DF) using a 1:20 scale model and a VU "Luximetro 9" daylight factor meter and comparing the results with calculated values for the minimum level of illumination needed for comfort in each room in the building.

Energy Performance Evaluation

In winter, the aim is to use the prevailing north-west winds to help transfer solar energy collected in the south part of the building through ducts to the rooms facing north. In summer, north-south ventilation is used to cool the building, especially at night when the ventilation rate is increased. Opening louver windows at night further supports night-time cooling. Manually-controlled external shades prevent over-heating in daytime.

The performance evaluations showed that the average temperatures achieved in the building month-by-month throughout the year will be as shown in Figure 10.

The total annual energy requirements will amount to 800,000 kWh (2,880,000 MJ). 39% of this is provided by solar energy. (This is used for winter heating and natural lighting throughout the year.) 24% is provided by wind energy for ventilation. 17% is provided by internal gains. (These are high because the building has a high level of occupancy.) The remaining 20% is provided by auxiliary energy.

A month-by-month breakdown of the solar gains and auxiliary heating and cooling energy is given in Figure 11.

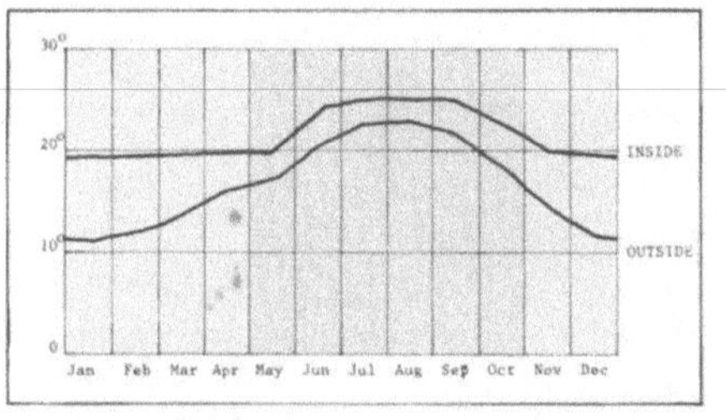

Key: cold months
warm months

Figure 10. Average temperatures achieved in the building compared with the average external temperatures

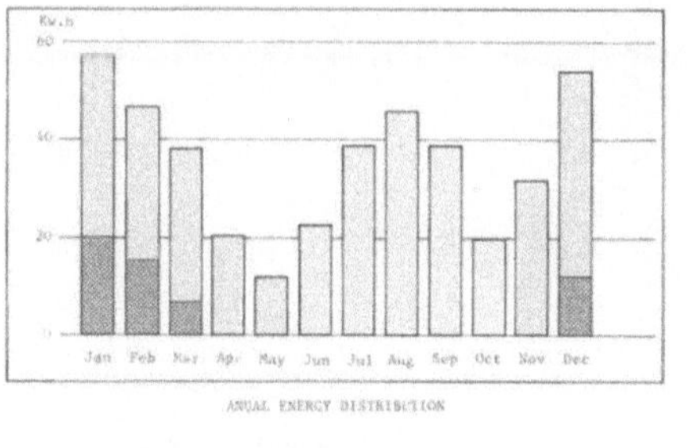

Key: solar gains
auxiliary heating
auxiliary cooling

Figure 11. Breakdown of solar gains and auxiliary heating and cooling requirements

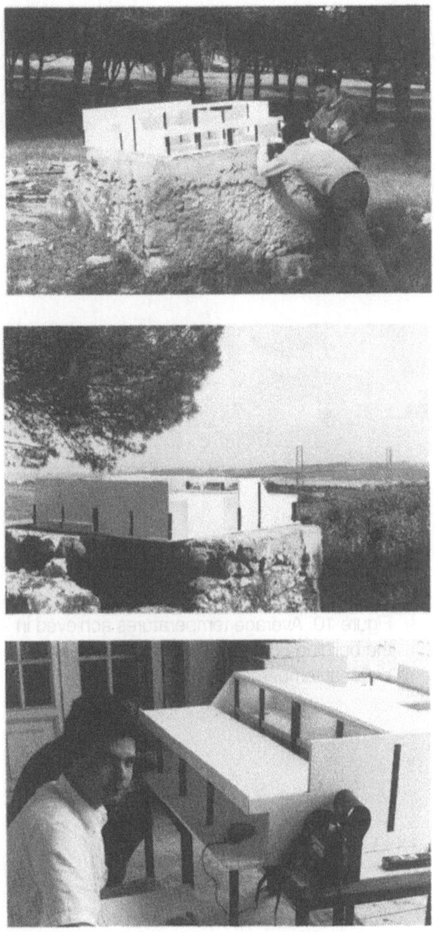

Figure 12. Photographs of daylighting tests

Daylighting Analysis

The initial daylighting tests on the 1:20 scale model were made at the Alto da Ajuda site. These were followed by evaluations conducted at Philips Lighting Division with the help of a lighting expert. The studies were carried out 1 m from the floor at 12 noon to maximize both natural lighting and the discomfort caused by glare and overheating, etc.

The daylight factors achieved in the ground-floor and first-floor rooms of the south-facing blocks of the model are shown in Figure 13.

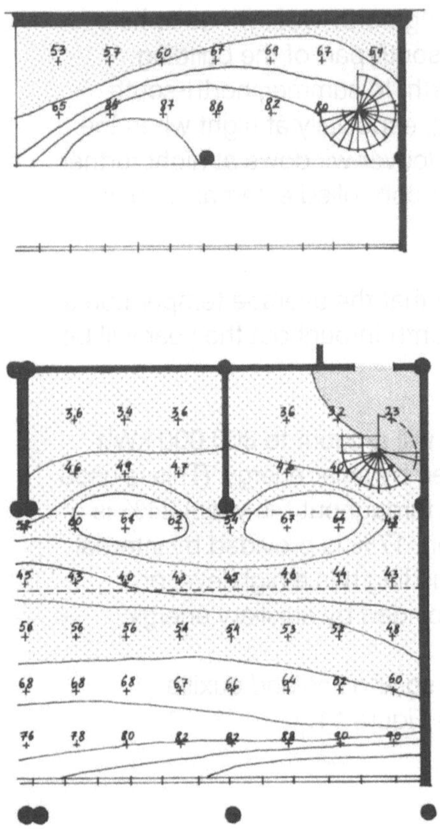

Ground-floor south-facing

Key: DF < 3%
DF 3-6%
DF > 6%

Figure 13. Daylight factors in ground- and first-floor rooms of south-facing studio blocks of model

Comparing these test results with the calculated minimum comfort values led to the conclusion that the experimental DF values were too high. To overcome this, it was decided that the vertical windows on the south wall of the south block should be fitted with full height reflective external venetian blinds to reflect light to the side or to the ceiling. The south-facing roof lights should be protected with external blinds. The equivalent glazing on the north block should be protected with similar external blinds but angled to reflect the winter sunlight back through the glazing. The natural lighting of the studio service area under the balcony should be improved by addition of a window to allow light to enter from the patio. This window will also aid cross ventilation in summer.

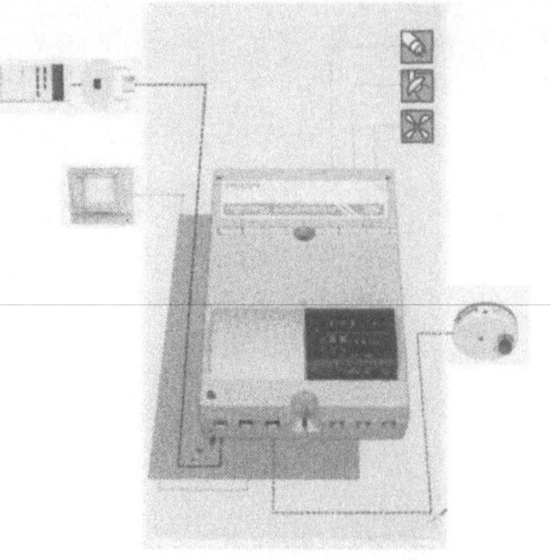

Figure 14. Local photocells control luminaires and lighting levels in different parts of the building .pa

Heating and Lighting Control Systems

The auxiliary heating and lighting systems can be manually operated but there are also automatic control systems. A computerized light switching system developed by Philips Lighting Division is programmed for day and night, summer and winter use. Sensors connected to local photocells allow the lighting to be switched off automatically when there is sufficient natural light in different parts of the building (see Figure 14). The system also controls motorized blinds according to lighting level and internal temperatures.

BUILDING 2000 Building 2000 brochures are published by Directorate General XII of the Commission of the European Communities to show how design studies can help architects and other building designers use passive solar principles to the best effect to produce attractive energy-efficient buildings. Each brochure describes studies carried out with the support of the Commission during the design phase of one of thirty-six non-domestic buildings in the EC Member States. The studies were on such topics as daylighting, heating, cooling, ventilation, comfort, control systems and urban design. They were carried out with the help of acknowledged European experts in these fields and drew heavily on lessons learned and techniques developed through the Commision's research and development programme on solar energy applications to buildings.

Commission of the European Communities/Directorate-General for Science, Research and Development

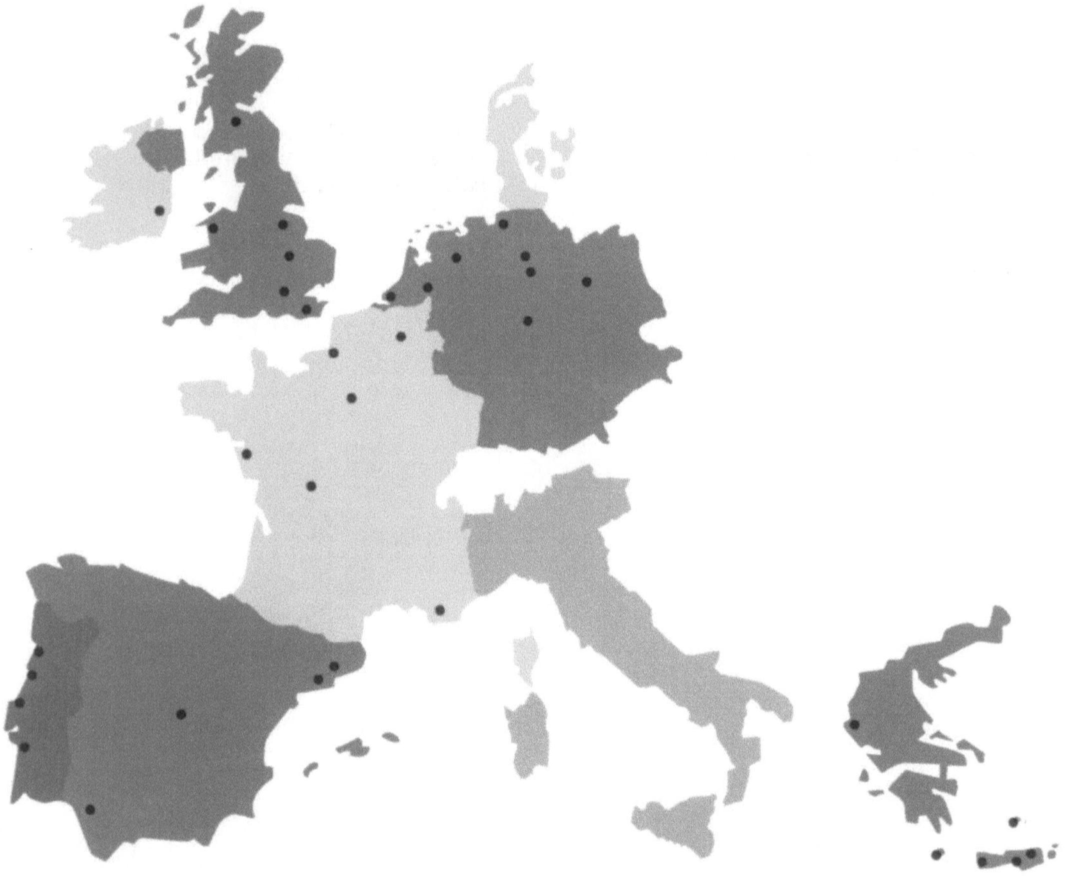

List of Design Team Participants and Advisers

Client
Universidade Téchnica de Lisboa,
Alameda Lomto Antonio dos Capucha
no. 1
1100 Lisboa
Portugal

Architecture
Cabinete das Novas Instalaçóes
(GNI) da Univeridade Téchnica de
Lisboa
R. Alexandre Herculano, 57 - 3°
1200 Lisboa
Portugal

Passive Solar Architecture
GNI and Research Group on
Environmental Science,
Faculdade de Arquitectura de Lisboa
Largo Academia Nacional de Belas
Artes
1200 Lisboa
Portugal

Building 2000 Adviser
J. Owen Lewis
Energy Research Group
School of Architecture
University College Dublin
Richview
Clonskeagh
Dublin 14
Ireland

This set of **Building 2000** brochures illustrates how architects and other building designers can successfully apply passive solar principles to produce energy-efficient buildings.

BUILDING 2000
Participants

Project Director
Theo C. Steemers

Coordinator
Cees den Ouden

Technical Steering Committee
Dean Hawkes
Nick Baker
Alex Lohr
Jean P. Lepoivre

Regional Liaison Agents
(D) Jörn Behnsen
(E) Vicente Sifre
(F) Michel Raoust
(GB) Alan Hildon
(GR) Matheus Santamouris
(P) Eduardo Maldonado

Further information or copies of the brochures can be obtained from prof. ir. Cees den Ouden, EGM Engineering BV, P.O. Box 1042, 3300 BA Dordrecht, The Netherlands.

BUILDING 2000

Commission of the European Communities

- Medical school building for University of Ioannina's new campus.
- Central atrium with carefully-designed roof, shading devices and ventilation provides attractive meeting and circulation space which is comfortable throughout the year and reduces heating and cooling requirements of surrounding building.
- Laboratory/office blocks heated in cold season by means of direct gain, ventilation air preheated in atrium and heat pump units fed with air preheated in atrium.

SCHOOL OF MEDICINE

IOANNINA/GREECE

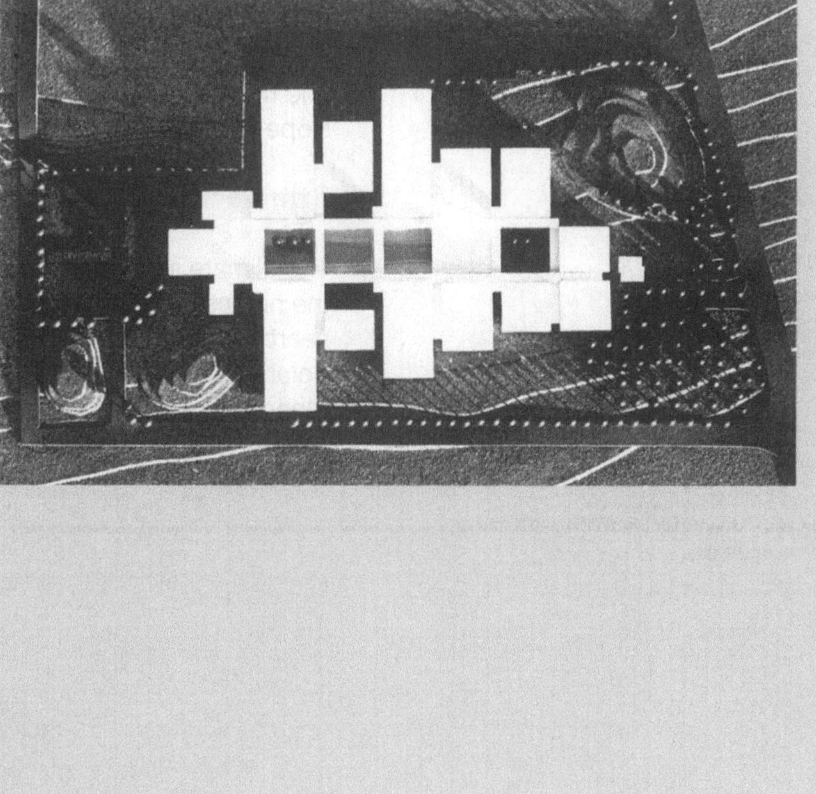

Building 2000 is a series of design studies illustrating passive solar architecture in buildings in the European Community.

ISSUE

L 4

FEB 1991

2

PROJECT DESCRIPTION

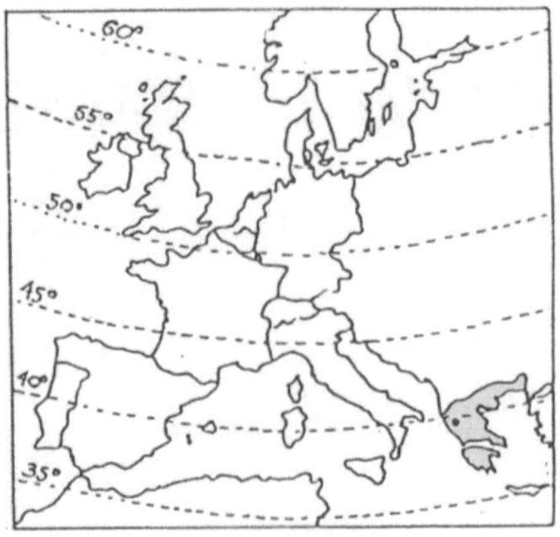

Figure 1. Location of Ioannina

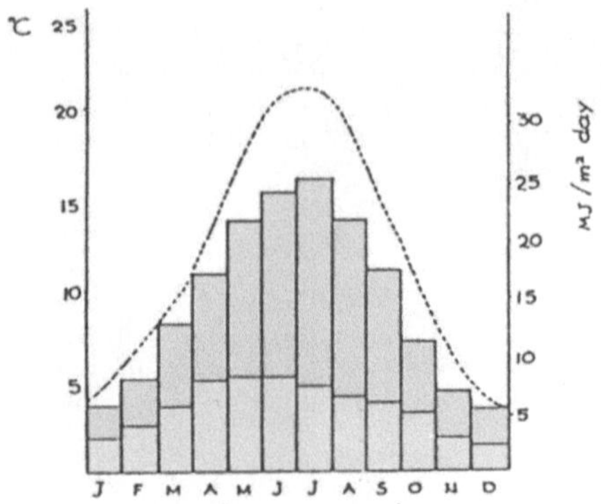

Key: external temperature
diffuse irradiation
------ direct irradiation

Figure 2. Mean solar irradiation and external temperature data

Sunshine hours	2252
Degree days (base 18 ° C)	2065
Annual global irradiation on horizontal plane	
in kWh/m^2	1524
(in MJ/m^2	5486)

Table 1. Some key climate data

Building Type

This is a new medical school building containing classrooms, laboratories, a library, auditorium and staff offices. It was commissioned by the University of Ioannina following an architectural competition.

Location

The building is being constructed on the new University of Ioannina campus between the main part of the campus (which is under construction to the south west) and the university teaching hospital (which is being built to the north). The campus is about 10 km to the south of the city of Ioannina (latitude 39 ° 42 ' N, longitude 20 ° 48 ' E) in north-west Greece (see Figure 1).

The medical school site is 500 m above sea level and has a 5% slope from west to east. The north-south axis is nearly flat.

Site Microclimate

The climate in Ioannina is among the coldest in Greece. It has one of the highest levels of rainfall. It can also be very humid because of the nearby mountains and lake. Summers tend to be short and humid. Solar irradiation and temperature data are given in Figure 2. Some other key climate data are given in Table 1.

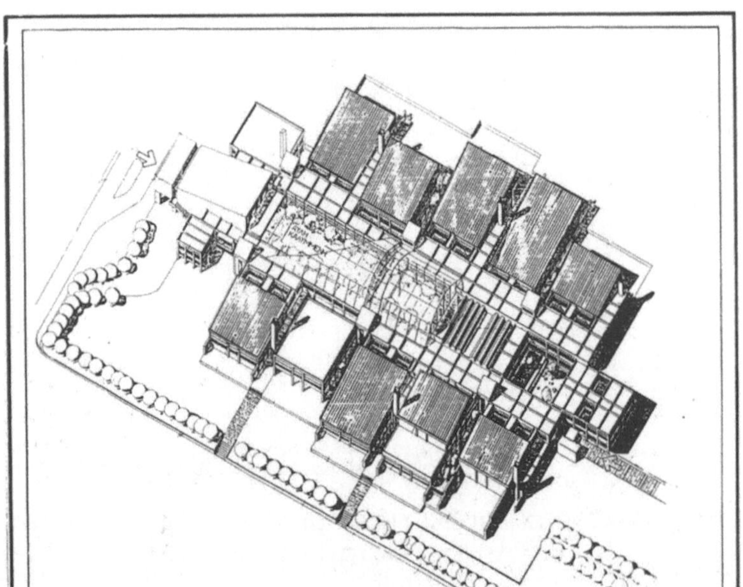

Figure 3. View from the north-east

Design and Construction Details

Because the complex is out-of-town, one of the aims was to create a building which is not merely a place of work but is also a warm, lively and interesting space pleasant to be in. It also needed to be capable of adapting to the local weather conditions, using passive solar features when appropriate, and providing students with meeting places which are sheltered from the cold and wet. The building also had to be simple and economical to construct and maintain but flexible in use and offer users high levels of comfort and quite sophisticated facilities.

The final design is shown in Figures 3-7. The main axis runs north-south, forming a link between the teaching hospital to the north and the roads and pathways to the main part of the campus to the south. The central covered courtyard (atrium) of the medical school is along this axis. Ten one- or two-storey blocks at right angles to it house the laboratories, research facilities and offices of the various medical departments (each of which is planned to be run independently) plus the main library, auditorium and classrooms. The total floor area is 23,000 m^2 and the building volume 70,000 m^3.

In order to keep construction and maintenance costs low, traditional materials have been used. The building has a concrete frame and 190 mm brick cavity walls insulated on the outside with 70 mm extruded polystyrol slabs. The insulation is protected on the vertical surfaces by a 90 mm hollow brick facing. The roofs over the heated parts of the building consist of 80 mm polystyrol foam insulation over a concrete slab; they are protected on the outside by two double layers of asphalt sheet covered with extruded perlite and cement paving. Some U-values are given in Table 2. The total construction cost is estimated to be 15 million ECU.

Walls	0.49
Roofs	0.41

Table 2. U-values of some building elements in W/m^2 K

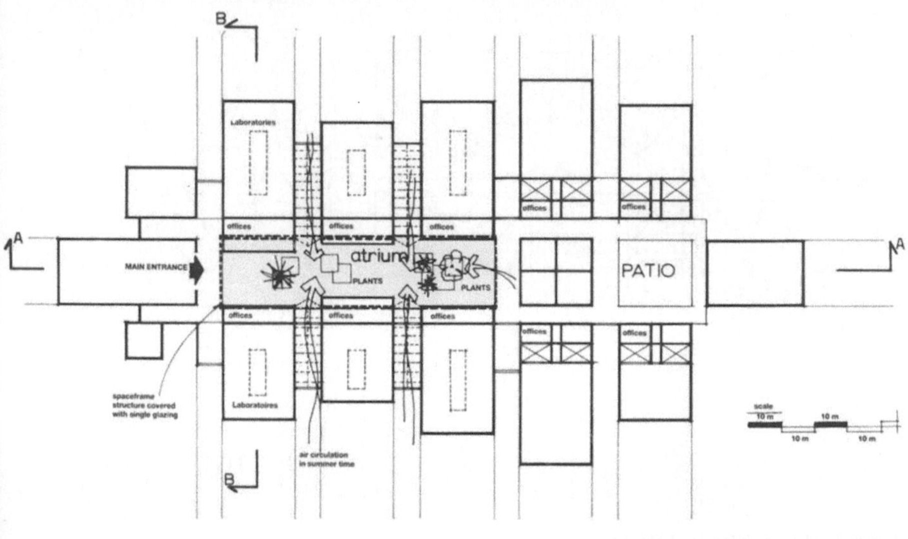

Figure 4. Ground floor plan

DESCRIPTION OF PASSIVE SOLAR FEATURES/COMPONENTS

Introduction

The building uses a number of passive solar and other energy- saving techniques as well as conventional heating and cooling systems in order to achieve maximum thermal and visual comfort as well as flexibility and energy-efficiency. Because the building is used mostly in autumn and winter, particular attention has been given to providing adequate heating in the cold part of the year.

Heat Pumps

So that the various parts of the building can be operated independently, easy-to-maintain air-to-water heat pumps have been installed in each of the departmental laboratory blocks, the library and the auditorium. The input air for these is preheated in the central atrium.

Passive Solar Features

Each building block is oriented towards the south, south-east or south-west to enable direct gain and ventilation systems to be used for heating and cooling the laboratory and classroom spaces. Rooflights help achievement of good daylighting levels throughout the building.

In addition, the courtyard running down the middle of the building has, by the addition of a roof, been transformed into a central atrium which plays a key role in the heating and cooling of the surrounding blocks and preheats the input air for the heat pumps. The atrium is also a comfortable and attractive space which serves as a useful meeting place for building users (see Figure 5).

Because of the costs of maintenance, the atrium is not provided with night- time insulation in winter. It does, however, have special shading devices and these, together with a good ventilation system, help keep the atrium and surrounding spaces cool in the summer months. Cross ventilation is achieved by opening windows on the sides of the atrium oriented in the direction of the prevailing breeze.

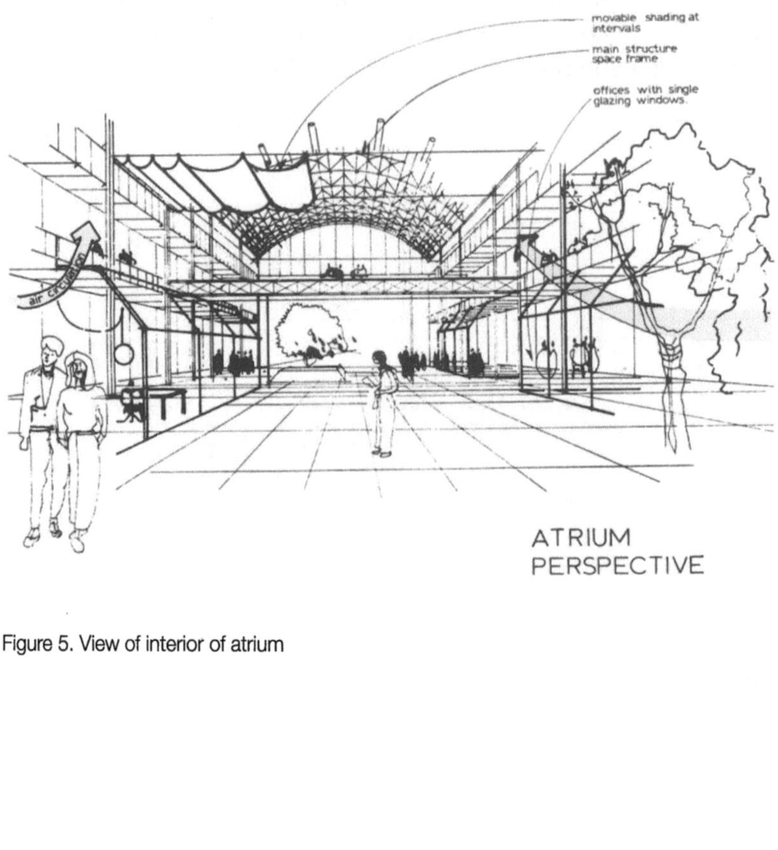

Figure 5. View of interior of atrium

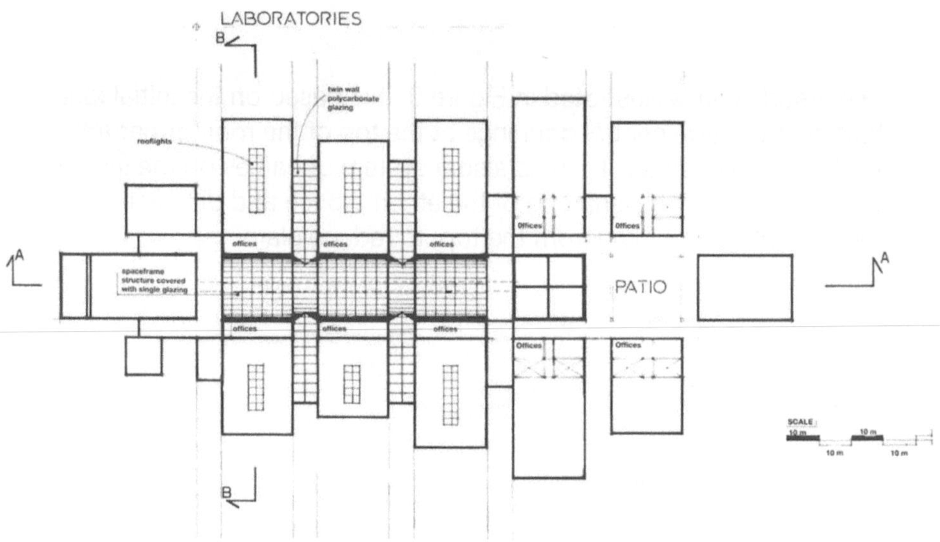

Figure 6. Plan of final design. (A discussion of the development of this design is given in the energy calculations section later in this brochure.)

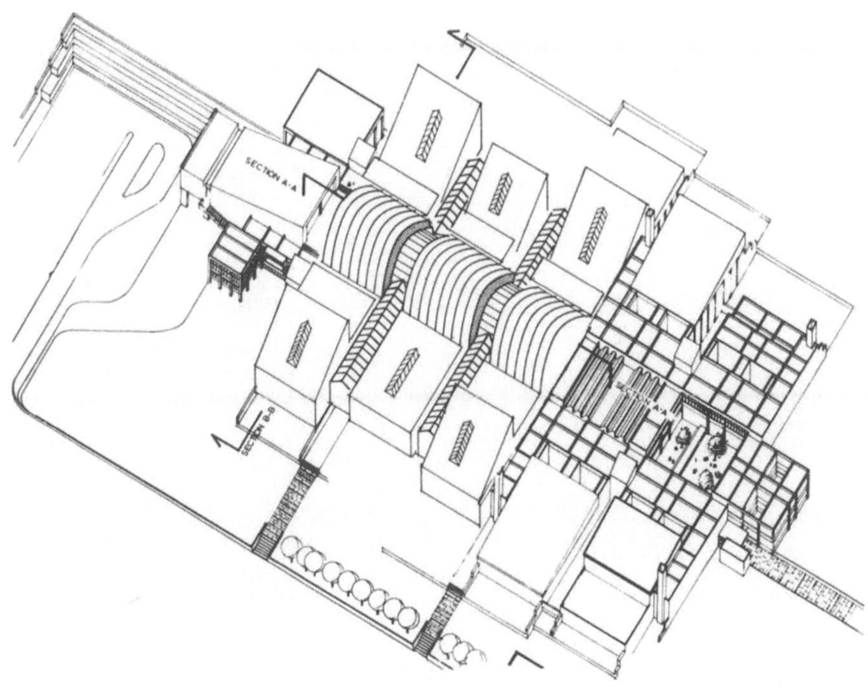

Figure 7. View of final design from north-east

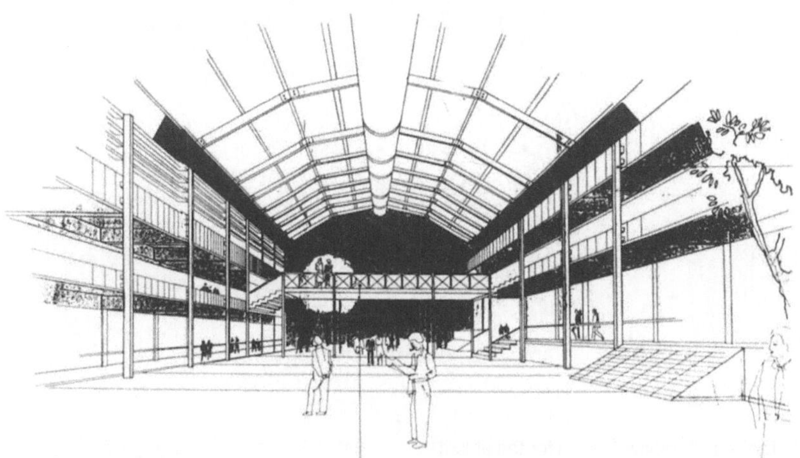

Figure 8. The preliminary design for the atrium

Auxiliary Heating

Auxiliary heating is provided by a conventional boiler system designed to meet the needs of the communal spaces and classrooms and serve as emergency back-up for the laboratories.

Control of Heating, Cooling and Lighting Systems

Because the building contains a number of interacting heating, cooling and lighting systems, a microprocessor-based energy management system is used to ensure that the right strategy is chosen at any one moment for optimizing comfort levels and energy savings.

ENERGY CALCULATIONS PERFORMED AND DESIGN TOOLS USED

Introduction

Design of the atrium was carried out with the help of thermal performance evaluations using computer simulation models, fluid modelling ventilation studies and daylighting tests. The prime aim was to develop a space which was comfortable throughout the year without the use of auxiliary heating and cooling and which reduced the heating requirements of the rest of the building during the cold season when the medical school is heavily occupied. Air from the atrium is to be used as preheated input for the heat pump units in the nearby laboratory blocks.

Preliminary Design Work

A number of initial designs were prepared all of which involved covering the central courtyard with a transparent polycarbonate roof to create what is essentially a giant greenhouse with a volume of 30,000 m^3 (see Figure 8). The roof was capable of being shaded in summer by movable exterior or interior shading devices. The atrium was naturally ventilated through openings on the east and west sides exposed to the prevailing summer breezes. The thermal performance of the preliminary designs was evaluated using a simulation model and an approximate calculation method involving a series of combinations of mean internal temperatures and air change rates. These preliminary studies showed that the central space and circulation areas round the perimeter of the atrium could be thermally comfortable in winter without the need for auxiliary heating. Moreover, the atrium could reduce the heating needs of the surrounding building by some 15- 20%. The studies showed, however, that the atrium temperature in these preliminary designs could reach uncomfortable levels in spring and summer.
It was thus decided to work on two further design alternatives in order to develop an atrium suitable for the Mediterranean climate.

The Two Alternative Designs

The first design is illustrated in Figure 9. It is based on the initial idea but has two large central openings at the top of the roof for natural ventilation in summer. There is also a system of baffles on the inside to be used for partial shading of the atrium space and diffusion of the natural light coming from the top, to reduce glare.

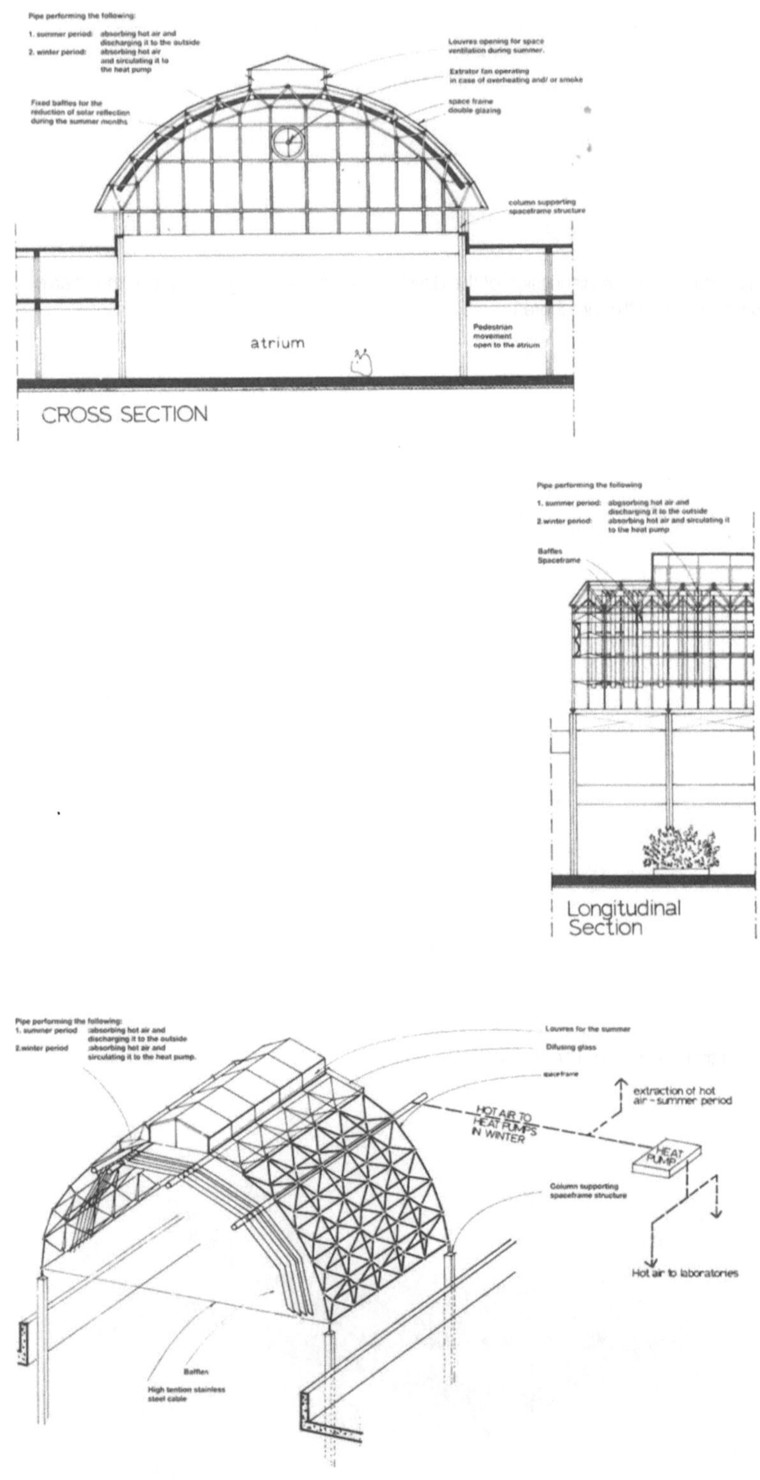

Figure 9. The first alternative design for the atrium

The second design (see Figure 10) consists of a series of shed roofs with sawtooth-shaped double-glazed vertical south-facing surfaces and opaque surfaces sloped to the north constructed of 50 mm polyurethane sandwich panels. Both designs have the same curved roof space achieved with a metallic space-frame system.

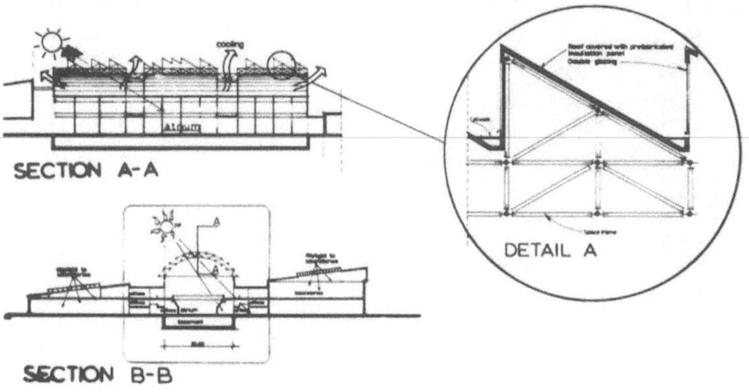

Figure 10. The second alternative design for the atrium

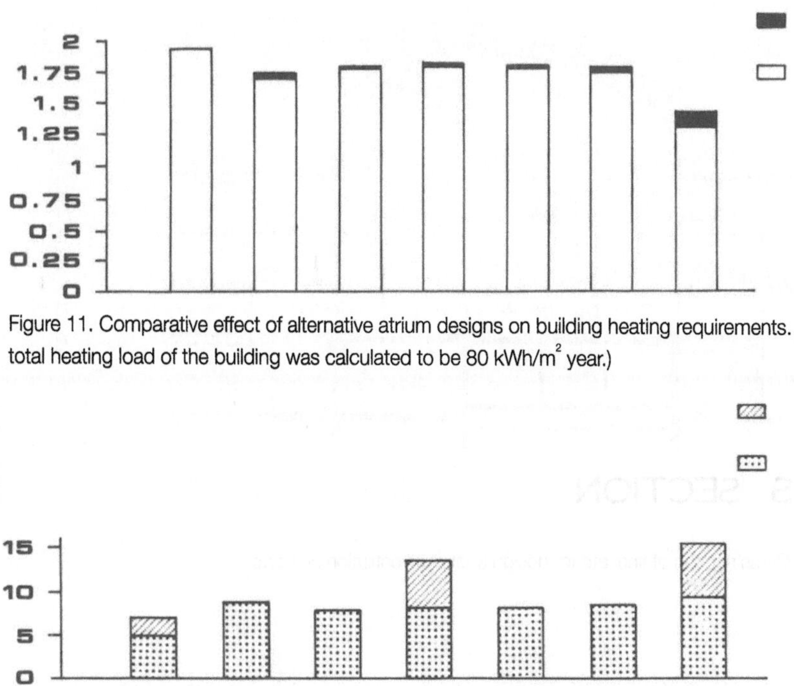

Figure 11. Comparative effect of alternative atrium designs on building heating requirements. (The total heating load of the building was calculated to be 80 kWh/m^2 year.)

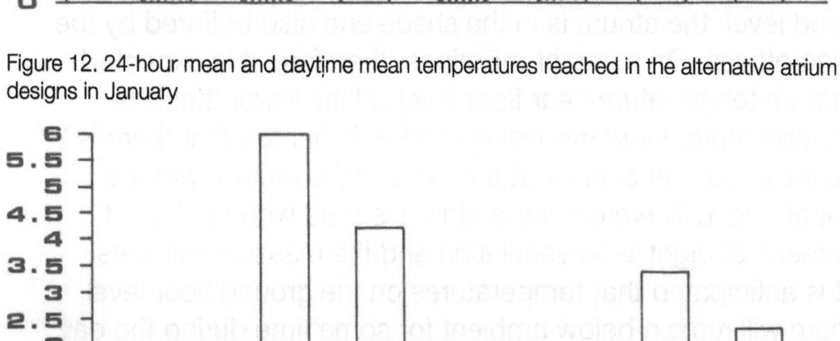

Figure 12. 24-hour mean and daytime mean temperatures reached in the alternative atrium designs in January

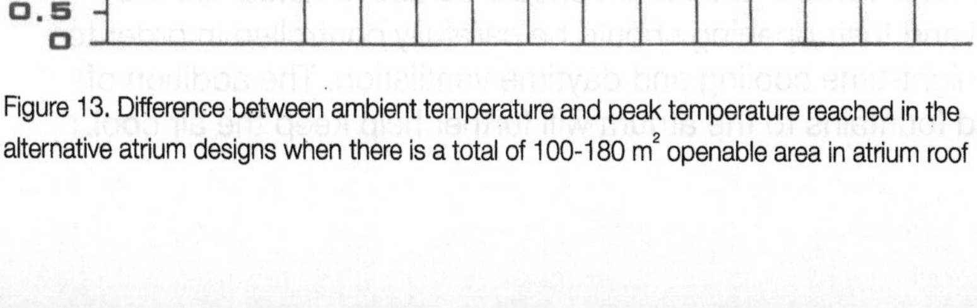

Figure 13. Difference between ambient temperature and peak temperature reached in the alternative atrium designs when there is a total of 100-180 m^2 openable area in atrium roof

Thermal Performance of the Two Alternatives

The results of the thermal performance evaluations on the two designs (and variations on them) are given in Figures 11-13. These show the comparative effect of the different atrium designs on the building heating requirements, 24- hour mean and daytime mean temperatures reached in the atrium in January, and the differences between peak summer atrium temperatures and external temperatures.

Both designs reduce the heating requirements of the surrounding building by some 6-12%. The higher end of the range is achieved if the ventilation air for the offices adjacent to the atrium is preheated in the atrium. (It is assumed that the amount of preheating is proportional to the area of office wall in contact with the atrium.)

It was concluded that, of the two roof types, the second alternative (the sawtooth design) is preferable because it requires less glazing for a similar winter performance and solar gains through the south-facing vertical glazing are easier to control in summer. Double glazing of the atrium roof has a minimal effect on energy requirements or comfort and is not justifiable in costs terms. Moreover, for the atrium to reach acceptable temperatures in winter and make a useful contribution to heating the adjacent offices, it is necessary to reduce the overall size of the atrium space and increase the thermal contact between the atrium and the offices.

Final Design

As a result of the thermal performance evaluations, it was decided:

(1) to adopt the sawtooth-type atrium roof with single-glazed operable windows and twin-wall polycarbonate glazing for the horizontal parts of the walkways in the main atrium area;

(2) to extend the circulation corridors and parts of the office blocks into the atrium in order to improve thermal contact and the daylighting of the offices;

(3) to eliminate the patios originally proposed for the office blocks (see Figure 4) as they increase fabric heat losses and are not essential for daylighting;

(4) to glaze over parts of the spaces between the office/laboratory blocks with twin- wall polycarbonate glazing of U- value 3.5 W/m^2 K. The small amount of glazing required provides thermal buffering for a large area of building envelope. As a result of doing this, large openings in the rooms overlooking these subsidiary atria will be required for daylighting and ventilation.

A cross-section of the final design of the central atrium is given in Figure 15. The plan and the view from the north-east are given in Figures 6 and 7.

Ventilation of Final Design

For the flows of ventilation air in the building to be as shown in Figure 15, the temperature of the air at the top of the atrium must be higher than that of the outside air. Positioning movable shades horizontally under the roof barrel will ensure that high temperatures are reached by retaining incoming solar gains. This is expected to provide the necessary temperature difference to induce convective flows at the top levels of the atrium.

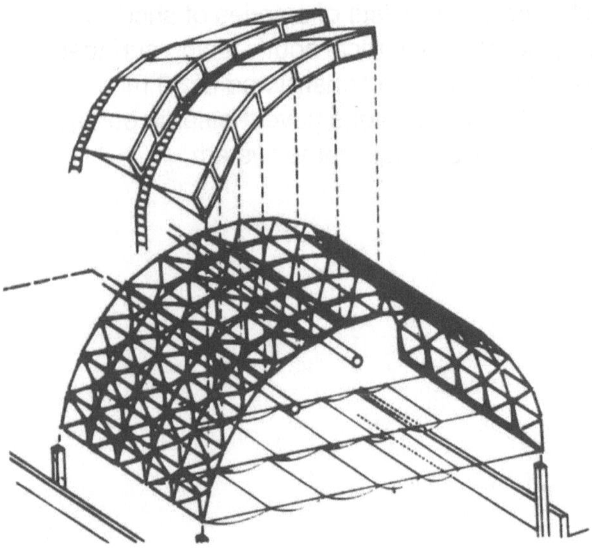

Figure 14. The final roof structure.

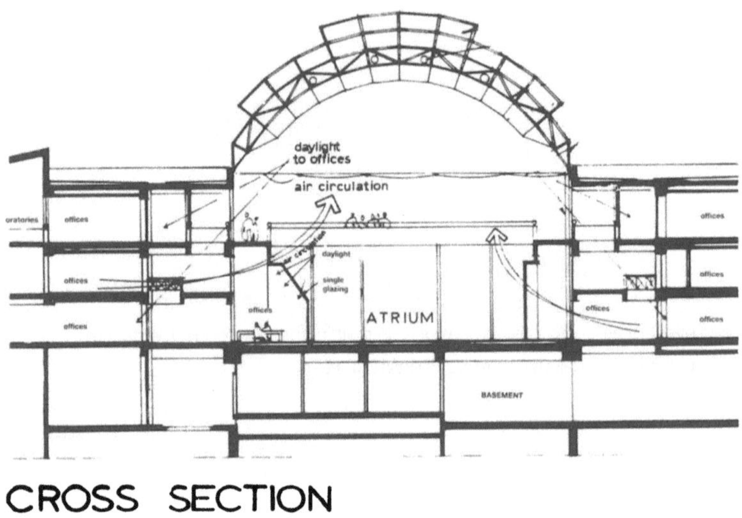

Figure 15. Cross section of final atrium design showing ventilation air flows.

Near ground level, the atrium is in the shade and also buffered by the surrounding offices. On summer mornings, therefore, it is expected that the atrium temperature near floor level will be lower than the outdoor temperature. Fluid modelling studies indicated that there is little or no mixing of the cool air at the base of the atrium with the warmer air above it. Therefore if the atrium is filled with cool air at night by means of night-time ventilation and the massive surfaces cooled, it is anticipated that temperatures on the ground floor level of the atrium will remain below ambient for some time during the day provided the pool of cool air is not disturbed by ventilation air flows. Thus, low level vents in the atrium should be above actual ground floor level and their opening should be carefully controlled in order to optimize night-time cooling and daytime ventilation. The addition of plants and fountains to the atrium will further help keep the air cool.

Shading and Daylighting in Final Design

The atrium glazing will be shaded by movable canvas shades hung at intervals below the roof. There does not need to be a continuous layer, as the sun patches will tend to be in fairly narrow bands at intervals corresponding to the geometry of the sawtooth roof.

To improve daylighting, the opaque parts of the atrium roof will be painted white and the floor will be a light colour to reflect daylight. Daylighting analysis showed that even under adverse weather conditions the lighting level in the atrium is generally satisfactory. The only slight exception is in the southern part where daylight factors are lowest. The variation of lighting levels along the north-south axis can be reduced by the use of light baffles. The total lighting load of the complete building is estimated to be 25 kWh/m^2 year.

Thermal Performance of Final Design

Some results of the thermal performance evaluation of the final atrium design are shown in the columns marked "recommendations" in Figures 11-13. The final design will reduce the heating requirements of the offices by 32- 40% for 1-0 air changes per hour between the atrium and the outside. These figures assume that the ventilation air for the offices adjacent to the atrium is preheated in the atrium.

Because of the reduction to its volume and increase in heat gains from the adjoining spaces, the atrium temperature will be higher in winter. As indicated in Figure 12, the evaluations showed that during the daytime in January (the coldest month of the year), the mean atrium temperature will be 15.5 ° C. This is perfectly acceptable for an indoor circulation space. During the rest of the year, the atrium will have a higher mean temperature.

The total cooling load of the building amounts to 7 kWh/m^2 year for the laboratory/office blocks and 30 kWh/m^2 year for the auditorium. The cooling contribution of the atrium (via shading and night ventilation) is estimated to be 16-20%.

With the sawtooth roof configuration, the sun patches in the atrium will tend to be at floor level. By using the movable shades to stop the sunshine reaching the lower parts of the atrium and opening some 182 m^2 of vent in the atrium roof, in summer the peak temperature in the atrium can be kept to within 2.3 ° C above ambient (see Figure 13). Controlled ventilation will maintain temperatures in the lower parts of the atrium (i.e. the ground floor and the walkways) below this.

GENERAL DESIGN GUIDELINES/POINTS OF INTEREST RESULTING FROM THIS PROJECT

General Rules for Passive Solar Architecture in the Mediterranean Region

The Mediterranean climate, which is quite different from that of northern Europe, is characterized by moderate heating requirements in winter and considerable cooling needs in summer. Passive solar components used in projects in the Mediterranean area, therefore, should be specifically addressed to the local climate and mode of living.

Traditionally, buildings in this part of the world were designed and constructed to that they responded naturally to the temperate climate with its mild winters and moderately warm summers. Auxiliary heating was limited or non-existent and yet the buildings remained comfortable. More recently, however, building design and practice and modern standards of living have made the use of mechanical heating and cooling devices commonplace. Climate factors have ceased to be taken into account in building design. This has caused problems over energy supply and brought pollution to the Mediterranean region which, in turn, has led to an interest in energy conservation and passive solar features.

The realization that energy conservation and passive solar techniques developed for northern climates are not necessarily applicable to the Mediterranean area has led to the search for new solutions. These had to be appropriate for buildings requiring moderate heating in winter and considerable cooling in summer. They also had to be able to cope with rapid changes in the weather. The aim has been to develop an architecture which increases human comfort without raising energy consumption and uses the natural techniques found in vernacular architecture.

Examples of traditional climate-responsive features which it is appropriate to use today include correct siting and landscaping at community and individual building level, organization of buildings round courtyards, use of suitable materials and construction techniques, use of adjustable openings to control ventilation and daylighting, location of transitional spaces between the interior of the building and the outside world to provide areas with microclimates adapted to daily and seasonal needs and, finally, designing different parts of the building so that they are suited to use at different times of the day or year.

Points of Interest Resulting from This Project

As a result of the experience gained in this particular project, it can be concluded that, in the Mediterranean region:

1. The presence of thermal inertia in a building is of fundamental importance. It evens out temperature swings throughout the year and determines the nature of the summer-time cooling system.

2. Shading devices are also very important because they control overheating and glare. A combination of permanent fixtures on the outside of the building and adjustable shades on the inside should be used. The correct mix will depend on the building size and type and the ability of the users to control the shades. In the Ioannina building, fixed shading devices were used on the outside of the north-facing inclined sides of the atrium shed roofs and retractable canvas shades were hung at intervals below the roof inside the atrium. The aim is to reduce overheating in spring, summer and early autumn.

3. Each space should be well ventilated in a way which is appropriate to its use. The need may vary with the season and time of day according to the pattern of Mediterranean life. Natural ventilation can be achieved by appropriate choice of building materials and components, the geometry of the spaces and facing openings towards the prevailing winds.

4. To improve thermal and visual comfort, light-coloured walls and roofs should be used as much as possible.

5. Finally, introduction of passive solar components into a building should not be treated in an inflexible way or as a remedy for all types of problems. Each building should be considered individually and passive solar features introduced into the overall design concept so that the best possible balance of thermal and visual comfort is achieved and the final building meets the occupants' practical and aesthetic needs.

Human Aspects of the Ioannina Building

The final Ioannina design seems to be suited to a university building in the climate conditions of northern Greece. It is expected not only to be energy efficient but also to have a comfortable noise- and pollution-free environment which encourages students and faculty to become involved in energy conservation research. It is the first public building in Greece where energy conservation and passive solar features have been used in combination and as such will serve as an example to designers and public authorities in the area and, indeed, throughout the country.

BUILDING 2000 Building 2000 brochures are published by Directorate General XII of the Commission of the European Communities to show how design studies can help architects and other building designers use passive solar principles to the best effect to produce attractive energy-efficient buildings. Each brochure describes studies carried out with the support of the Commission during the design phase of one of thirty-six non-domestic buildings in the EC Member States. The studies were on such topics as daylighting, heating, cooling, ventilation, comfort, control systems and urban design. They were carried out with the help of acknowledged European experts in these fields and drew heavily on lessons learned and techniques developed through the Commision's research and development programme on solar energy applications to buildings.

Commission of the European Communities/Directorate-General for Science, Research and Development

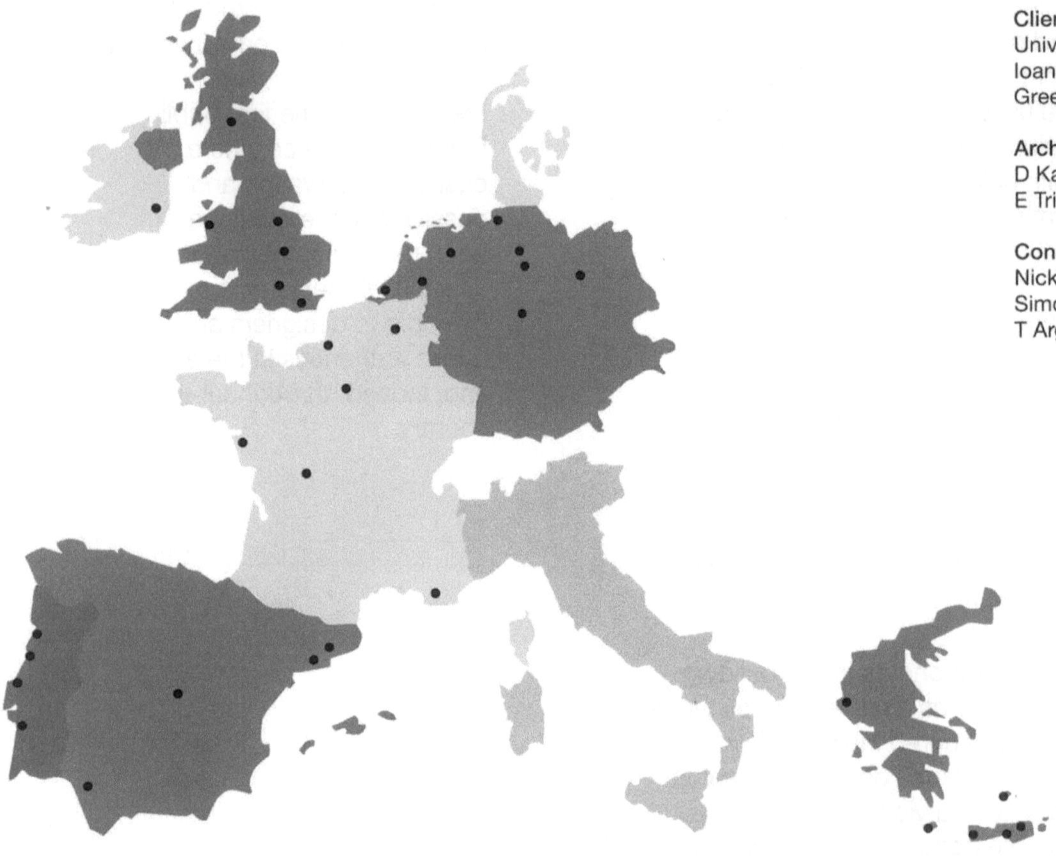

List of Design Team Participants and Advisers

Client
University of Ioannina
Ioannina
Greece

Architects
D Kataropoulos
E Triantis

Consultants
Nick Baker
Simos Yannas
T Argyriou

This set of **Building 2000** brochures illustrates how architects and other building designers can successfully apply passive solar principles to produce energy-efficient buildings.

BUILDING 2000
Participants

Project Director
Theo C. Steemers

Coordinator
Cees den Ouden

Technical Steering Committee
Dean Hawkes
Nick Baker
Alex Lohr
Jean P. Lepoivre

Regional Liaison Agents
(D) Jörn Behnsen
(E) Vicente Sifre
(F) Michel Raoust
(GB) Alan Hildon
(GR) Matheus Santamouris
(P) Eduardo Maldonado

Further information or copies of the brochures can be obtained from prof. ir. Cees den Ouden, EGM Engineering BV, P.O. Box 1042, 3300 BA Dordrecht, The Netherlands.

BUILDING 2000

Commission of the European Communities

- Research complex in south of France
- Layout and building design aim to provide comfortable indoor and outdoor work spaces throughout the year
- Traditional measures of cooling, shading and heating include use of sunspaces, venetian blinds, shelter belts of trees and water pools and streams
- New passive solar techniques include diode walls and enthalpy collectors

GEOSCIENCES RESEARCH LABORATORY

AMPUS/FRANCE

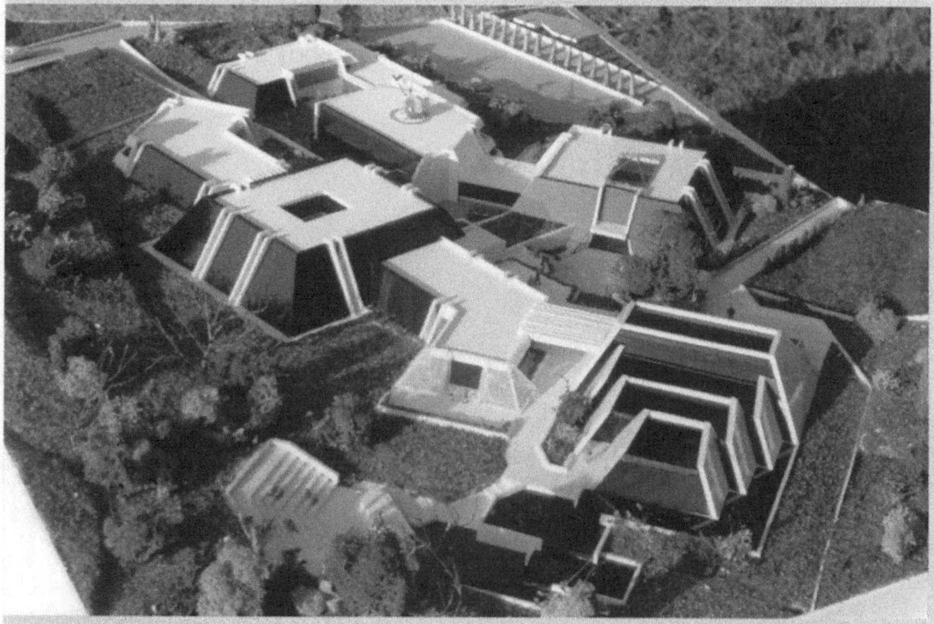

Building 2000 is a series of design studies illustrating passive solar architecture in buildings in the European Community.

ISSUE L 5

FEB 1991

PROJECT DESCRIPTION

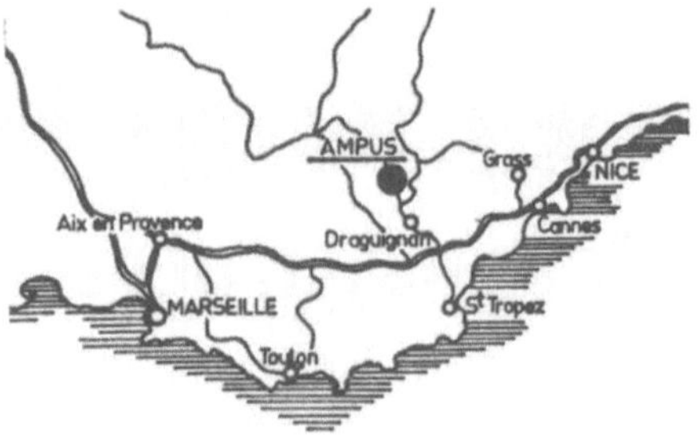

Figure 1. Location

Degree days	1980
Sunshine hours	
Oct-Apr	1400
Jan-Dec	2820
lobal irradiation on a horizontal plane	
Oct-Apr in kWh/m²	2417
(Oct-Apr in MJ/m²	8700)
Jan-Dec in kWh/m²	1056
(Jan-Dec in MJ/m²	3800)

Table 1. Some key climate data. (Source: Meteorologie Nationale Francaise)

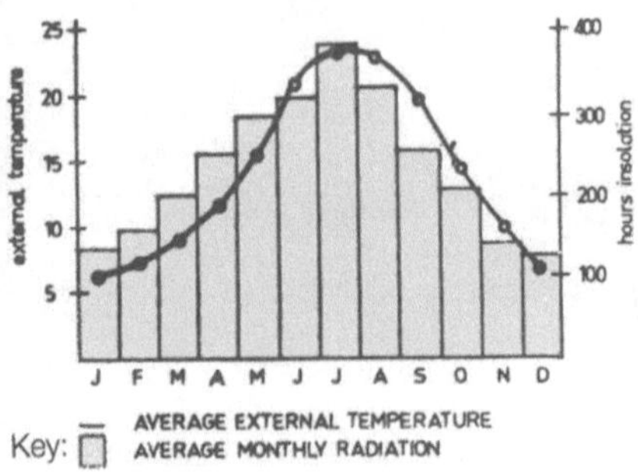

Key:

Figure 2. Average monthly solar irradiation and ambient temperature data

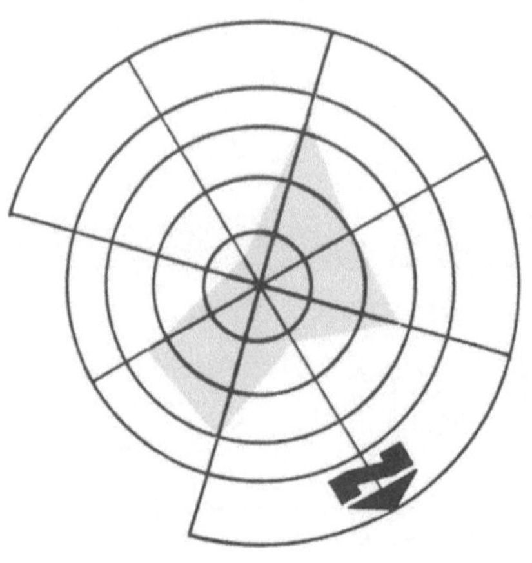

Figure 3. Prevailing winds

Building Type

This is a research complex for use by 150 students, 50 temporary assistants, 30 research staff and 20 service personnel of the Institut Méditerranéen des Géosciences (IMG).

IMG is a federation of organizations in the Provence-Alpes-Côte d'Azur region of France which are concerned with the earth sciences and associated data processing and communications operations. It carries out scientific research in collaboration with several local universities - Nice, Toulon, Aix, Marseille and Avignon. It has achieved financial independence by providing a number of services.

Location

The complex is on a 30,000 m site 650 m above sea level at Ampus. This is an hour's drive from Nice or Marseille and is in the high-technology belt between Toulon and Sophia-Antipolis.

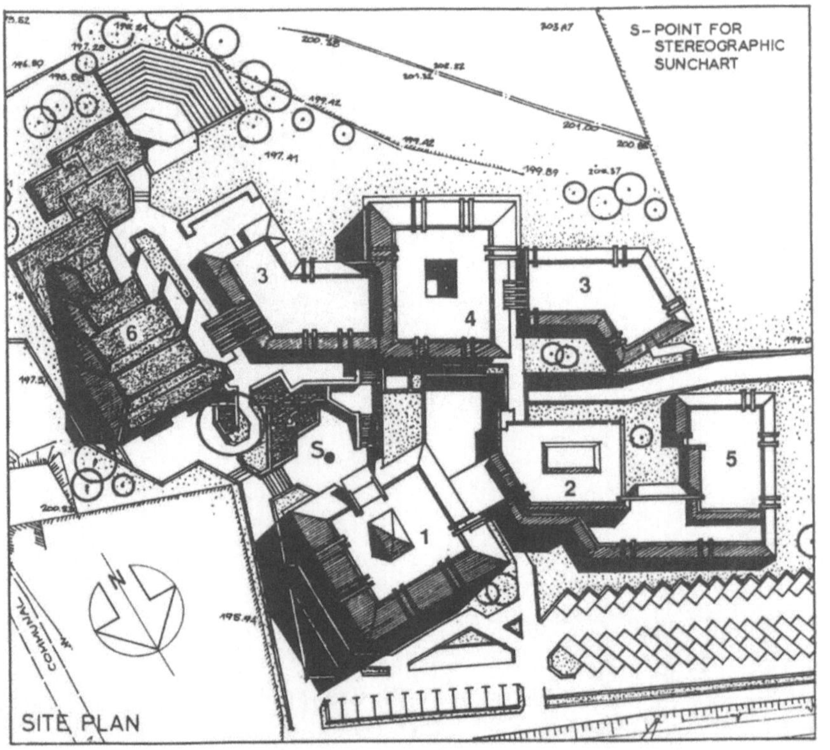

Key:
1. reception, scientific centre
2. research centre
3. research units
4. training centre
5. library, cafeteria
6. conference hall

Figure 4. Site plan

Site Microclimate

Ampus has record sunshine levels (there are 300 sunny days a year) but the weather is less clement there than at places on the coast. The dominant winds are: the mistral, a cold, dry wind from north to south; the marin, a hot wind from the south (i.e. the sea); and the tramontane, a hot dry wind from north to south. Some climate data for the site are given in Table 1 and Figures 2 and 3.

The site is not overshadowed by neighbouring buildings.

Design and Construction Details

The complex (which has an estimated cost of 9 M ECU) has a floor area of 6,820 m^2 and a volume of 18,840 m^3.

The layout (see Figure 4) takes into account key features of the site - the slope, the best views over the surrounding area and the means of access - all of which face north.

The natural difference in levels suggested that the complex should be divided into two distinct parts, linked above ground by a bridge and below ground by storage areas (see Figures 5, 6 and 7).

Pedestrian and vehicular movement is separated. The centre of pedestrian traffic is a square next to the entrance of the main building.

An amphitheatre for summer use is integrated into the landscape.

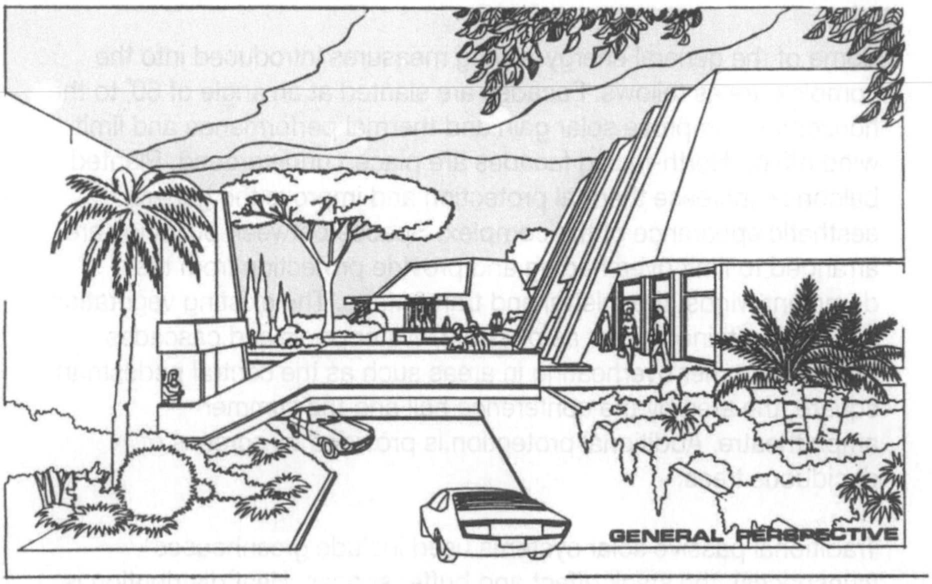

Figure 5. General perspective

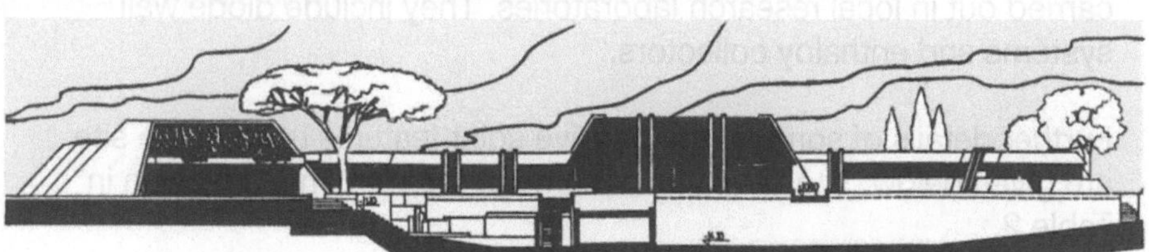

Figure 6. North elevation

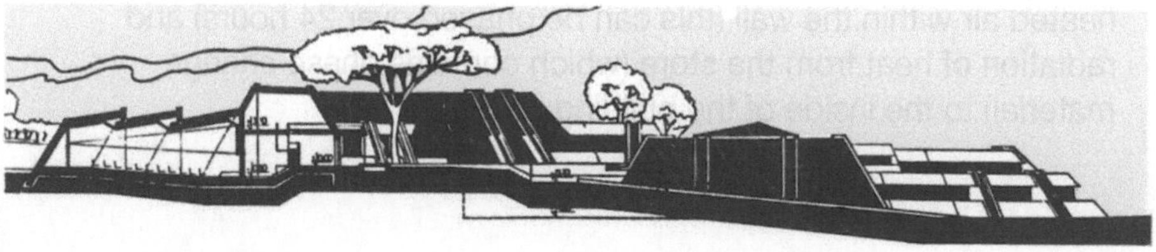

Figure 7. West elevation

DESCRIPTION OF PASSIVE SOLAR FEATURES/COMPONENTS

roof	0.28
external walls	0.47
diode walls	0.55
south-facing windows	5.50
north-facing windows	3.20

Table 2. U-values of some building elements (W/m² K)

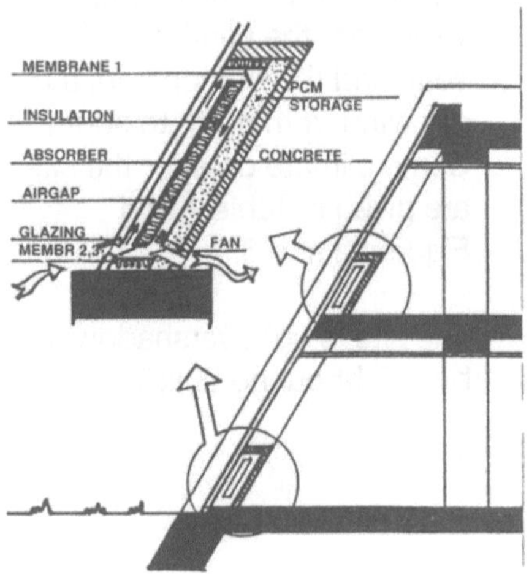

Figure 8. Diode wall system

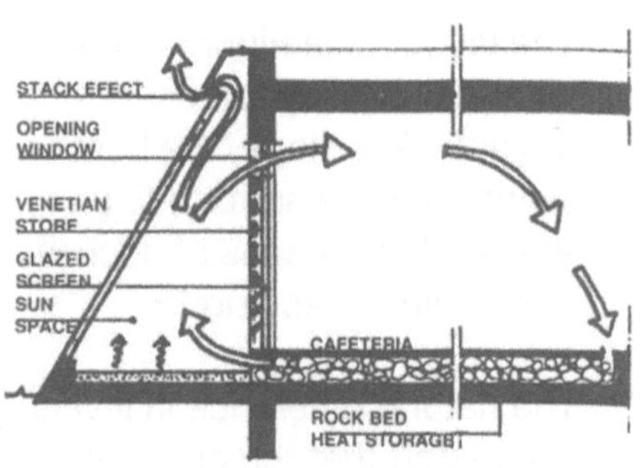

Figure 9. Greenhouse system

The aims of the project are to express the work of the Institut through the geometric form of the complex and to use various complementary passive solar systems to achieve optimal comfort levels throughout the year and reduce overall energy consumption by 60%. It is hoped that the complex will provide a permanent demonstration of solar technologies adapted to the climate conditions of the south of France.

Some of the general energy-saving measures introduced into the complex are as follows. Facades are slanted at an angle of 60° to the horizontal to improve solar gain and thermal performance and limit wind effect. North-facing facades are placed underground. Planted balconies increase thermal protection and improve the general aesthetic apearance of the complex. Spaces between buildings are arranged to limit overshading and provide protection from the dominant winds, the mistral and tramontane. The existing vegetation has been retained as far as possible. Water pools and cascades prevent summer overheating in areas such as the central pedestrian square, the area by the conference hall and the summer amphitheatre. Additional protection is provided by screens of deciduous trees.

Traditional passive solar systems used include greenhouses (sunspaces), the stack effect and buffer spaces. Heat distribution is provided by a low temperature underfloor system. All the collector zones and patios serve as thermal stores. Glazed areas are equipped with blinds to reduce direct solar gains.

Other passive solar techniques used in the complex result from work carried out in local research laboratories. They include diode wall systems and enthalpy collectors.

Further details of some of the passive solar features used at the site are given below. The U-values of some of the elements are given in Table 2.

Diode Walls

The diode wall system shown in Figure 8 is used on all south- facing facades (i.e. those in facilities 1, 3, 4 and 5 of Figure 4) to provide heating in winter. The system involves entry of fresh air into the wall, heating the air by contact with the absorber, redistribution of the heated air within the wall (this can be phased over 24 hours) and radiation of heat from the store (which contains phase change material) to the inside of the building.

Greenhouses

The greenhouses illustrated in Figure 9 are located on the west pyramid patio, the laboratory patio and the cafeteria corners (i.e. in 1, 3 and 5 of Figure 4). In summer, they make use of the thermal stack effect to provide cooling. In winter, the heated air is transferred to the north part of the complex where it is stored in a pebble bed.

The 'Jardinieres'

The "jardinieres" shown in Figure 10 are on the north- and north-west-facing facades (i.e. in 2, 3 and 5 in Figure 4). They create buffer spaces angled at 60° and increase insulation. They also improve the building aerodynamics and prevent unnecessary turbulence.

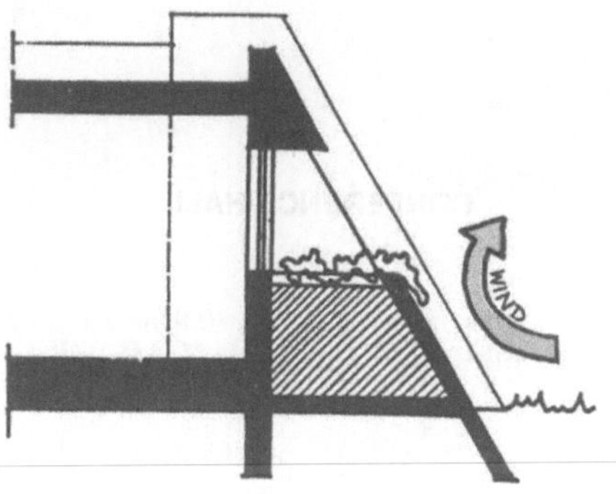

Figure 10. "Jardiniere"

Venetion Blinds

Some of the bay windows and greenhouses are equipped with venetian blinds with hollow blades containing phase change material. They have a 6 hour thermal storage capacity to prevent overheating and smooth out solar gains over the working day.

Enthalpy Collectors

The solar system used in the conference hall is new and involves use of enthalpy collectors whereby a stream of water passes over the roof. The system not only provides economical passive cooling in summer but also constitutes a very attractive architectural feature. Furthermore, it creates a confortable external environment in the nearby amphitheatre. The system is illustrated in Figure 11 and involves: circulation of low-temperature water (5-10° C); solar gain in warm weather; use of the air enthalpy in poor weather; transfer of solar gains by a water-water heat pump; heat storage.

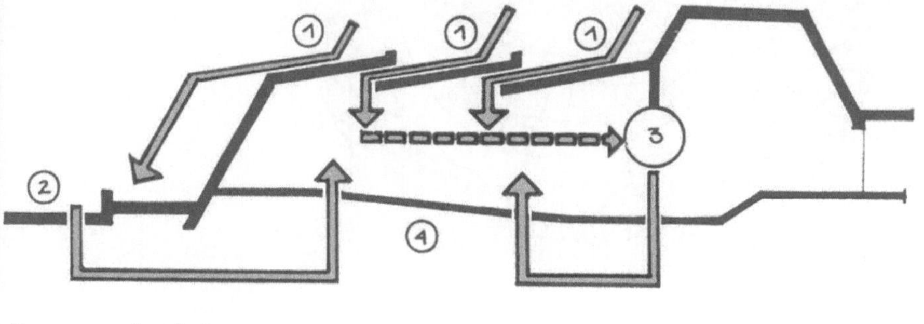

Key:
1. collector
2. pool storage
3. storage/central core
4. underfloor distribution

Figure 11. The solar system in the conference hall

ENERGY CALCULATIONS PERFORMED AND DESIGN TOOLS USED

Thermal Performance

The thermal performance of the complex was calculated with and without the benefit of the solar heating and ventilation systems. It was found that, without the solar systems, the energy load of the conference hall would amount to 240,000 kWh/year. Inclusion of the enthalpy collector system would reduce this by 50% to 120,000 kWh/year (see Figure 12). Together, the other buildings on the site would have an energy requirement of 540,000 kWh/year (i.e. 100 kWh/m^2) without the solar systems. With the solar systems the load would be reduced to 300,000 kWh/year (i.e. 55 kWh/m^2) - see Figure 13. Overall, therefore, the annual energy requirements of the whole complex would be reduced from 780,000 kWh to 420,000 kWh - a saving of 46% (see Figure 14).

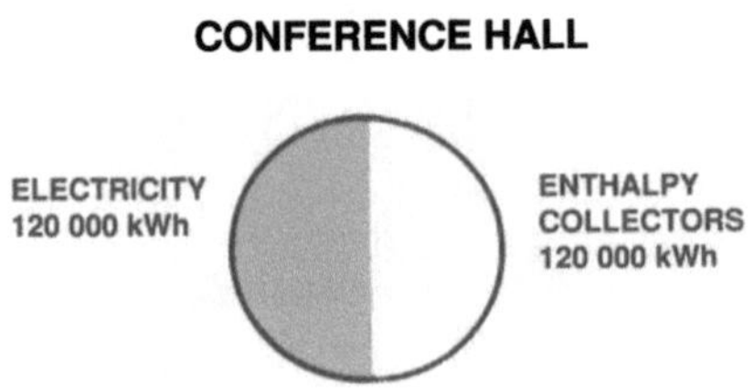

Figure 12. Annual energy requirements of conference hall with enthalpy collector system

Stereographic Sunchart

During the design stages, the stereographic sunchart illustrated in Figure 15 was used in conjunction with the site layout plan (Figure 4) to give an idea of the availability of solar radiation at particular locations, shading, etc.

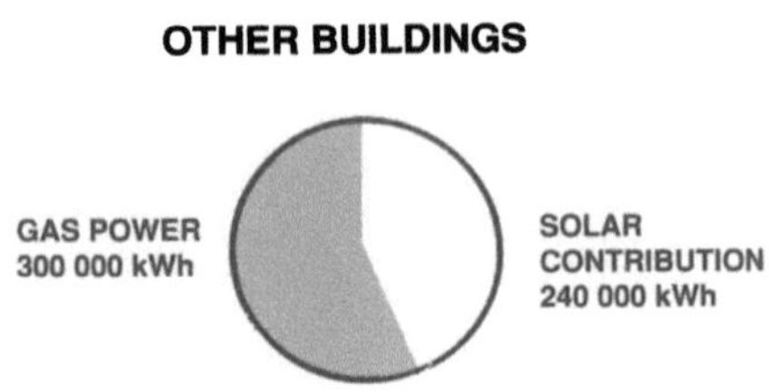

Figure 13. Annual energy requirements of other buildings after inclusion of solar systems

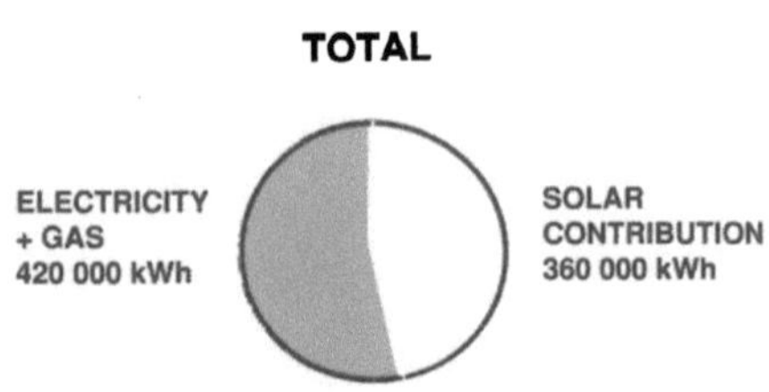

Figure 14. Annual energy requirements of whole complex after inclusion of solar systems

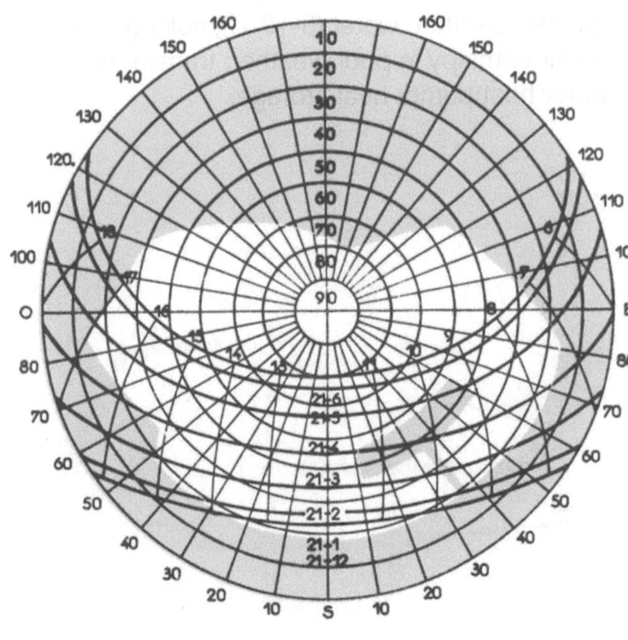

Figure 15. Stereographic sunchart

Daylighting

The penetration of daylight into the buildings in winter and summer days is shown in Figure 16. There is protection in summer and deep penetration in winter. The pyramidal form of the buildings helps optimize penetration while keeping the central patio in shadow, even in summer.

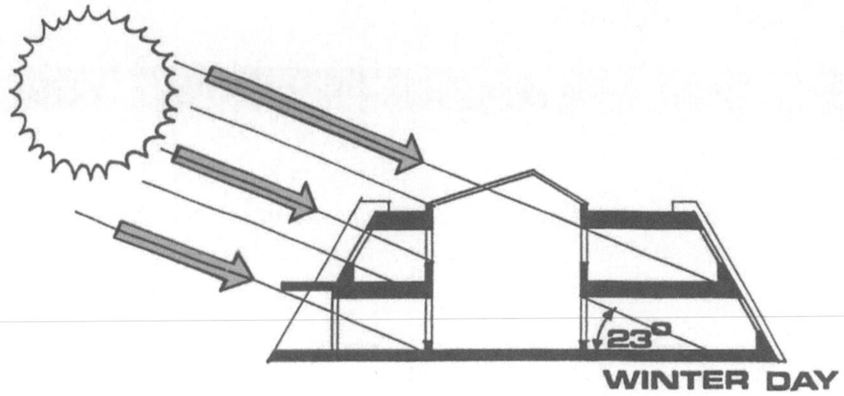

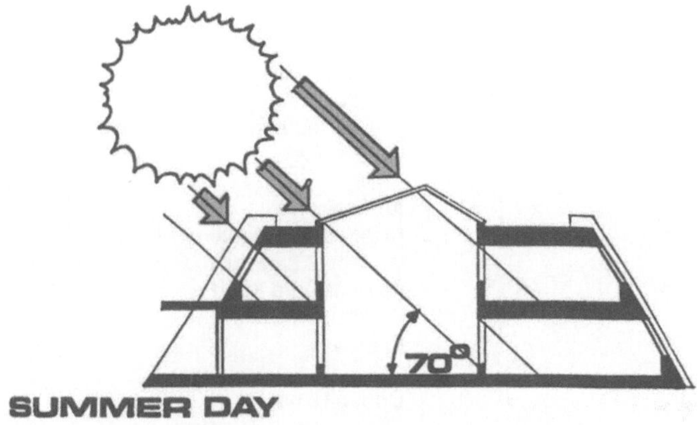

Figure 16. Penetration of daylight into the buildings in winter and summer

GENERAL DESIGN GUIDELINES/POINTS OF INTEREST RESULTING FROM THIS PROJECT

Serious delays to the building programme, which have nothing to do with the architects, have hindered realization of this project. The principal aim of the design studies, besides saving energy, has been achievement of improved comfort in the indoor and outdoor work areas. Overheating has been prevented by use of venetian blinds containing phase change material in the offices and water circulation in the conference hall. Outside, various water cascades, pools and protective belts of trees have been introduced to create a comfortable environment in summer, especially in assembly areas. In winter, the shape of the buildings optimizes collection and storage of solar gains. The pyramidal form of the taller buildings provides minimum resistance to wind and allow optimum daylight penetration.

Participation in the CEC Building 2000 project enabled the architects overcome daylighting problems and led them to modify their original plans by omitting a central store containing phase change material and giving more emphasis to passive solar systems.

BUILDING 2000

Building 2000 brochures are published by Directorate General XII of the Commission of the European Communities to show how design studies can help architects and other building designers use passive solar principles to the best effect to produce attractive energy-efficient buildings. Each brochure describes studies carried out with the support of the Commission during the design phase of one of thirty-six non-domestic buildings in the EC Member States. The studies were on such topics as daylighting, heating, cooling, ventilation, comfort, control systems and urban design. They were carried out with the help of acknowledged European experts in these fields and drew heavily on lessons learned and techniques developed through the Commision's research and development programme on solar energy applications to buildings.

Commission of the European Communities/Directorate-General for Science, Research and Development

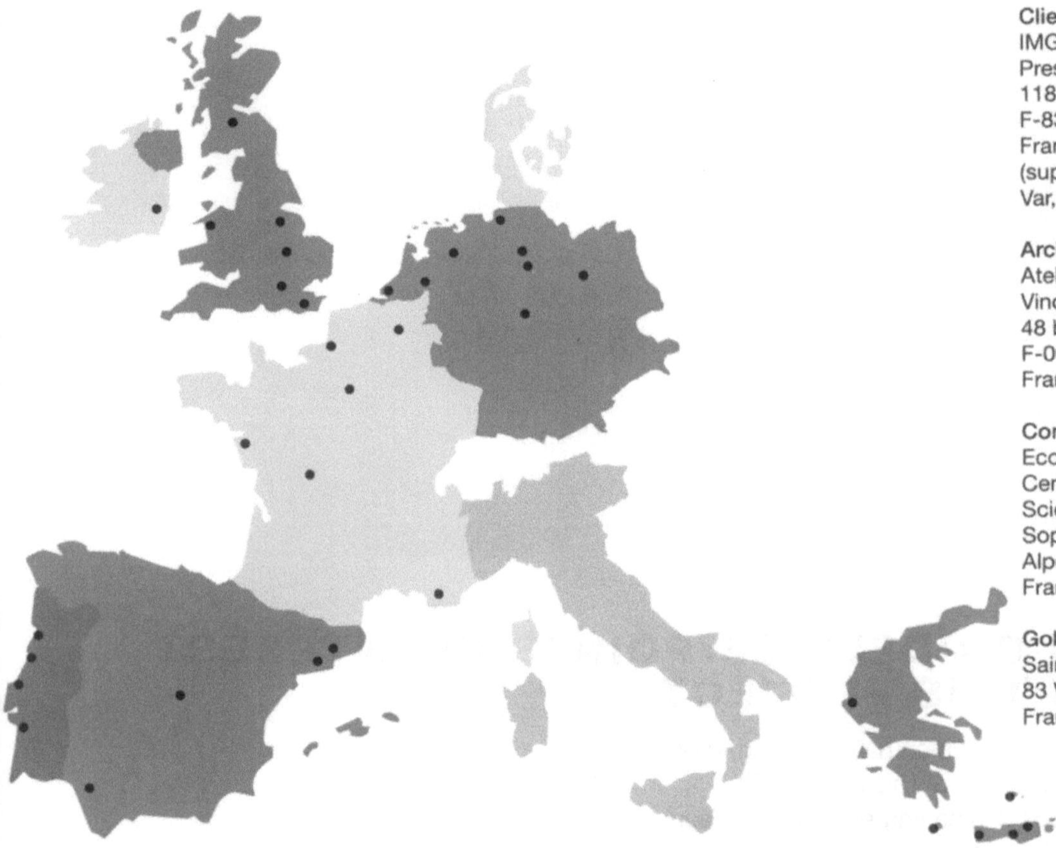

List of Design Team Participants and Advisers

Client
IMG (Professor Roger D Casanova, President)
118 avenue Julien Cazelles
F-83300 Draguignan
France
(supported by the Conseil General du Var, France)

Architect
Atelier d'Architecture VIP
Vincent Z Zarzycki/J P et I Lehembre
48 boulevard Foch
F-06600 Antibes
France

Consultants
Ecothermic Laboratory
Centre Nationale de Recherche Scientifique
Sophia-Antipolis
Alpes Maritimes
France

Golfe Ingénierie
Sainte-Maxime
83 Var
France

This set of Building 2000 brochures illustrates how architects and other building designers can successfully apply passive solar principles to produce energy-efficient buildings.

BUILDING 2000 Participants

Project Director
Theo C. Steemers

Coordinator
Cees den Ouden

Technical Steering Committee
Dean Hawkes
Nick Baker
Alex Lohr
Jean P. Lepoivre

Regional Liaison Agents
(D) Jörn Behnsen
(E) Vicente Sifre
(F) Michel Raoust
(GB) Alan Hildon
(GR) Matheus Santamouris
(P) Eduardo Maldonado

Further information or copies of the brochures can be obtained from prof. ir. Cees den Ouden, EGM Engineering BV, P.O. Box 1042, 3300 BA Dordrecht, The Netherlands.

Commission of the European Communities

BUILDING 2000

SPORTS
AND
EDUCATIONAL
CENTRES

Building 2000 is a series of design studies illustrating passive solar architecture in buildings in the European Community.

VOLUME 1 / **KLUWER ACADEMIC PUBLISHERS**

BUILDING 2000

Commission of the European Communities

- Renovation of Caneton indoor public swimming pool
- New roof with increased thermal insulation and solar collecting surface
- Preheat of fresh air by passage through solar roof
- Reduction of ventilation heat losses
- Addition of solar water heating system

INDOOR PUBLIC SWIMMING POOL

CESSON/FRANCE

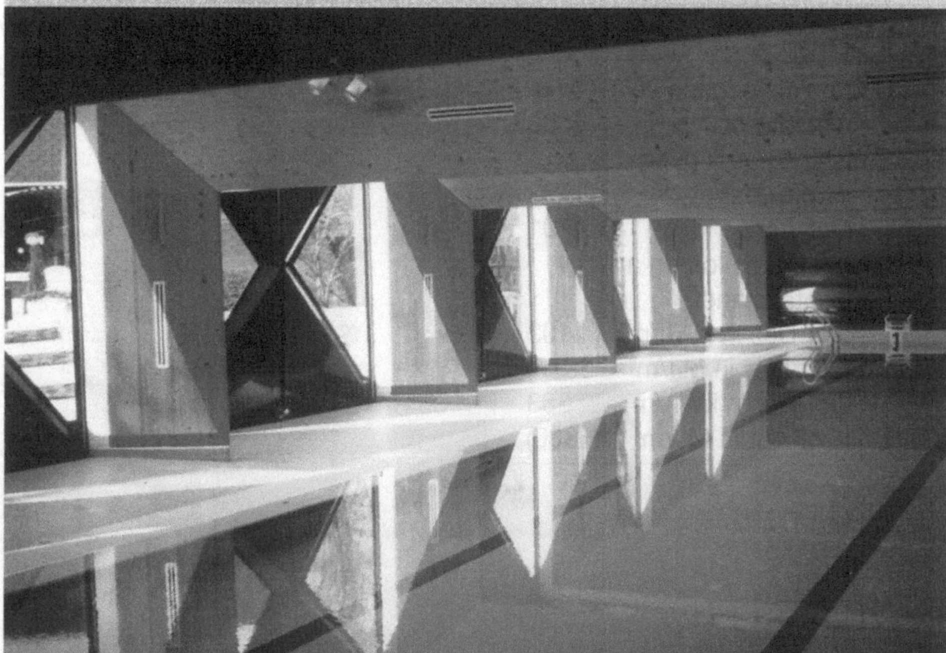

Building 2000 is a series of design studies illustrating passive solar architecture in buildings in the European Community.

ISSUE

E 1

FEB 1991

PROJECT DESCRIPTION

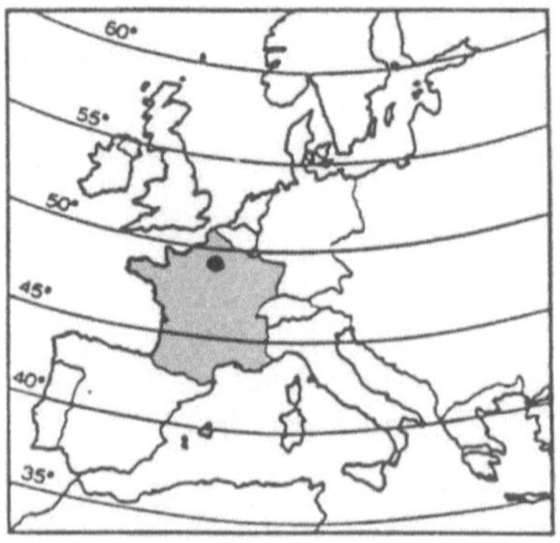

Figure 1. Location

Degree days (base 25° C) over	
whole year	5000
Sunshine hours over	
whole year	1793
Oct-Apr	701
Global irradiation on a horizontal plane	
Jan-Dec in kWh/m²	1103
(Jan-Dec in MJ/m²	3970)
Oct-Apr in kWh/m²	383
(Oct-Apr in MJ/m²	1380)

Table 1. Some key climate data

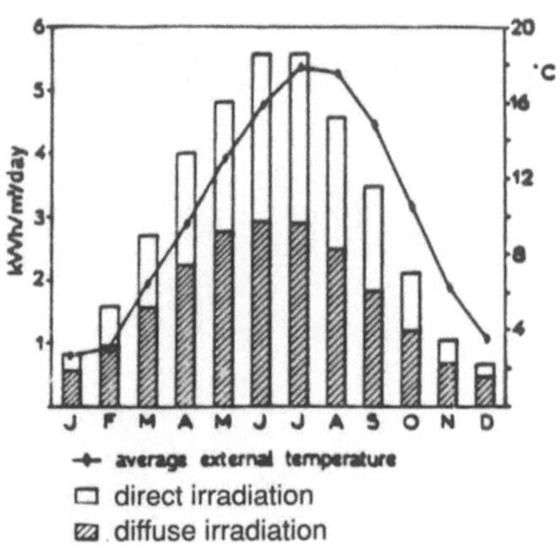

Figure 2. Average ambient temperatures and solar irradiation data

Building Type

This project concerns the renovation of a Caneton-type indoor public swimming pool which had fallen into disrepair. The original pool (see Figure 3) was typical of the 200-250 Caneton pools built throughout France over the past fifteen years as part of a French government programme aimed at encouraging people to take up swimming. In all, around a thousand indoor pools were built in this programme using a total of five designs chosen through a design competition held in 1969.

A large number of the pools are now in a state of considerable degradation and the aim of this project is to demonstrate on one pool a system of renovation which is capable of easy replication in the other Caneton pools. It involves replacement of the roof and installation of a new solar heating system.

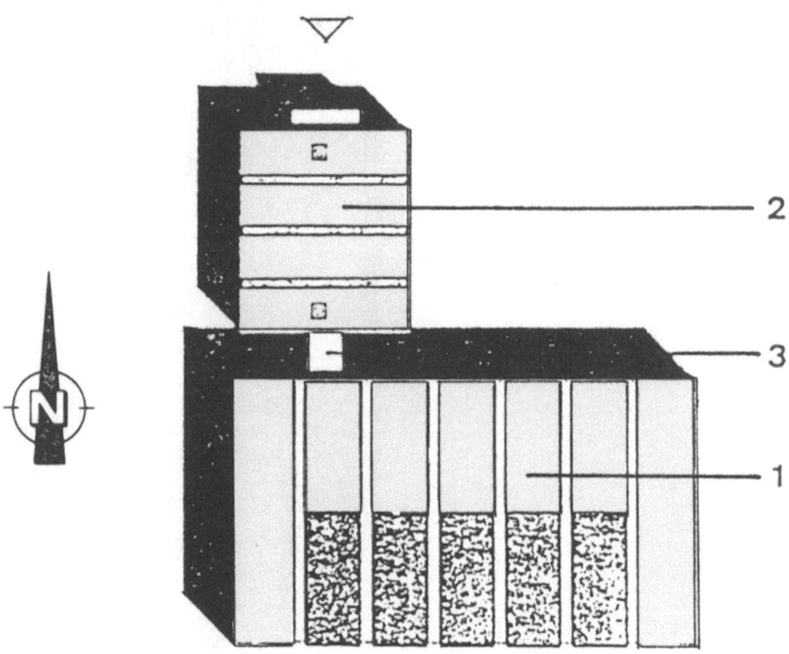

Key: 1. Swimming pool hall
2. Building containing entrance hall and changing rooms
3. Passage with footbaths

Figure 3. A Caneton indoor swimming pool

Location

The demonstration pool is located in the Cesson region of Paris (Figure 1) at a latitude of 48° 33' and an altitude of 77 m.

Site Microclimate

The site is not overshadowed by other buildings. Some climate data are given in Table 1 and Figure 2.

Design and Construction Details

The original facility at Cesson had an area of 925 m^2 and a volume of 3475 m^3. It consisted of a well-insulated swimming pool hall (34 m x 19.5 m) containing a 25 m x 10 m pool and a reception building (12.3 m x 14.7 m) with entrance hall and changing rooms.

The building frame was made of six self-supporting post-and-beam structures of triangular cross-section.

The renovated system was developed using the results of studies carried out through the Building 2000 programme. A scheme of the final design is shown in Figure 4. To achieve it the existing roof, apart from the support structure, is dismantled and a new roof constructed. This has been designed to:

- capture solar radiation;
- have good thermal insulation properties (see Table 2);
- allow good natural lighting of the pool;
- be comfortable acoustically.

In addition, solar water heaters are installed.

The approximate cost of the renovation is FF 1.5 million (250,000 ECU).

U-values in W/m^2 K	
of triple-glazed surface	2.60
of opaque surface	0.35

Table 2. U-values of the new roof

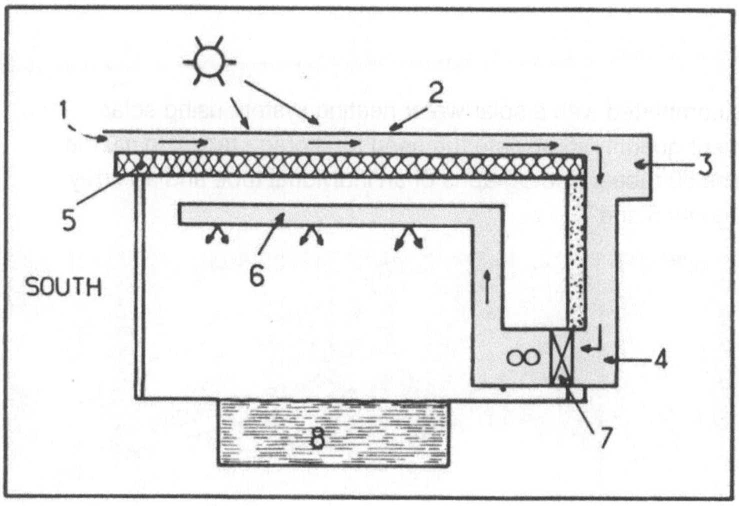

Key:
1. Fresh air
2. Aluminium panels
3. Inlet duct
4. Heated fresh air
5. Insulation
6. Air supply duct
7. Heating coil
8. Pool

Figure 4. The renovated Caneton swimming pool system

DESCRIPTION OF PASSIVE SOLAR FEATURES/COMPONENTS

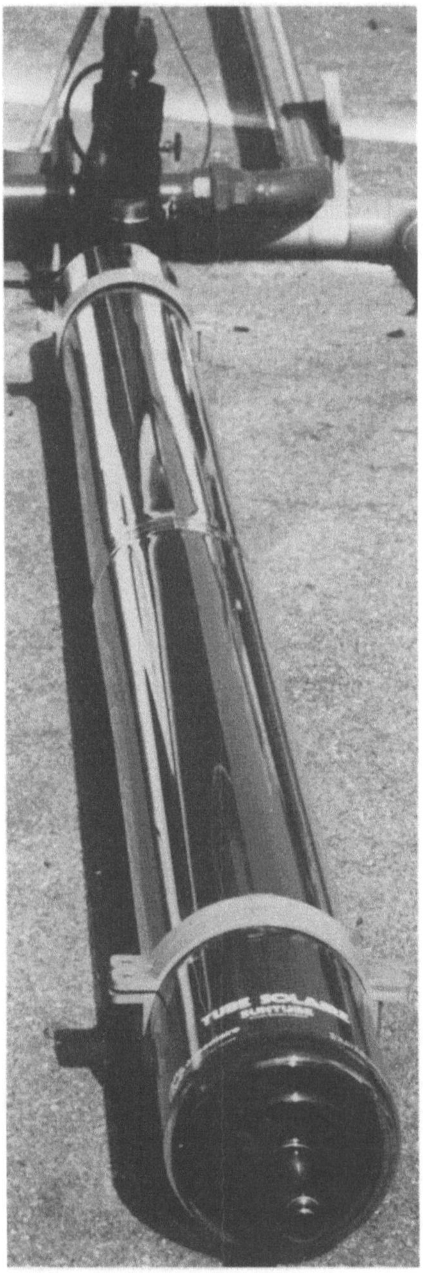

Figure 5. Photograph of a solar tube

Design Principles

The renovated system was designed to take into account the fact that nearly 40% of the energy used to run the original building was taken up in conditioning the ventilation air. A further 5% was used to provide hot water. Hot water consumption is highest in summer because it is a function of pool occupancy.

New Roof Structure

Part of the new roof structure is triple glazed. In the remainder, insulating sandwich panels consisting of 80 mm polyisocyanurate foam with a waterproofing surface on top and a ribbed galvanized steel sheet on the under-side are fixed to detachable cross girders. The under-side serves as a vapour barrier. Aluminium alloy or steel roof panels (black on top) are mounted above the sandwich panels in such a way that air can circulate through a 50 mm space between the two. These panels cover an area of approximately 500 m^2, providing additional waterproofing of the roof and serving as a solar collecting system which warms the air passing beneath. A profile of the roof surface is shown in Figure 6.

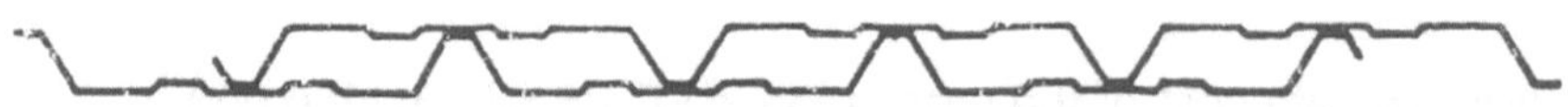

Figure 6. Profile of the roof surface

Solar Water Heaters

The system is completed with a solar water heating system using solar tubes in sufficient quantity to obviate the need for storage tanks. In all, the system contains 50 tubes. Photographs of an individual tube and an array are given in Figures 5 and 7.

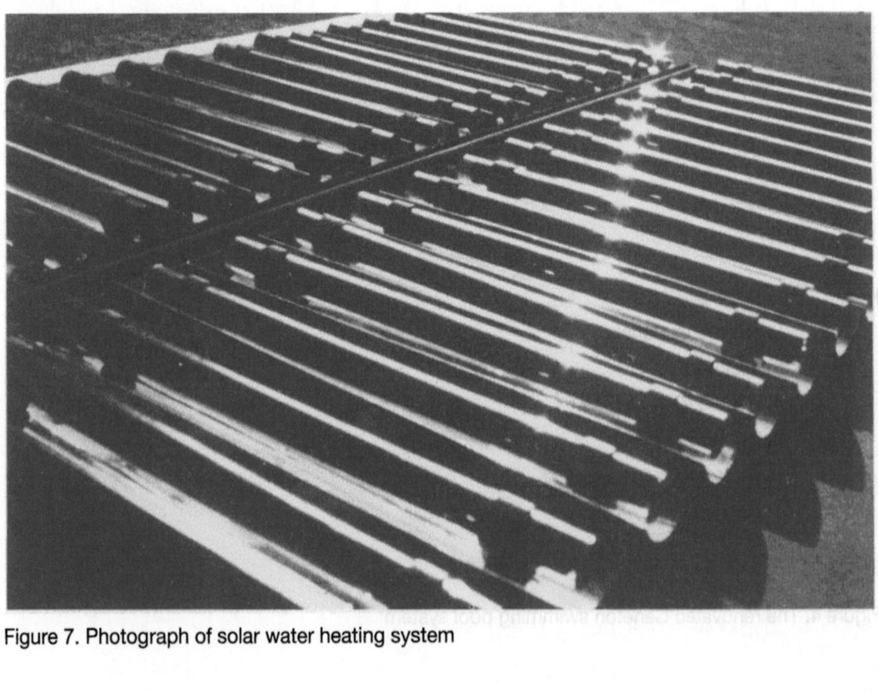

Figure 7. Photograph of solar water heating system

ENERGY CALCULATIONS PERFORMED AND DESIGN TOOLS USED

Aim of Energy Calculations

The details of the renovated system were developed with the help of thermal performance calculations. The aim was to create a design which was not only energy-efficient but also had appropriate values of those parameters which determine thermal comfort conditions in an indoor pool building. Examples of the latter are relative humidity, air temperature and temperature of surfaces inside the building - parameters which are very different from those which create comfort in the home.

Computer Model

The thermal performance calculations were carried out using the PISCOV computer model of a covered swimming pool.

Input of Meteorological Data

In carrying out the studies, meteorological data for the city of Trappes was used.

Building Parameters

To conduct the simulation studies, values of certain parameters had to be input into the model to describe the building before and after renovation. The parameters for the original building are given in Table 3.

Input of various parameters for the renovated building showed the following:

- increasing the insulation of the roof and the glass walls of the hall had the effect of:
  - increasing the mean radiant temperature (raising inside surface temperature);
  - increasing the inside wall temperature (reducing risks of inside condensation);
  - making it possible to increase internal relative humidity (improving comfort of the swimmers when they are wet and saving energy by reducing the flow of fresh air and evaporation).
- reducing the area of roof glazing in order to compensate for the increased area of glazing in the south wall had the effect of:
  - improving the winter heat balance (solar gains from a south- facing vertical wall are greater than those through a horizontal wall);
  - reducing overheating in summer and warm weather.
- dehumidifying with fresh air at a flow rate which takes into account the relative humidity inside the building had the effect of:
  - increasing comfort by increasing relative humidity in winter;
  - saving energy by reducing amount of heated air and pool evaporation.

Cont. on page 6

U-values in W/m² K	
fixed roof	0.86
translucent mobile roof	3.5
doors on south side	
opaque section	1.0
translucent section	5.6

Dehumidification by fresh air was at fixed flow rate of 14,000 kg/hour (approximately 12,000 m³ /hour)

There was no pool cover

Temperature was regulated by ambient air sensor

Coefficient of heat loss through walls, roof and floor as ratio of internal volume (G1) was 0.83 W/m³ K

Table 3. Some parameters for the original building

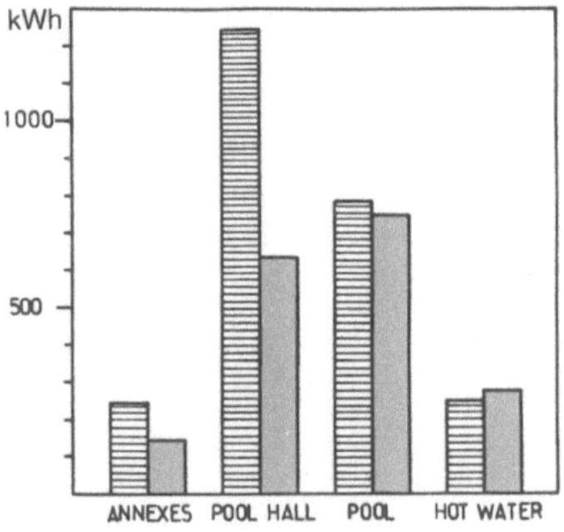

Figure 8. Daily heat losses (kWh) in December and April after renovation

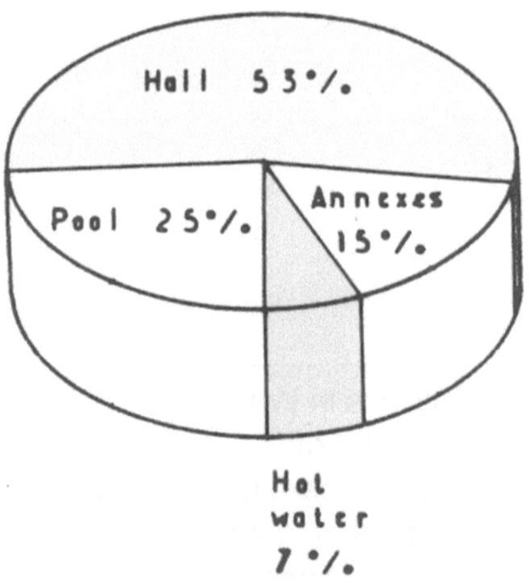

Figure 9. Breakdown of energy savings after renovation. (Total savings: 545,400 kWh a year.)

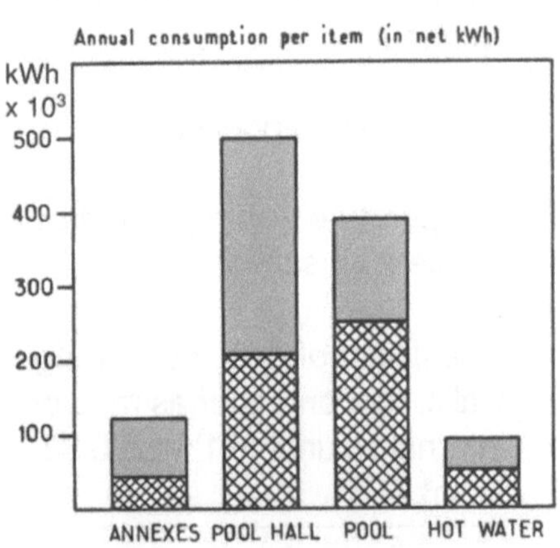

Figure 10. Annual energy consumption (net kWh x 10^3) before and after renovation

Cont. from page 5

- allowing settings of inside temperature and relative humidity to vary as function of ambient temperature had the effect of:
  - saving energy by reducing the temperature difference between the hall and the outside air, reducing the fresh air flow and evaporation during warm weather;
  - maintaining comfort conditions (raising humidity when the inside temperature falls);
  - preventing condensation on inside of the cold exterior walls (lowering relative humidity of air inside the building and raising the indoor temperature during cold weather).
- using a pool cover at night saved energy.
- maximum fresh air flow was 14,000 kg/hour (approximately 12,000 m/hour).
- the coefficient of heat loss through walls, roof and floor as a ratio of internal volume (G1) was 0.55 W/m³ K.

Results of Thermal Performance Calculations

Some results of the energy calculations are given in Table 4 and Figures 8, 9 and 10.

Figure 8 shows the daily heat losses of the different parts of the building - pool hall, annexes (i.e. the reception building), the pool itself and the hot water system - after renovation.

Table 4 and Figures 9 and 10 show energy consumption before and after renovation. The total energy requirements of the original building amounted to 1.1 x 10^6 kWh a year. Renovation reduced this by nearly 50% to just over 0.5 x 10^6 kWh a year. Over half the savings were made in the pool hall, 15% in the annexes, 25% in the pool itself and 7% in the hot water system.

The 500 m solar collecting roof surface was calculated to recover 65,000 kWh energy a year - about 12% of the total savings made.

	Annual consumption (kWh) before renovation	after renovation	Net savings (kWh)	% savings
Annexes	121,000	40,700	80,300	66.36
Pool hall	499,400	208,100	291,300	58.32
Pool	389,400	250,600	138,800	35.64
Hot water	91,000	56,000	35,000	38.46
Total	1,100,800	555,400	545,400	49.54

Table 4. Energy consumption before and after renovation

GENERAL DESIGN GUIDELINES/POINTS OF INTEREST RESULTING FROM THIS PROJECT

The recent interest in saving energy has led to the development of some complex techniques which require considerable capital outlay. Design and construction of systems using these techniques are often poor because the technology is not always fully understood. Further, the completed systems are often incorrectly run by unqualified operators, with bad results.

To avoid this, simple passive solar techniques have been chosen for renovation of the Caneton pool at Cesson which are reliable to operate and produce consistent results. The lessons learned through the Building 2000 studies enabled systems to be specified which make effective use of solar energy in the form of heat and light. It is expected that, as these techniques are used in pools in the future, the problems encountered in the old covered pools will be avoided and there will be improved user comfort and reduced energy consumption.

BUILDING 2000 Building 2000 brochures are published by Directorate General XII of the Commission of the European Communities to show how design studies can help architects and other building designers use passive solar principles to the best effect to produce attractive energy efficient buildings. Each brochure describes studies carried out with the support of the Commission during the design phase of one of thirty-six non-domestic buildings in the EC Member States. The studies were on such topics as daylighting, heating, cooling, ventilation, comfort, control systems and urban design. They were carried out with the help of acknowledged European experts in these fields and drew heavily on lessons learned and techniques developed through the Commission's research and development programme on solar energy applications to buildings.

Commission of the European Communities/Directorate-General for Science, Research and Development

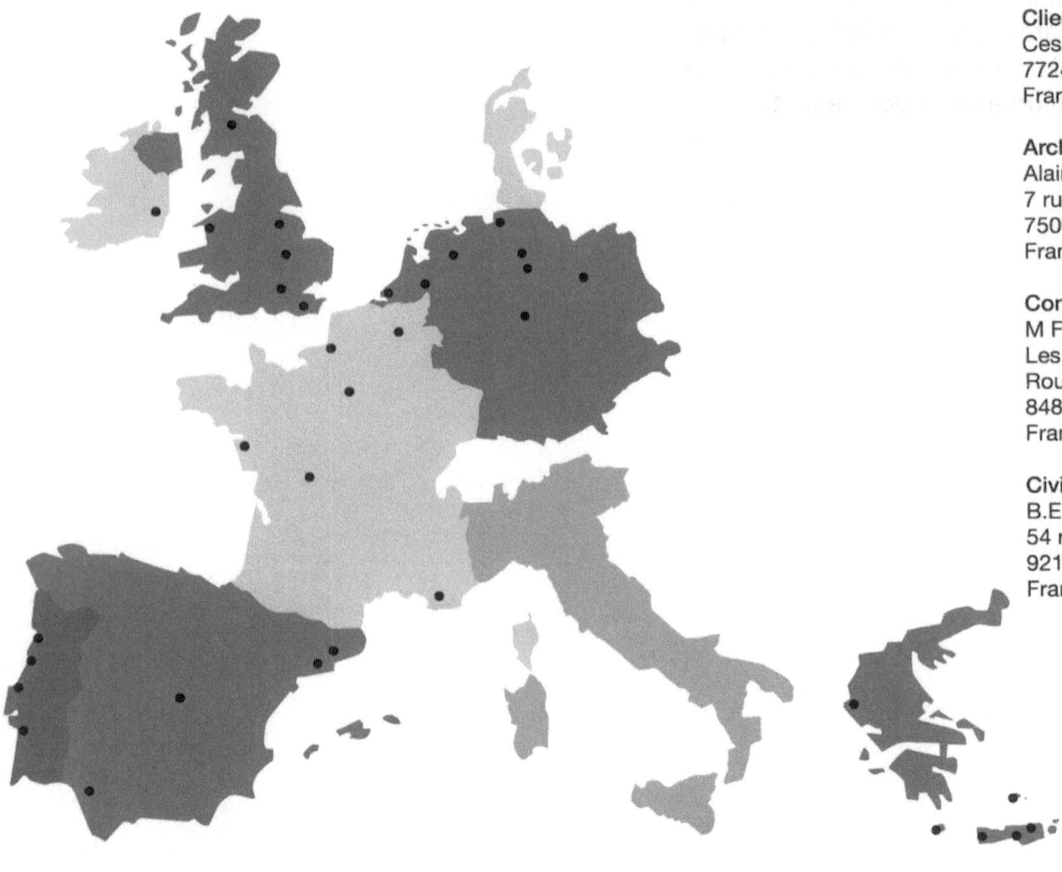

List of Design Team Participants and Advisers

Client
Cesson Town Council
77240 Cesson
France

Architect
Alain Liebard
7 rue d'Argenteuil
75001 Paris
France

Consultant
M Farnallier
Les Grandes Terres
Route des Princes d'Orange
84850 Travaillan
France

Civil Engineer
B.E.T. Genirie
54 rue Carves
92120 Montrouge
France

This set of **Building 2000** brochures illustrates how architects and other building designers can successfully apply passive solar principles to produce energy-efficient buildings.

BUILDING 2000 Participants

Project Director
Theo C. Steemers

Coordinator
Cees den Ouden

Technical Steering Committee
Dean Hawkes
Nick Baker
Alex Lohr
Jean P. Lepoivre

Regional Liaison Agents
(D) Jörn Behnsen
(E) Vicente Sifre
(F) Michel Raoust
(GB) Alan Hildon
(GR) Matheus Santamouris
(P) Eduardo Maldonado

Further information or copies of the brochures can be obtained from prof. ir. Cees den Ouden, EGM Engineering BV, P.O. Box 1042, 3300 BA Dordrecht, The Netherlands.

BUILDING 2000

Commission of the European Communities

- Educational complex containing teaching and gymnasium facilities.
- Separate, appropriately oriented, buildings for the different activities.
- Passive solar gains in winter through south-facing facade and solar roofs.
- Summer cooling through radiative cooling and night-time ventilation.
- Daylighting through solar roofs and south-facing facades.
- Passive solar features provide 86% energy requirements compared with similar, conventional buillding.

EDUCATION CENTRE

MADRID/SPAIN

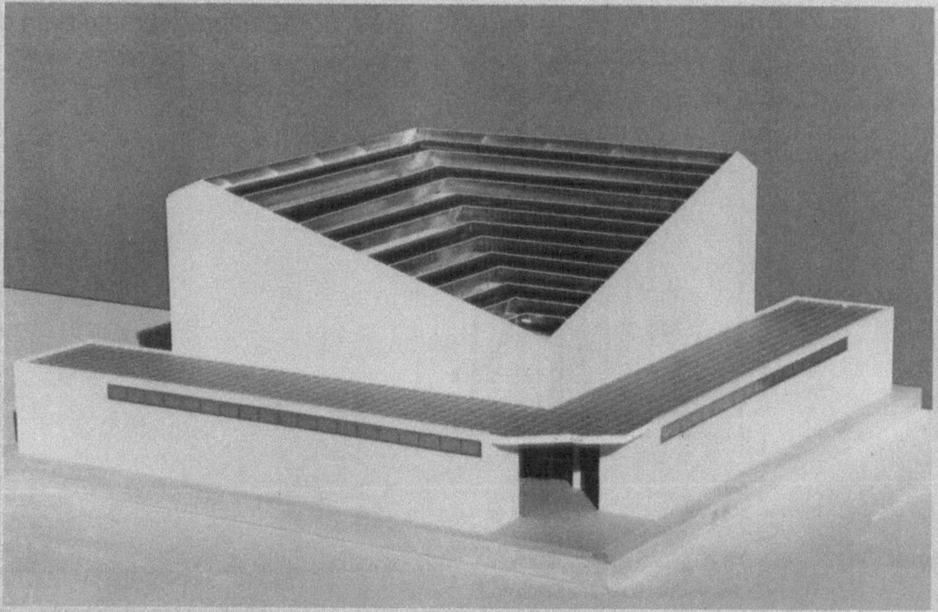

Building 2000 is a series of design studies illustrating passive solar architecture in buildings in the European Community.

ISSUE

E 2

FEB 1991

PROJECT DESCRIPTION

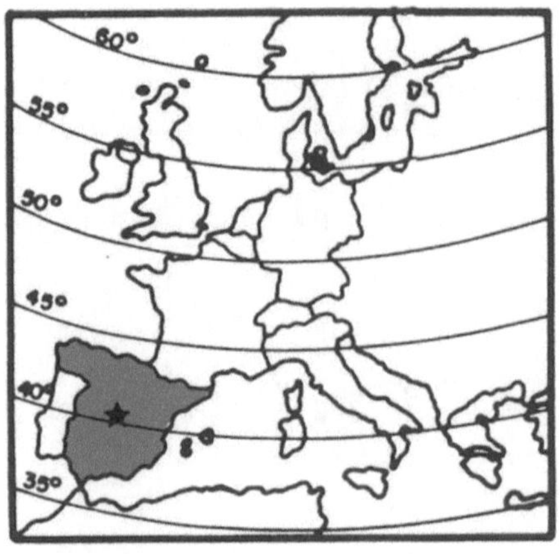

Figure 1. Location of Madrid

Building Type

This project is concerned with an educational centre containing teaching and sports gymnasium facilities.

Location

The building is on a site 667 m above sea level in Madrid, Spain (latitude 40 ° 25 ' N, longitude 3 ° 41 ' W) - see Figure 1.

Site Microclimate

Mean daily solar irradiation and temperature data are given in Figure 2. Additional climate data can be found in Table 1.

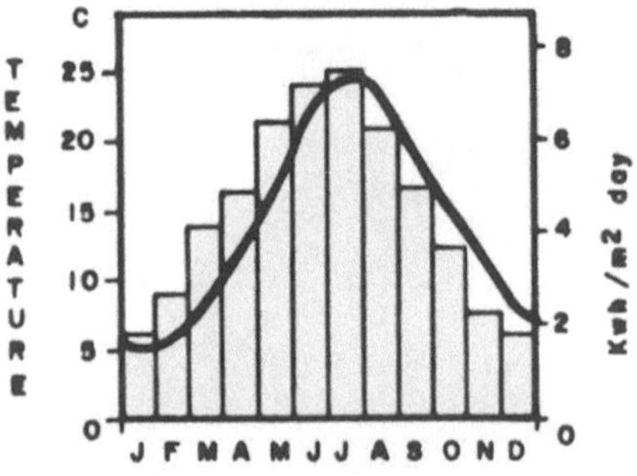

Key: — external temperature solar irradiation on horizontal plane

Figure 2. Mean daily solar irradiation and temperature data

Mean daily temperature	
Oct-Apr	9 ° C
Jan-Dec	13.9 ° C
Sunshine hours	
Oct-Apr	1260
Jan-Dec	2869
Degree days (base 18 ° C)	1910

Table 1. Some key climate data

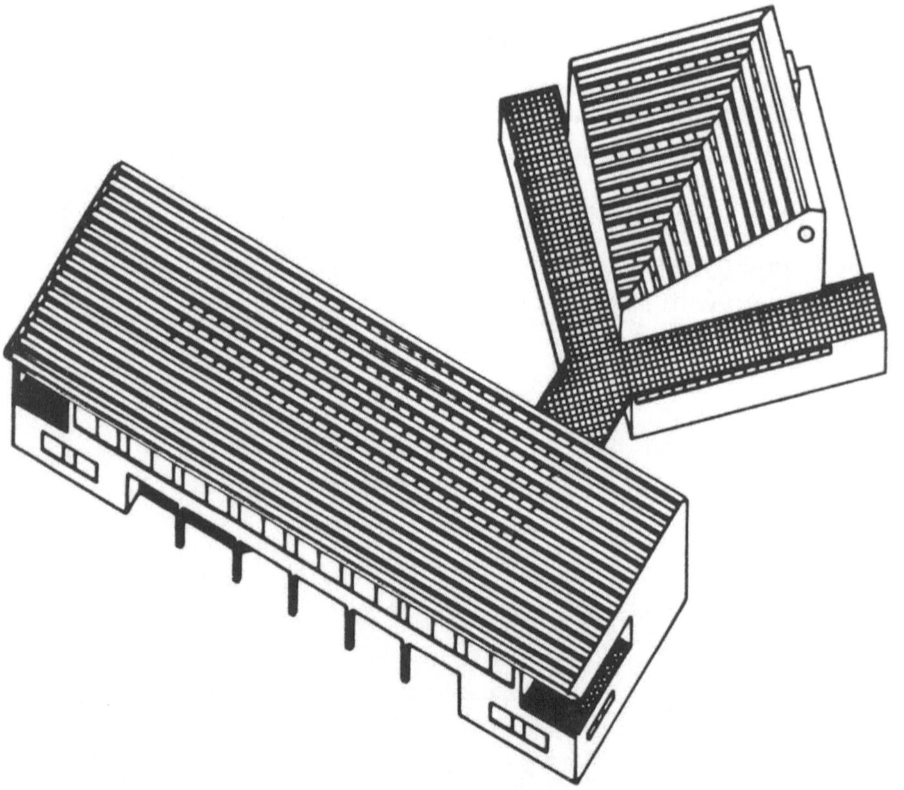

Figure 3. View of the educational centre from the south-east

Design and Construction Details

The centre consists of two linked units - a 1600 m^2 teaching block and a 400 m^2 sports gymnasium (see Figures 3-5). The teaching block is to the south of the gymnasium and oriented east-west to maximize the use of solar energy in winter. The east and west facades have small areas and few openings. Orientation of the gymnasium is such that the wind is deflected away from the teaching block.

A ramp and heavy east and west walls provide a large thermal mass in the teaching block.

Both units have the solar roofs described in more detail below. Solar energy or coolth collected in the roofs is stored in water stores, one in the gymnasium and one in the teaching block. The water is passed through tubes embedded in the floors to heat or cool the building.

The building has a metal structure made of 5 m x 8 m modular units. The external walls are of heavy brick, insulated on the outside. Perimeter insulation prevents heat transfer between the building and the ground. U-values of some building elements are given in Table 2.

Floors	0.75
Walls	
opaque areas - south	1.2
opaque areas - other	0.5
transparent areas	2.6
Roofs opaque areas	0.5
transparent areas	2.6

Table 2. U-values (in W/m^2 K) of some building elements

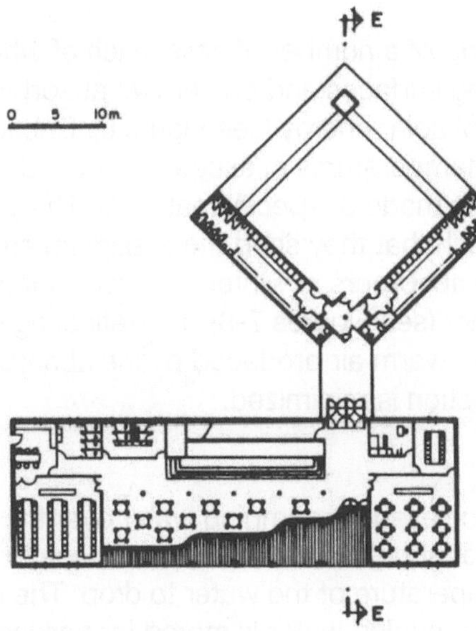

Figure 4. Plan of the education centre showing the teaching block to the south of the gymnasium

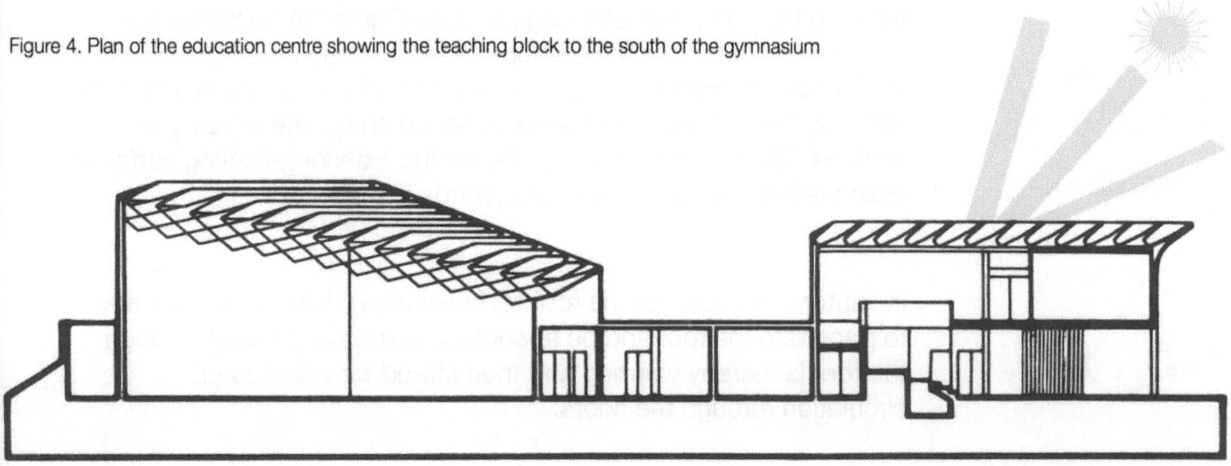

Figure 5. North-south section through E-E

DESCRIPTION OF PASSIVE SOLAR FEATURES/COMPONENTS

Introduction

A number of passive solar features have been incorporated into the complex to ensure that it is thermally and visually comfortable, that it can be operated with the minimum amount of auxiliary heating and that most of the lighting requirements can be met by daylight. Particular care has been taken over the orientation and siting of the different blocks. The south- facing wall of the teaching block is glazed to maximize solar gains in winter and daylight in summer. In the gymnasium (where the maximum acceptable temperature is 24 ° C) the walls have few openings in order to prevent overheating in summer; daylighting is achieved through the roof. Both blocks of the complex have solar roofs. These act as a heat collector in winter and as a cooling device in summer and permit entry of daylight throughout the year. Additional cooling in summer is achieved by means of night ventilation.

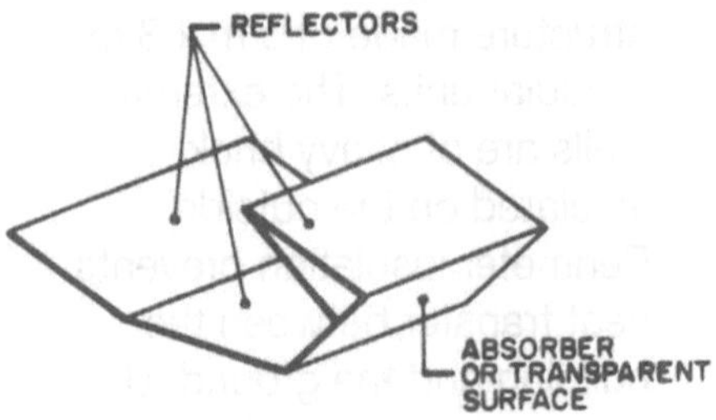

Figure 6. The reflecting and absorbing surfaces of a solar roof cell

Solar Roof

The solar roof is made up of a number of cells, each of which contain several reflecting surfaces and one or two absorbing surfaces for collection of solar energy (see Figure 6). Because the absorbers never reach temperatures in excess of 100 ° C, it is not necessary for them to be made of special materials. The angle of the reflecting surfaces is such that they send the maximum amount of solar radiation onto the absorbers in winter but prevent it from reaching them in summer (see Figures 7-9). The reflecting surfaces also serve to contain the warm air produced by the absorbers so that heat loss by convection is minimized.

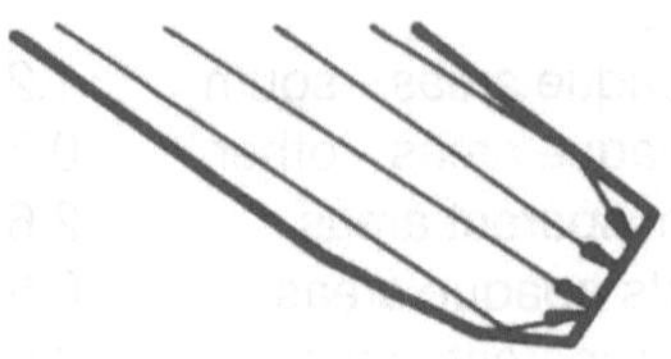

Figure 7. The absorption of solar radiation in a solar roof cell in winter

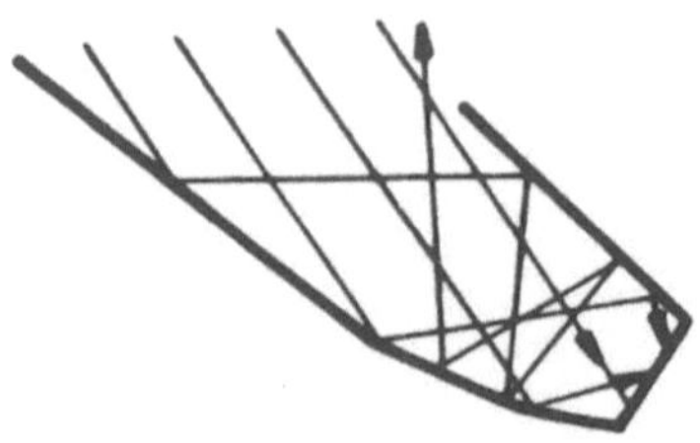

Figure 8. The reflection and absorption of solar radiation in a solar roof cell in spring

Mode of Operation in Summer

In summer, cooling is achieved by pumping water over the reflecting surfaces of the solar roof at night. Radiative heat losses to the cool night sky cause the temperature of the water to drop. The roof temperature falls and the cooled water is stored for passage through tubes in the floors the following day (see Figure 10).Daylighting is provided by entry of diffuse solar radiation through the transparent roof areas.The angles of the various roof surfaces are such that only about 2-3% of the incident solar radiation enters the building in summer. Most of the sun's rays fall on the exterior reflecting surfaces which reflect the radiation back towards the sky.

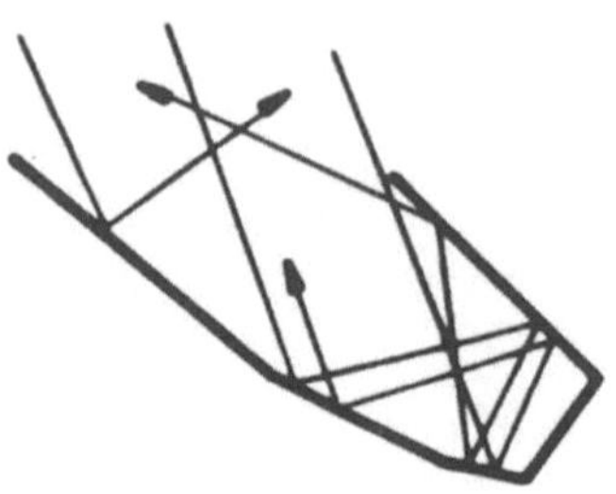

Figure 9. The reflection of solar radiation in a solar roof cell in summer

Operating Mode in Winter

In winter, the angles of the roof surfaces allow direct solar radiation to penetrate the roof and be absorbed. The water passing through the roof is thereby warmed and then stored for subsequent circulation through the floors.

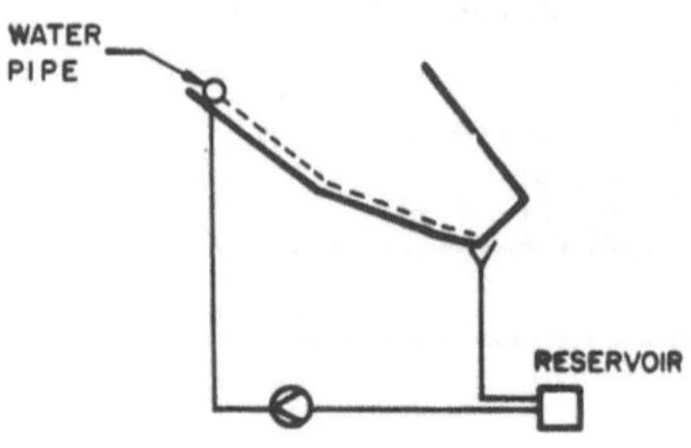

Figure 10. Storage of water cooled or heated in solar roof

ENERGY CALCULATIONS PERFORMED AND DESIGN TOOLS USED

Introduction

Several studies of the daylighting and thermal performance of the complex were carried out during the design stages. The daylighting of the gymnasium through the solar roof was tested using a scale model. The amount of solar energy collected by the solar roof was calculated. The cooling achieved by both radiative cooling to the night sky of water passing over the solar roof and night ventilation was determined using various formulae. Finally, the total annual fuel savings achieved from the passive solar measures were estimated.

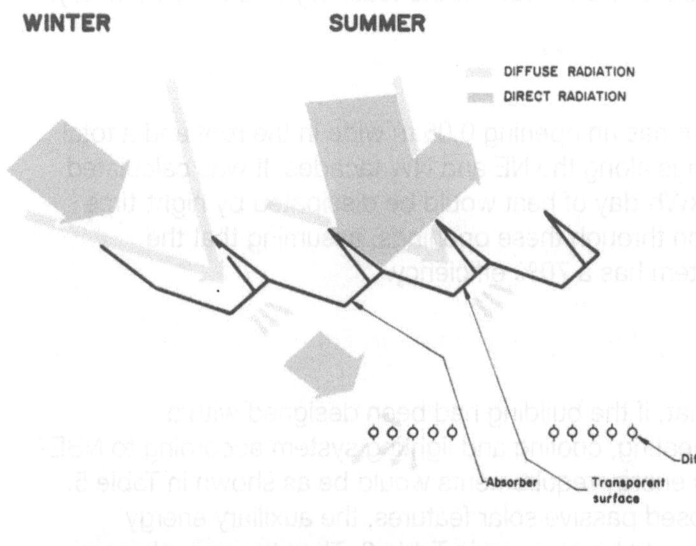

Figure 11. Daylighting through solar roof

DAY	WITHOUT DIFFUSER		WITH DIFFUSER	
	D.F.	LUX	D.F.	LUX
21 - 12	25 %	12.750	≤ 20 %	≤ 10.200
21 - 1	20 %	11.500	≤ 16 %	≤ 9.200
21 - 2	16 %	10.600	≤ 13 %	≤ 8.000
21 - 3	12,5 %	10.000	≤ 10%	≤ 8.000
21 - 4	8,5 %	3.600	≤ 7%	≤ 2.900
21 - 4	4 %	3.500	≤ 3%	≤ 2.800
21 - 5	3 %	2.800	≤ 2,5 %	≤ 2.200
21 - 6	2 %	2.500	≤ 1,6%	≤ 2.000

Table 3. Average daylight factors (DF) and lux levels achieved in gymnasium with and without installation of diffuser below solar roof

Gymnasium Daylighting Tests

The daylighting of the building through the solar roof is illustrated in Figure 11. The lighting performance of the gymnasium solar roof with and without the installation of an adjustable diffuser below the transparent surface was determined using a 1:50 scale model. (The aim of the diffuser was to overcome glare problems.) The tests were carried out under clear sky and cloudy conditions on several winter, spring and summer days. Photometric cells were used to measure the lighting levels 1.2 m above the floor. The results are given in Table 3. 3.

month	MJ/m²	kWh/m²
January	4.35	1.21
February	5.02	1.39
March	9.71	2.70
April	3.97	1.10
May	0.07	0.02
June	0.07	0.02
July	0.02	0.01
August	1.06	0.29
September	2.24	0.62
October	8.06	2.24
November	3.45	0.96
December	2.56	0.71

Table 4. Average daily solar energy collected by solar roof

	kWh	kWh/m²
Space and water		
Heating	90,000	45
Cooling	50,000	25
Lighting	6,400	3.2
Total	146,400	73.2

Table 5. Annual energy requirements of building if conventionally heated, cooled and lit

	kWh	kWh/m²
Space and water		
Heating	20,000	10
Cooling	1,000	0.5
Lighting	600	0.3
Total	21,600	10.8

Table 6. Annual energy requirements of building with passive solar features

Collection of Solar Energy by Solar Roof

The average daily solar energy collected by a solar roof month-by-month throughout the year is given in Table 4.

Summer Cooling by Night-Time Radiation

As indicated earlier, both the gymnasium and teaching block are partially cooled in summer by circulating water over the reflecting surface of the solar roof during the night hours so that heat is lost from the water to the cold night sky. It was estimated that on a clear night the gymnasium roof would be able to dissipate 180 kWh/day of heat by this means and the roof of the teaching block 360 kWh/day.

Summer Cooling by Night Ventilation

The gymnasium has an opening 0.05 m wide in the roof and a total of 40 m openings along the NE and NW facades. It was calculated that some 33 kWh/day of heat would be dissipated by night-time cross ventilation through these openings, assuming that the ventilation system has a 70% efficiency.

Overall Energy Savings by Passive Solar Features

It was found that, if the building had been designed with a conventional heating, cooling and lighting system according to NBE-CT-79, then its energy requirements would be as shown in Table 5. With the proposed passive solar features, the auxiliary energy requirements would be as given in Table 6. Thus the annual energy savings due to the passive solar features amount to 124,800 kWh (62.4 kWh/m²) - 86% of the total energy requirements of the comparable conventional building (see Figure 12).

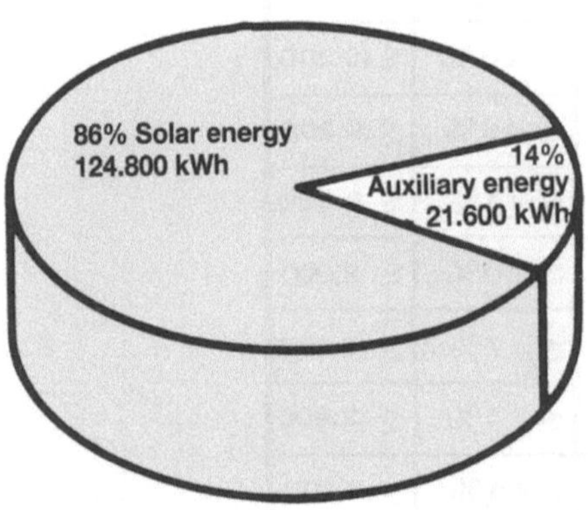

Figure 12. Breakdown of energy supply for building with passive solar features

GENERAL DESIGN GUIDELINES/POINTS OF INTEREST RESULTING FROM THIS PROJECT

The main feature of this building, the solar roof, was designed following a study of an energy-efficient building in the Mediterranean with a roof area approximately equal to the wall area. This showed that the amount of solar radiation incident on the roof was twice that falling on the vertical walls and led to the idea that the best way of avoiding overheating in warm climates would be to lessen the amount of radiation collected by the roof.

The solar roof works as following.

Operation of Roof in Summer

It reflects most of the incident solar radiation back to the sky.

Because it is made of light-weight material it reaches an equilibrium temperature quickly during the day or night.

Because it is composed of a number of surfaces it operates as if it were two roofs, one dealing with most of the incident solar radiation and the other receiving a minimum amount of energy.

Water is pumped over the roof at night so that the water temperature drops below ambient. This cold water is stored for later circulation through the floors of the building.

The transparent surfaces permit diffuse sunlight to pass through and daylight the building.

Operation of Roof in Winter

The maximum amount of solar energy is collected because the solar radiation falls directly onto the absorbers.

As the profile of the roof contains cavities heat loss is minimized.

The transparent surfaces allow direct and diffuse sunlight to reach the rooms.

BUILDING 2000 Building 2000 brochures are published by Directorate General XII of the Commission of the European Communities to show how design studies can help architects and other building designers use passive solar principles to the best effect to produce attractive energy-efficient buildings. Each brochure describes studies carried out with the support of the Commission during the design phase of one of thirty-six non-domestic buildings in the EC Member States. The studies were on such topics as daylighting, heating, cooling, ventilation, comfort, control systems and urban design. They were carried out with the help of acknowledged European experts in these fields and drew heavily on lessons learned and techniques developed through the Commision's research and development programme on solar energy applications to buildings.

Commission of the European Communities/Directorate-General for Science, Research and Development

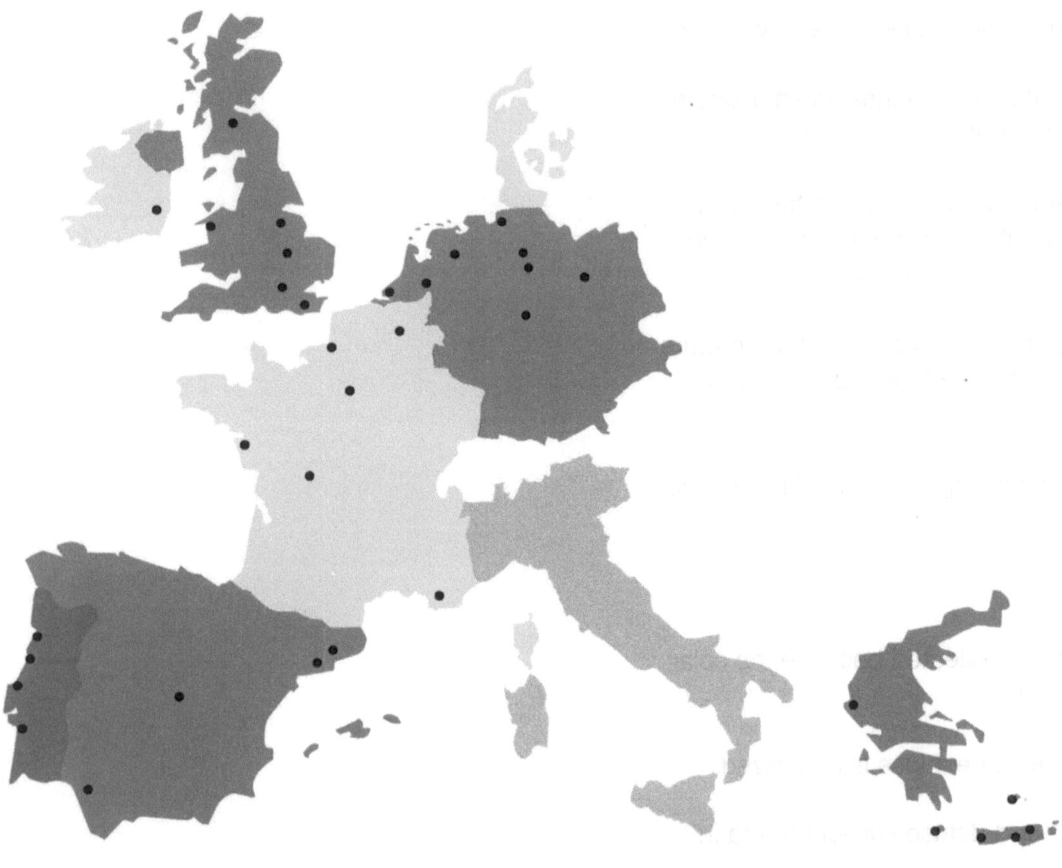

List of Design Team Participants an Advisers

Architects
Jose Antonio de la Fuente Cerezo
Angel Sanz García

Engineers
Eduard Morere
Miguel Angel Sanz

Project Manager
José Martinez Galán
Enercom S.A.
Fermin Caballero 65
28034 Madrid
Spain

CEC Consultants

General Design
Jaime Lopez Asiain (Bioclimatic Architecture Department, Sevilla Architecture University) and Jose Maria Cabeza Leinez (Architect)

Daylighting
Rafael Serra (Head, Bioclimatic Architecture Department, Barcelona Architecure University)

Ventilation
Nick Baker
Cambridge University
UK

Noise and Thermal Store Design
Cees den Ouden
EGM Engineering
The Netherlands

This set of **Building 2000** brochures illustrates how architects and other building designers can successfully apply passive solar principles to produce energy-efficient buildings.

BUILDING 2000 Participants

Project Director
Theo C. Steemers

Coordinator
Cees den Ouden

Technical Steering Committee
Dean Hawkes
Nick Baker
Alex Lohr
Jean P. Lepoivre

Regional Liaison Agents
(D) Jörn Behnsen
(E) Vicente Sifre
(F) Michel Raoust
(GB) Alan Hildon
(GR) Matheus Santamouris
(P) Eduardo Maldonado

Further information or copies of the brochures can be obtained from prof. ir. Cees den Ouden, EGM Engineering BV, P.O. Box 1042, 3300 BA Dordrecht, The Netherlands.

BUILDING 2000

Commission of the European Communities

- Study centre for National Centre for Alternative Technology containing teaching, study and research areas and living accommodation for sixteen students
- Arrangement of spaces and contruction type is responsive to building's occupancy mode
- Daylighting design is optimized to ensure minimum overall energy consumption
- Ventilation air is preheated in six circulation/sunspace areas which serve as buffer zones
- Auxiliary heating amounts to 15,400 kWh (55,400 MJ) a year. This represents an 87% saving on the heating requirements of a comparable building constructed to 1985 UK Building Regulations requirements

STUDY CENTRE

MACHYNLLET/WALES/UNITED KINGDOM

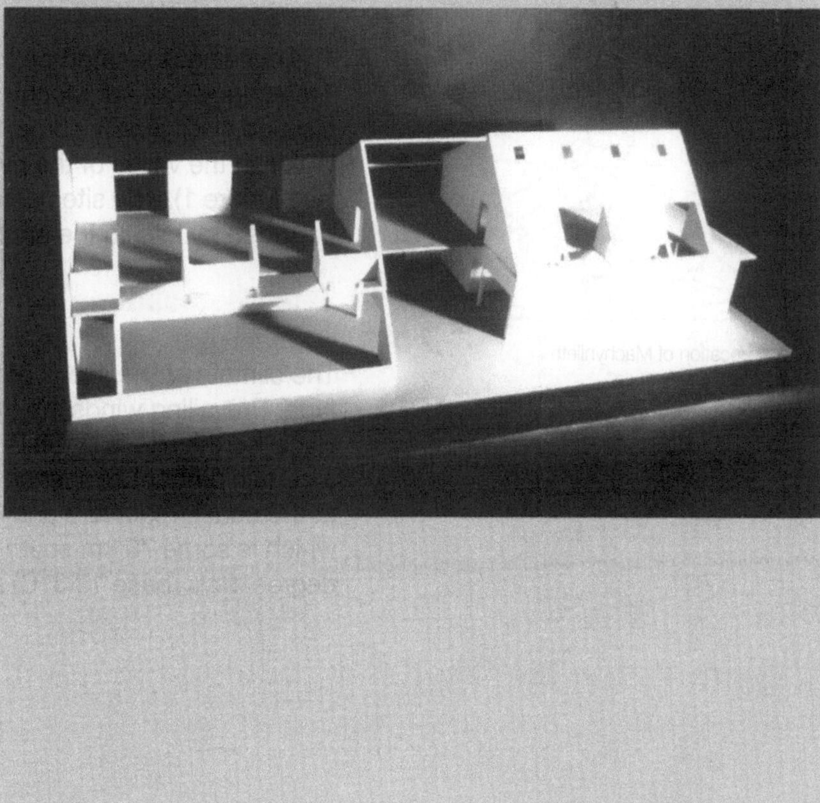

Building 2000 is a series of design studies illustrating passive solar architecture in buildings in the European Community.

FEB 1991

PROJECT DESCRIPTION

Building Type

This building is used by the National Centre for Alternative Technology for its teaching courses on low-energy building, self build and organic agriculture. It contains eight study-bedrooms and associated washing and eating facilities, a large teaching space, study areas, a library/reading room and a public demonstration and research room. The eight bedrooms are designed primarily for single occupancy but each contains space for an additional bed in what is termed a crog loft, i.e. a mezzanine deck above the storage and utilities area.

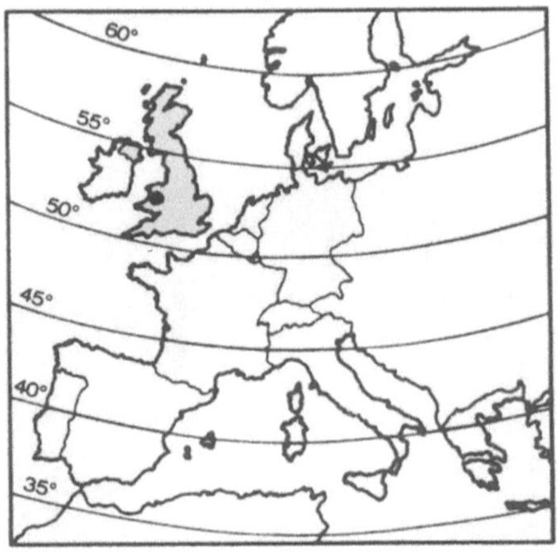

Figure 1. Location of Machynlleth

Location

The building is located on the National Centre for Alternative Technology's site at Machynlleth. This has been developed over a disused slate quarry some 15 km inland from the west coast of Wales in the valley of the river Dulas, a tributary of the river Dovey (see Figure 1). The site has an open aspect to the south-west but is bounded by hills to the south-east.

Site Microclimate

The climate at the site is of the temperate north-western maritime type. Prevailing winds are from the south-west and rainfall is high. Because the site is in a hilly area and some way inland it experiences a certain amount of shelter from the wind and has less sunshine than at the coast. Figures 3 and 4 give some climate data for Aberporth, which is some 70 km south-west of Machynlleth. There are 2342 degree days (base 18.3° C) a year at Aberporth.

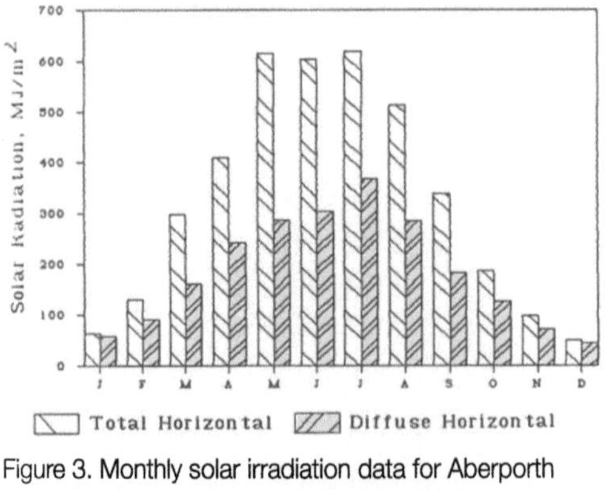

Figure 3. Monthly solar irradiation data for Aberporth

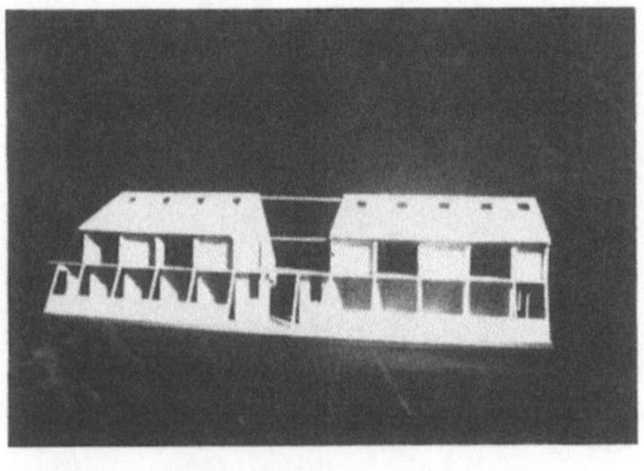
Figure 2. Model of building viewed from the south

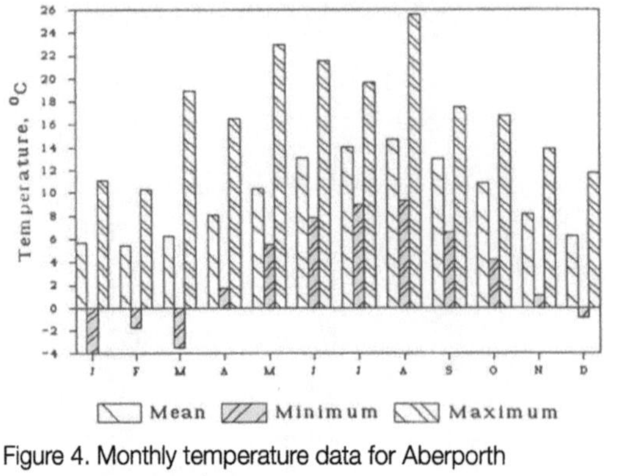

Figure 4. Monthly temperature data for Aberporth

Design and Construction Details

The building is on two storeys with total area 560 m² and total volume 1703 m³. It faces 10 ° west of south. The layout is linear, along an east-west axis, with work and living spaces on either side of a central sunspace/atrium (see Figures 2 and 6). The large teaching and study spaces are on the upper floor (Figure 7). Additional teaching spaces, the study-bedrooms and utility spaces are on the ground floor (Figure 8).

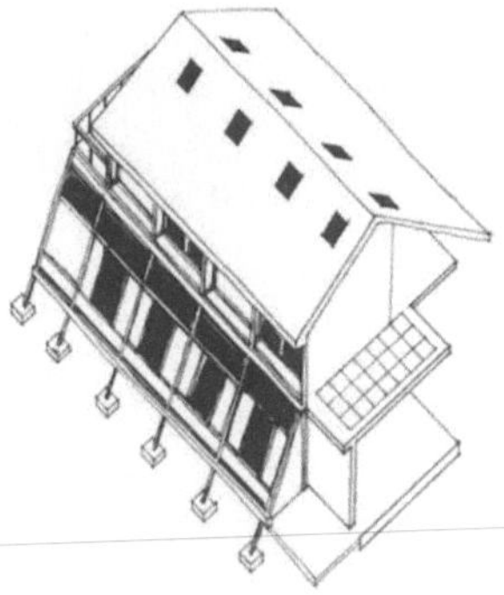

Figure 6. West half of building showing section through atrium and south elevation.

The large teaching spaces on the upper floor have a lightweight timber-frame structure. This is supported by the monolithic blockwork cross wall system which encloses the study-bedrooms, separating them from the teaching and utility spaces on the ground floor. This solid internal core of rooms is surrounded by circulation spaces, which are timber framed and provide the main points of entry.

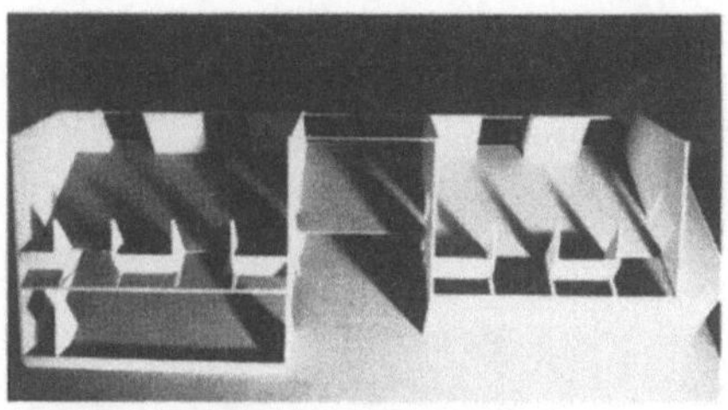

Figure 7. Model showing large teaching spaces on first floor

400 m² of the building's floor area (1325 m³ of the volume) is heated (see Figure 5). Heating is provided by an underfloor system on the ground floor accommodated in a 100 mm screed. The system uses heat from an existing hot water district supply ring main on the site. The screed is separated from the ground floor slab by 100 mm mineral fibre bats. The 150 mm cavity in the timber-frame walls is completely filled with mineral fibre bats. The roof has a total of 190 mm of mineral fibre in the form of a cross lattice of 140 mm and 50 mm. The U-values of the walls, roof and floor are given in Table 1. All windows are double glazed.

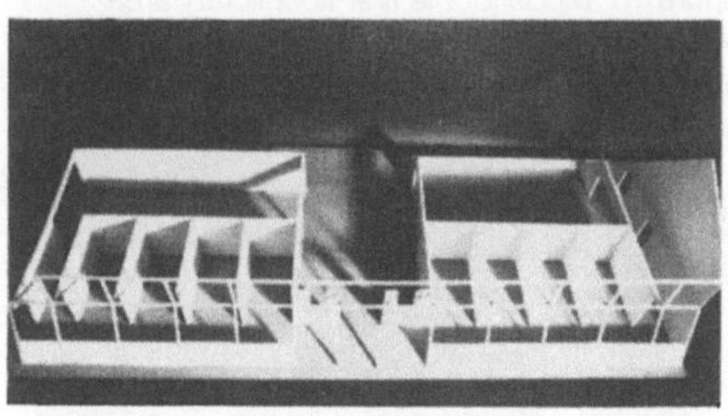

Figure 8. Model showing study-bedrooms on ground floor behind sunspace access corridor

The upper floor responds rapidly to direct solar gain in the morning and pre-heated warm air entering from the south-facing sunroom. The ground-floor construction, with its greater mass, accumulates heat during the day and is warm in the evening when the study-bedrooms are occupied. The circulation spaces (which are not heated) provide thermal buffering for indirect solar gain and preheated ventilation air. Because the buffer zones are necessary circulation spaces, their capital costs are not regarded as being attributable to the passive solar system even though they perform a key role in the operation of that system.

walls	0.26
roof	0.18
floor	0.32

Table 1. U-values (W/m² K)

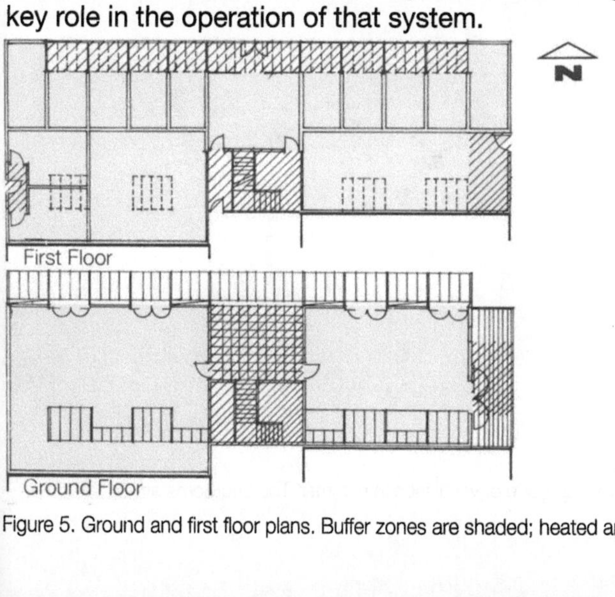

Figure 5. Ground and first floor plans. Buffer zones are shaded; heated areas are pink of coloured.

DESCRIPTION OF PASSIVE SOLAR FEATURES/COMPONENTS

Introduction

To produce an energy-efficient building which is comfortable for occupants, particular care was taken in the areas of daylighting and of ventilating the building with air which has been preheated by direct solar gain. These are described in more detail below.

Daylighting

The study centre adjoins existing buildings. Therefore, rooflights have been introduced along the whole length of the north side to light the northern ground floor rooms and the teaching rooms on the upper floor (see Figures 9 and 13). A glazed stairwell penetrates the core of the building and leads to the south-facing sunspace. It introduces side lighting for the study areas in the northern rooms on the ground floor. The south-facing corridor linking the sunspace to all the bedrooms, kitchen and bathrooms is also highly glazed. Thus all the circulation spaces have the maximum amount of daylighting. The teaching area and library on the first floor receive daylight from four sides and from rooflights placed to provide an even distribution of light with minimum glare. Figures 10 and 11 show models of the library with and without the high level rooflights.

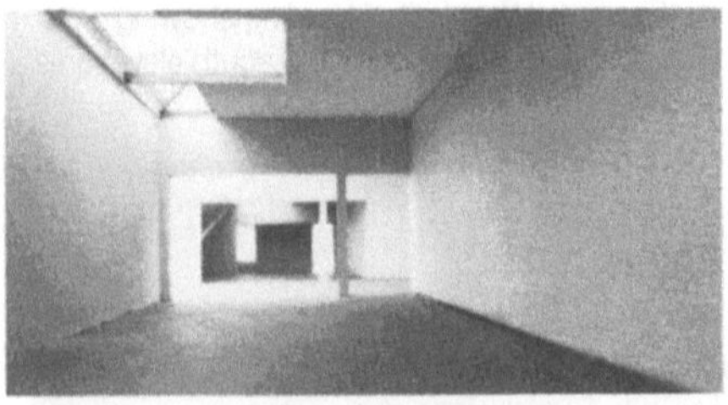

Figure 9. Model of north-lit ground floor spaces showing quality of lighting

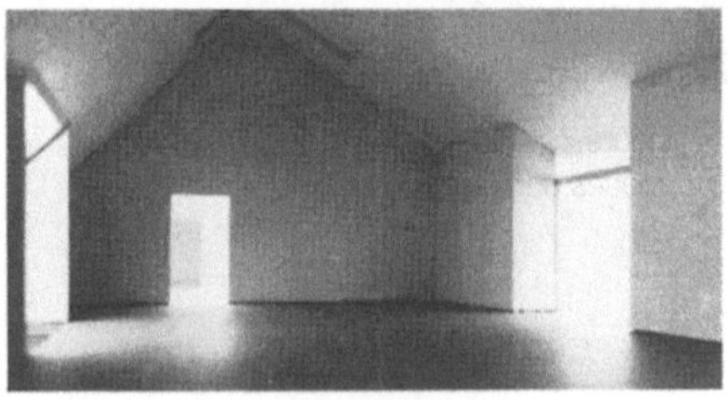

Figure 10. Model of the first-floor library area without rooflights. The designed lighting level is 300 lux.

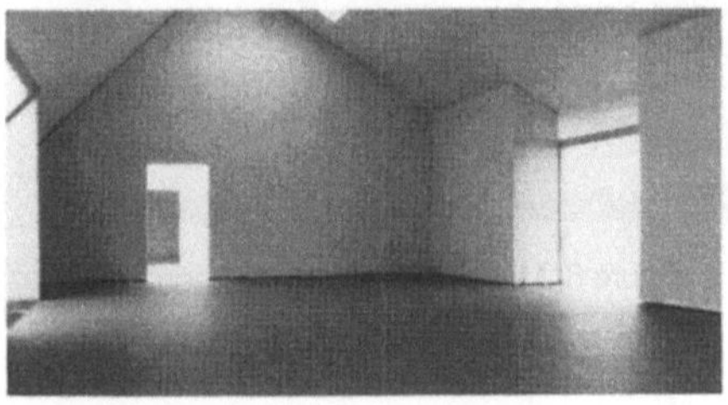

Figure 11. Model of the first-floor library with rooflights added to minimize glare and give a better overall light quality

Preheating of Ventilation Air

The six circulatory areas and sunspaces which serve as buffer zones reduce infiltration losses and in winter provide ventilation air which has been preheated by direct solar gains in the south-facing sunspaces to all the "dry" zones of the building (i.e. the teaching, study and bedroom areas) by means of re-circulation vents at high and low levels (see Figure 12). The "wet" zones (i.e. the kitchen, laundry and bathrooms) contain extract ventilators and, because of the negative pressure, draw in preheated air from the buffer spaces. The main buffer spaces face south and west so that the prevailing south-west winds will provide preheat ventilation from re-circulation with the buffers or direct supply from the buffers as necessary. The building is designed to have one air change per hour, half of which is expected to be provided by the preheated ventilation air. The mode of ventilation in summer is illustrated in Figure 13.

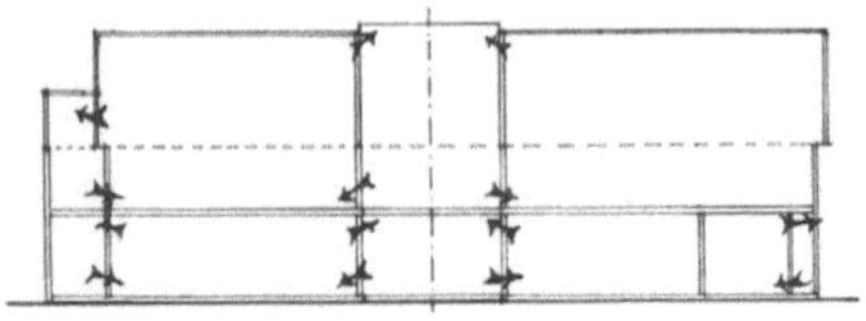

Figure 12. West-east section showing flow of preheated ventilation air in winter

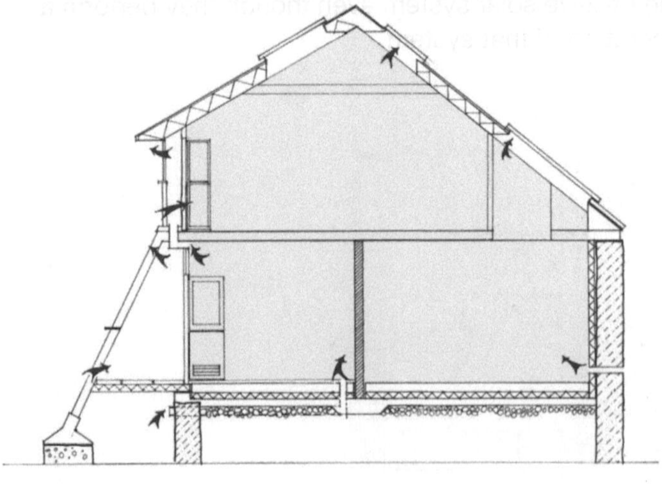

Figure 13. South-north section showing mode of ventilation in summer. The bedrooms are ventilated using solar chimneys.

ENERGY CALCULATIONS PERFORMED AND DESIGN TOOLS USED

Design Tools

To produce a building where daylighting and heating systems are optimized and the best possible use is made of preheated ventilation air, a number of studies were carried out during the design stages. A series of physical models were constructed and tested in the overcast sky facility at the Polytechnic of North London to evaluate daylighting availability, distribution and quality. Minimum and average daylight factors were calculated for all 24 rooms in the building using the DAYLIGHT computer program. Thermal performance evaluations were carried out with the SERI-RES simulation program using a 17-zone representation of the building and test reference year weather data for Aberporth. Daylight factors were input to SERI-RES to assess the annual lighting and heating energy conseqences of changing the fenestration. The effect on thermal performance of preheating ventilation air, changing the thickness of wall and roof insulation and altering the thermal mass of different parts of the building was also tested.

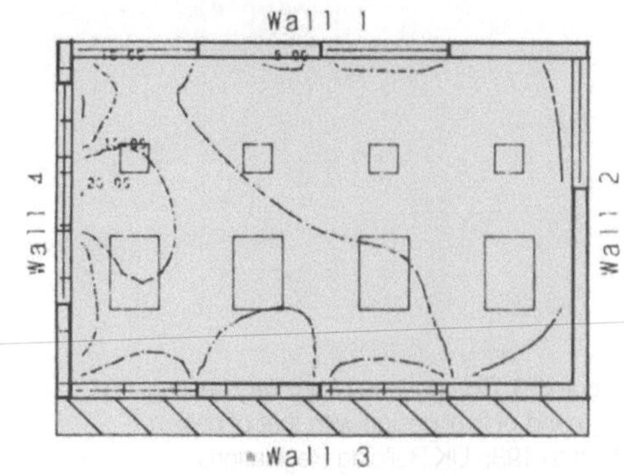

Figure 14. Daylight distribution in library. Example of output from DAYLIGHT program.

Results of Daylighting Studies

Some of the results of the physical modelling of daylighting are shown in Figures 9-11 and 22-25.

An example of the output from the DAYLIGHT program is given in Figure 14. This shows how daylight factor distribution in the library becomes more uniform by introduction of small northern rooflights.

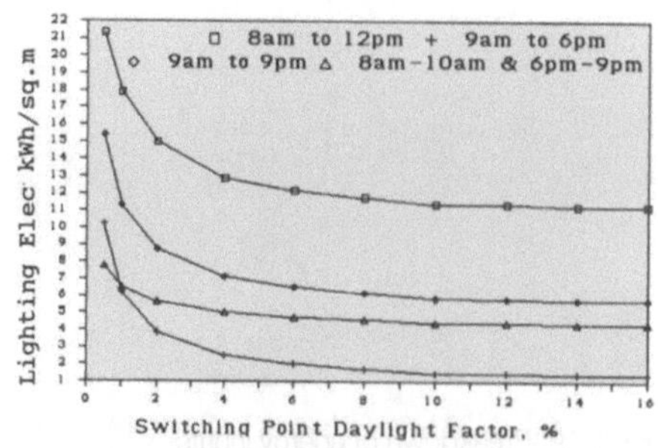

Figure 15. Effect of occupancy period on lighting energy use with lighting standard of 150 lux

Effect on Lighting and Heating Energy Use of Changing Fenestration

The effect of occupancy period on lighting energy use is shown in Figure 15.

In a building which contains public spaces occupied from 9 am to 6 pm, study areas used from 9 am to 9 pm and living accommodation occupied from 8 am to 12 pm, there are a range of demands for lighting and heating. Occupancy patterns have a great effect on the sensitivity of energy consumption to fenestration changes. Early evening use gives a potential for increasing window areas to offset electric lighting demand by daylighting. Continued use of the building into the night means that increased glazing leads to higher thermal losses and a net increase in energy consumption.

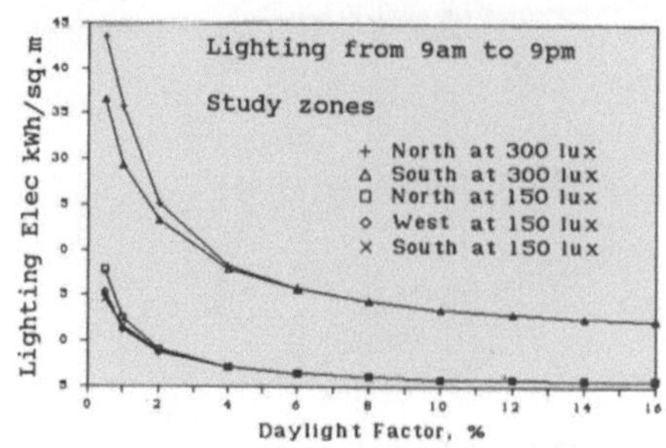

Figure 16. Effect of switching point daylight factor on lighting energy use

The sensitivity of lighting energy use to the daylight factor at the switching point is illustrated in Figure 16. The pattern was the same for all zones of the building. Energy consumption fell rapidly as the daylight factor rose from 2% to 4% and then levelled off. These initial savings were incorporated in the base case design.

The studies showed that lighting energy use was very dependent on lighting standards. 300 lux was the chosen standard for areas where visual tasks were to be carried out. 150 lux was used elsewhere. It took three times the lighting energy to achieve double the lighting standard.

Studies were carried out on all the heated zones to determine the effect on primary energy consumption of increasing the glazing by 20% and by 40%. Two cases were taken - double glazing using plain glass and double glazing with low-emissivity (low-E) glass (see Figure 17). In the calculations it was assumed that the primary energy requirements for the electricity used in the building are three times those for the space heating.

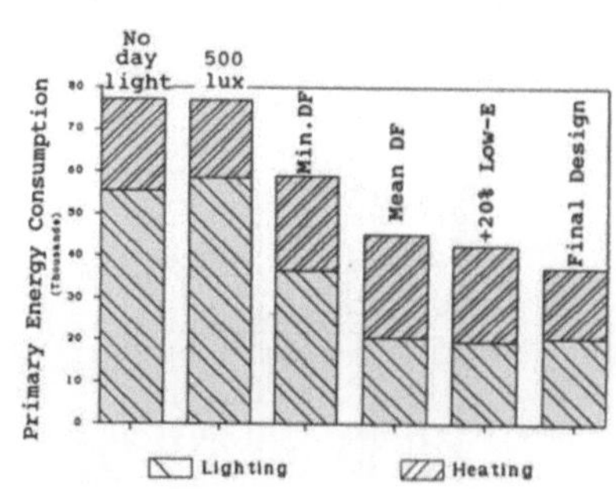

Figure 17. Lighting and heating energy use for selected cases.

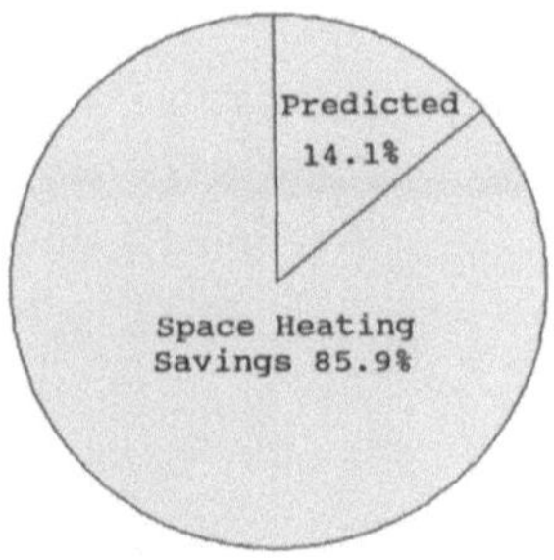

Figure 18. Comparison of auxiliary heating demand of final design with that of design built to 1985 UK Building Regulations requirements. (Latter would require 108,222 kWh (390,000 MJ) a year.)

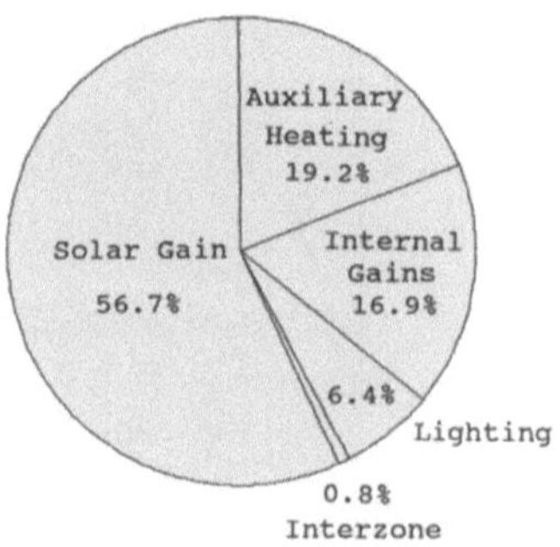

Figure 19. Breakdown of energy inputs during heating season. (Total energy input is 79,584 kWh (287,000 MJ) a year.)

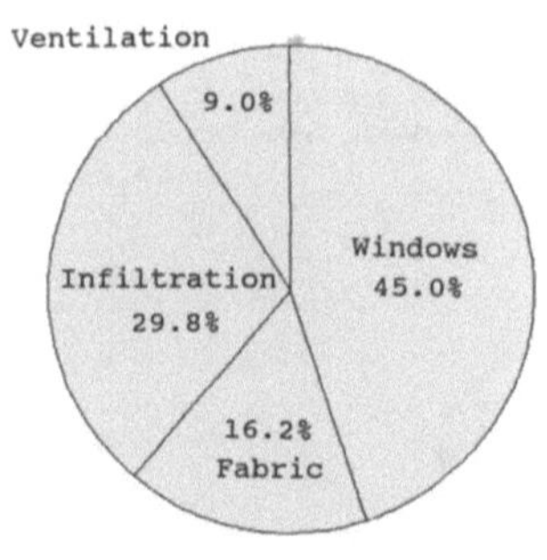

Figure 20. Breakdown of energy losses during heating season. (Total is 79,584 kWh (287,000 MJ)).

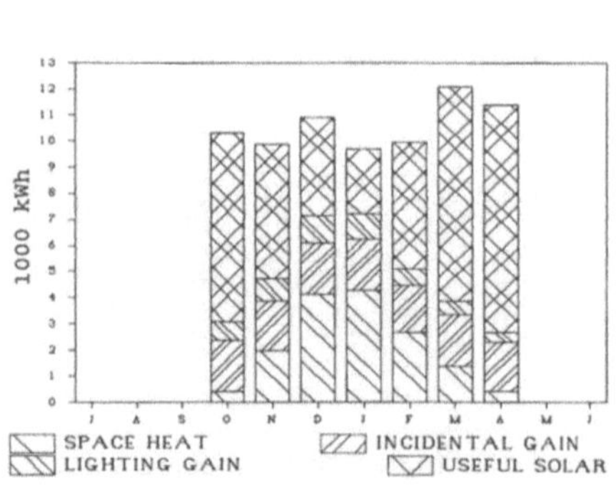

Figure 21. Monthly breakdown of energy inputs, including those of buffer spaces

Because manual switching patterns are complex the potential savings were investigated using automatic on/off switching. Two cases were taken - with a minimum daylight factor at the switching point and with an average daylight factor at the switching point. (In practice, switching is likely to take place when the daylight factor is between the two values and to be highly dependent on room layout and use.) It was found that the lighting energy use was greater when switching took place at the point of minimum daylight factor (see Figure 17). For plain double glazing, increasing the glazing area was found to reduce primary energy demand up to the point when the average daylight factor was approximately 8% in zones which are used in the morning and early evening. For zones which are not used after 6 pm or which are used very late at night, average daylight factors of 3% or 4% were found to produce optimum results. When the plain glass was replaced by low-E glass, the total energy consumption did not change much. The lower transmissivity of the low-E glass led to increased lighting load; this offset the savings in heating energy use. Increasing glazing area was found to reduce primary energy demand up to the point where the average daylight factor was 12% or more (compared with 8% with plain glass). However, the additional savings of around 2600 kWh (9360 MJ) a year made possible by low-E units did not justify the additional capital cost of 5500 ECU.

A manual switching system will never realize the lighting savings of greater fenetration predicted for an automatic system, although the heating savings will probably be the same. Therefore, in the Machynlleth building (which has manual switching), the window sizes originally developed in the physical model studies were retained.

Effect on Heating Energy Use of Preheating Ventilation Air

In the study centre, the extensive use of thermal buffer spaces in combination with well-sealed windows ensures that a high proportion of ventilation air is preheated before it enters the heated zones. If 50% of the infiltration air is preheated, then annual space heating savings of 7800 kWh (28,080 MJ) are predicted - this amounts to around 200 ECU.

Effect on Heating Energy Use of Increasing Wall and Roof Insulation

The effect of increasing the insulation in the external walls and roof from a base level of 150 mm and 190 mm respectively was evaluated. On material costs alone, each 10 mm increase in the insulation in the walls and roof produced savings of 60 kWh and 106 kWh a year respectively - giving simple payback times of 57 and 33 years. This supported the original decision to fill the existing cavities fully but not make extensive additional provisions for insulation. The larger cavities found in timber frame buildings bring advantages in this respect.

Effect Of Changing Thermal Mass

The effect of increasing and reducing thermal mass throughout the building was tested and it was found that either option could lead to increased energy consumption. However, analysis of the individual zones showed that reducing the thermal mass in the northern ground floor rooms and increasing it in the southern ground floor rooms could lead to savings of around 500 kWh (1800 MJ) a year.

Overall Thermal Performance

The overall thermal performance of the final design is shown in Figures 18-21. The energy use for heating is 15,400 kWh (55,440 MJ) a year. This is equivalent to 27.5 kWh per square metre a year or 0.012 kWh (0.043 MJ) per square metre per degree day per year. This is comparable to heating a 20 square metre room with a 100 W light bulb for 4 hours a day.

GENERAL DESIGN GUIDELINES/POINTS OF INTEREST RESULTING FROM THIS PROJECT

General Design Guidelines

From an early design stage the planning of the building aimed to minimize space heating and lighting energy use. The following points were found to be useful, with potential for replication. The fast-response timber-frame structure on the first floor and the slow-response mass wall structure on the ground floor were directly in line with the timed use of the building. Corridors and other cirulation areas were moved to the perimeter of the building to avoid the need for daytime internal lighting. The multiple use of glazed spaces allowed solar gain circulation and ventilation pre-heat. The deep-plan spaces were given a mixture of daylighting sources to achieve better light quality and distribution.

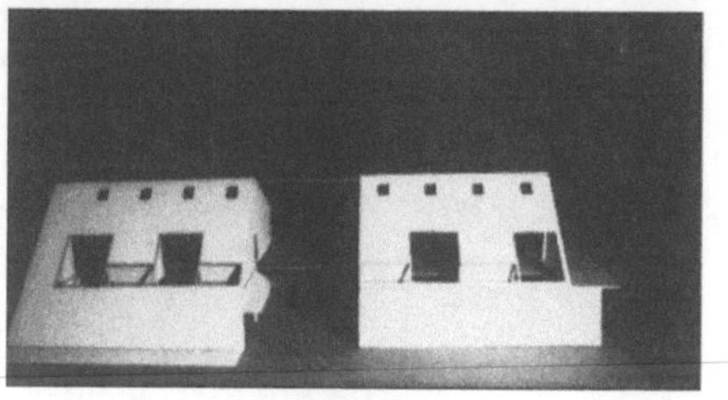

Figure 22. View of model from the north, showing rooflight arrangement

Modelling

The physical models proved very useful for planning purposes and lighting visualization but could be altered easily as the design evolved. The computer daylight model provided rapid results. It was found that dynamic thermal modelling is necessary if the multiplicity of factors contributing to overall energy use was to be taken into account. The main points arising from the modelling study were as follows. (Fuller details are available in a separate technical report.) The greatest energy savings from daylighting came when daylight factors of 2%-4% were reached. Large savings resulted from avoidance of high lighting standards where they are not needed. In non-office accommodation, summer evening daylighting made an important contribution to energy savings. Ventilation air is one of the remaining areas of major heat loss in a well-insulated building; preheating this in the buffer spaces provided large energy savings. Low-E double glazing was found to reduce space heating demand but was not economically viable because of its low light transmittance.

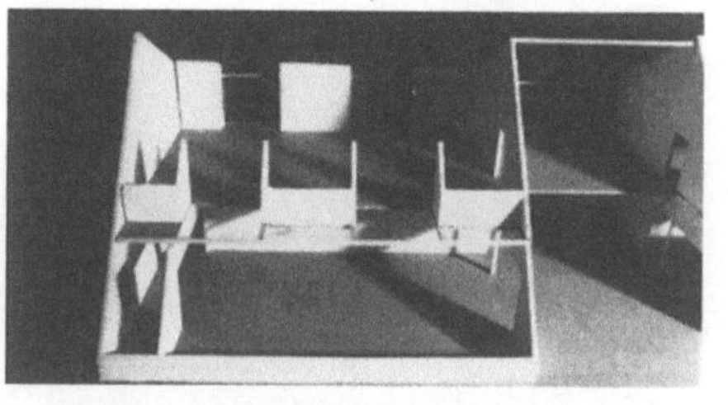

Figure 23. Model with north face removed to show how larger openings light ground and first floor

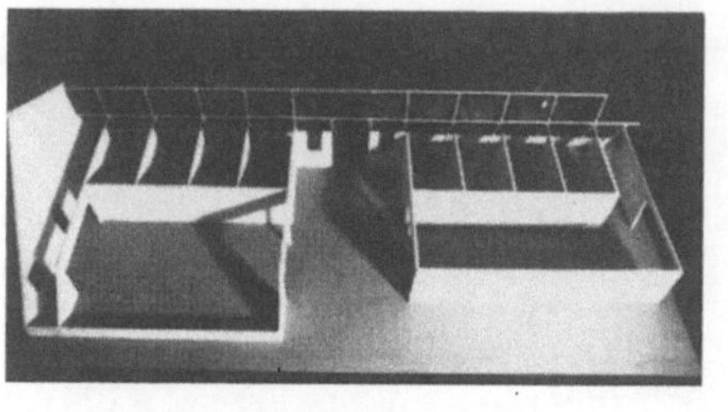

Figure 24. North side of ground floor model, showing building halved by atrium

The Building 2000 Studies

The decision to test the Machynlleth building performance through the Building 2000 programme to see if the main principles of construction and organization of spaces were justified was made early in the design process. The results of the thermal modelling showed that the basic ideas were valid but some alterations to the glazing and mass would be advantageous. The work proved that funds should be made available to test designs at as early a stage as possible using first daylighting studies and then thermal modelling. To optimize the design, sensitivity studies should be carried out.

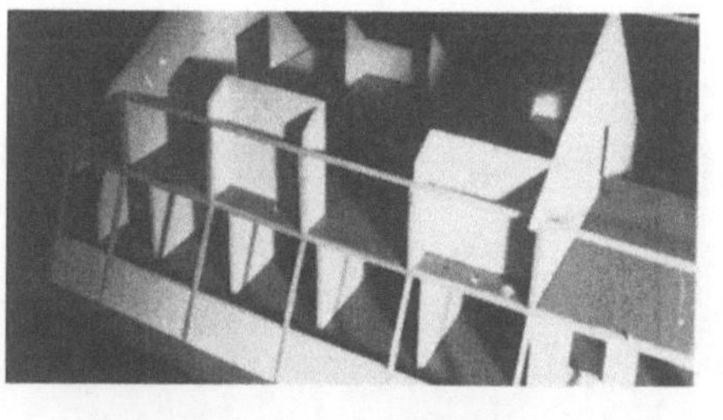

Figure 25. Detail of model showing solar chimneys for summer ventilation of bedrooms

At the beginning of the Building 2000 programme the designers had identified a number of points which needed investigation. The advantage of the programme was that immediate feed-back was available from the co-ordinators. This enabled ideas not worth pursuing in the Machynlleth climate (e.g. rock storage) and complex features to be eliminated quickly so that attention could be focussed on two important issues, daylighting and construction.

BUILDING 2000 Building 2000 brochures are published by Directorate General XII of the Commission of the European Communities to show how design studies can help architects and other building designers use passive solar principles to the best effect to produce attractive energy-efficient buildings. Each brochure describes studies carried out with the support of the Commission during the design phase of one of thirty-six non-domestic buildings in the EC Member States. The studies were on such topics as daylighting, heating, cooling, ventilation, comfort, control systems and urban design. They were carried out with the help of acknowledged European experts in these fields and drew heavily on lessons learned and techniques developed through the Commision's research and development programme on solar energy applications to buildings.

Commission of the European Communities/Directorate-General for Science, Research and Development

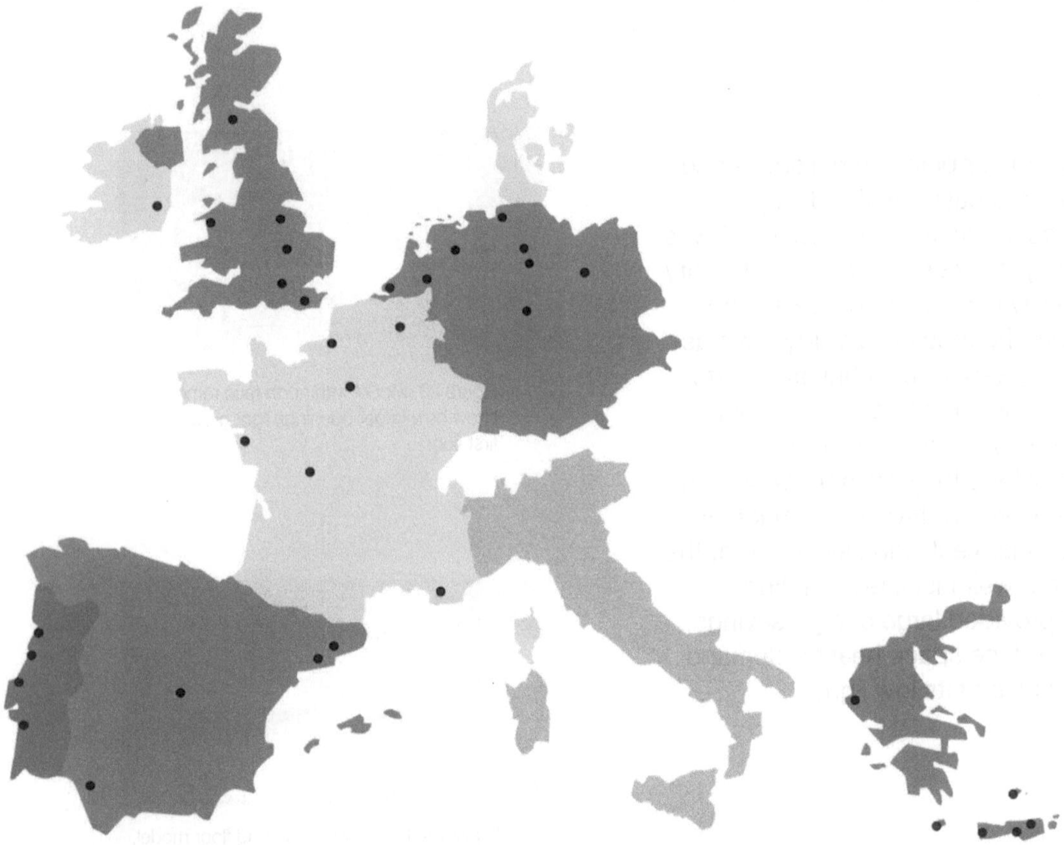

List of Design Team Participants and Advisers

Client
National Centre for Alternative Technology
Machynlleth
Powys, Wales
(represented by Pat Borer)

Architect
Yvonne Dean
Department of Environmental Design
Low Energy Architectural Research U
Polytechnic of North London
Holloway Road
London N7 8DB

Environmental Science and Design
Mike Wilson
Department of Environmental Design
Low Energy Architectural Research U
Polytechnic of North London
Holloway Road
London N7 8DB

Energy Consultants
Mike Wilson and Jon Walker
Department of Environmental Design
Low Energy Architectural Research U
Polytechnic of North London
Holloway Road
London N7 8DB

Acknowledgements:
To part-time diploma students at PNL (course tutor, John Osbourn) who conducted model building and daylighting studies.

This set of Building 2000 brochures illustrates how architects and other building designers can successfully apply passive solar principles to produce energy-efficient buildings.

BUILDING 2000 Participants

Project Director
Theo C. Steemers

Coordinator
Cees den Ouden

Technical Steering Committee
Dean Hawkes
Nick Baker
Alex Lohr
Jean P. Lepoivre

Regional Liaison Agents
(D) Jörn Behnsen
(E) Vicente Sifre
(F) Michel Raoust
(GB) Alan Hildon
(GR) Matheus Santamouris
(P) Eduardo Maldonado

Further information or copies of the brochures can be obtained from prof. ir. Cees den Ouden, EGM Engineering BV, P.O. Box 1042, 3300 BA Dordrecht, The Netherlands.

BUILDING 2000

Commission of the European Communities

- Renovation of existing cottages and development of new building to form research and educational facilities for permaculture institute.
- Renovated building contains attached greenhouses which provide direct solar gains, preheated ventilation air and space for crop growing.
- Design for new building was developed by comparing performances of designs containing wrap-around greenhouses and atria.

RESEARCH AND EDUCATIONAL CENTRE

STEYERBERG/ FEDERAL REPUBLIC OF GERMANY

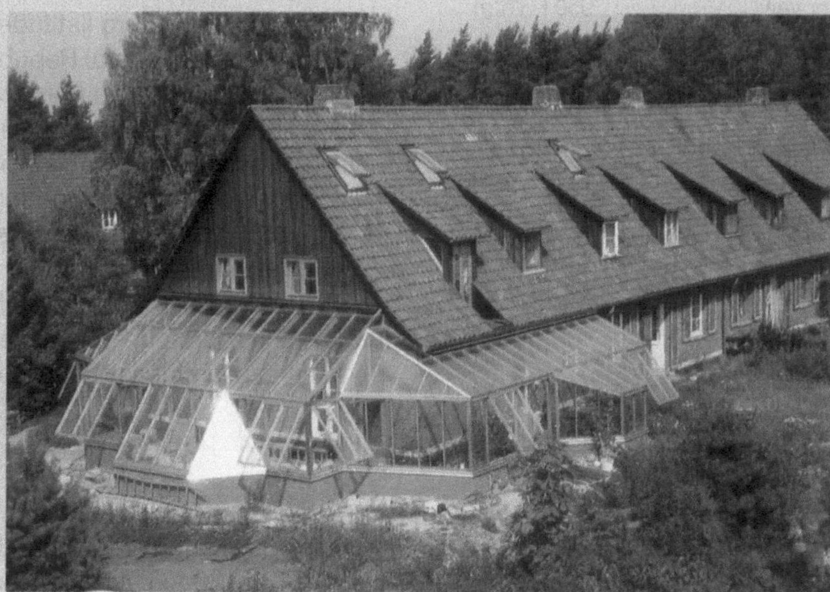

Building 2000 is a series of design studies illustrating passive solar architecture in buildings in the European Community.

ISSUE

E 4

FEB 1991

PROJECT DESCRIPTION

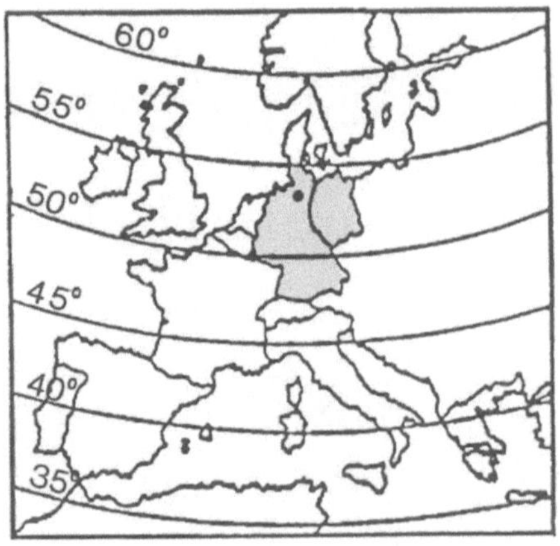

Figure 1. Location of Steyerberg

Sunshine hours per year	1600
Mean global irradiation per year	950 kWh/m^2
Sun as % of daylight time	
Nov-Feb	>25%
Mar, Apr and Oct	35%
May-Sep	45%
Mean temperature	
Jan	1.5 ° C
Jul	17.4 ° C
Annual average	9.1 ° C
Frost-free days per year	176
Rainfall per year	630mm
Relative humidity	85%

Table 1. Some climate data

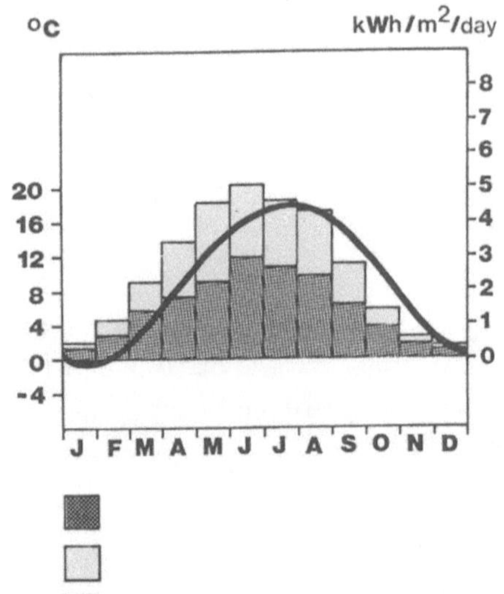

Figure 2. Mean solar irradiation and temperature data

Building Type

This project concerns the development of two separate buildings which together form a research and education centre for the Permaculture Institute of Europe. One building was formed by renovation of two existing cottages. The other is a new construction. Both buildings will house offices, seminar rooms and accommodation for staff, visiting lecturers, researchers, students and apprentices.The development is being carried out in two phases. Renovation of the cottages (Phase 1) took place over the period 1988-89. Construction of the new building (Phase 2) is expected to begin in 1991 and will take into account the results of the design studies described in this brochure and the lessons learned from constructing and monitoring the Phase 1 Building.

Location

The two sites, which are 600 m apart, are some 3 km outside the old village of Steyerberg (latitude 52 ° 3 ' N, longitude 9 ° 3 ' E) in Lower Saxony in the Federal Republic of Germany - see Figure 1. They are about an hour's drive from Hannover and two hours from Bremen and Hamburg.The sites are in the sandy hills of the west bank of the river Weser in the southern part of the county of Nienburg. The cottages are part of a community originally constructed in 1939 (see Figure 3). The new building is on a 2.6 ha site being developed as an example of permaculture for Northern European climates (see Figure 4). This site is unobstructed from the south-west round to the east. There is a larch tree plantation to the south but it is far enough away to let in even the lowest winter sun.

Site Microclimate

The climate is typical of that found in regions near the coast in north-west Germany. Over a year there are 100 days when the temperature is at least 20 ° C. Additional climate data are given in Table 1 and Figure 2.

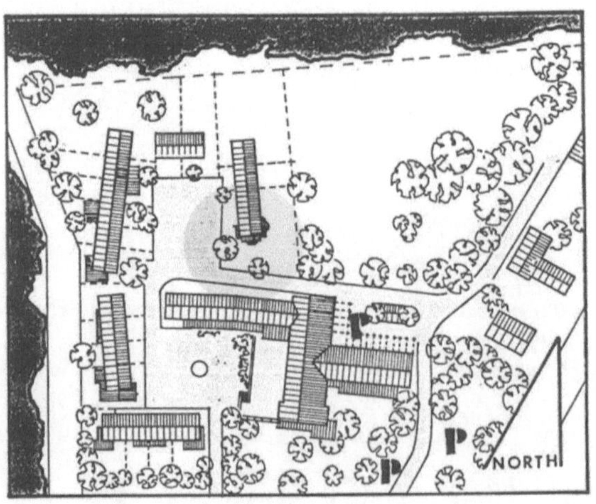

Figure 3. Site of the original cottages renovated in Phase 1 of the project

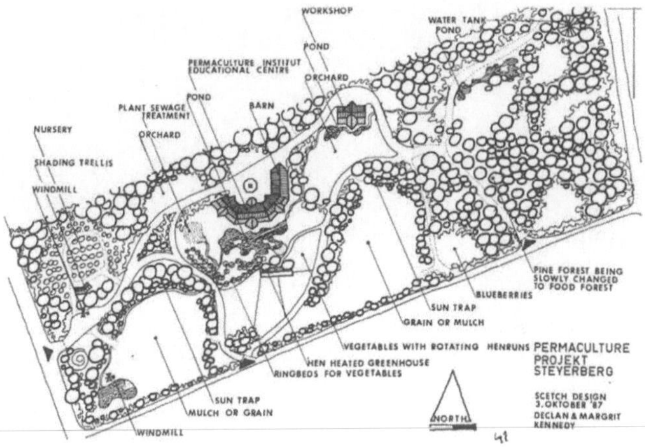

Figure 4. Site of the new building to be constructed as Phase 2 of the project

Design and Construction Details

Introduction

In developing the two buildings the objective has been to demonstrate the concepts of permaculture in a European cultural and social setting in a cold temperate climate. The term 'permaculture', originally formed from the phrase 'permanent agriculture', has now come to mean the planning and design of stable, self- supporting systems based on the principles of a sustainable ecology. It involves a holistic approach to life and includes the supply of wholesome food, energy and warmth and participation in good pursuits. The Permaculture Institute of Europe aims to support and develop these ideas by practicing them in different conditions and climates and disseminating the results of their work through seminars and workshops.

General Design Concepts

In both projects there are attached greenhouse or atrium areas for growing edible plants. The buildings and greenhouses are ventilated with pre-heated or cooled air to avoid temperature drops in winter and overheating in summer. The ventilation air for the main buildings is preheated in the greenhouses. In the Phase 1 building, the greenhouses are located on the south, east and west sides of the original cottages. In developing the Phase 2 building, the task has been to find out whether a building with a wrap-around greenhouse or one with an atrium would be the better choice for optimizing energy efficiency in the building and greenhouse areas. The buildings also incorporate the following ideas:- solar collectors to the south, with superinsulated water or magnesium tanks underground for long-term storage;- water-saving devices (e.g. humus toilets and used-water recycling systems) to save energy in treatment of waste water;- recycling of disposable wastes, preferably on site;- production of food (especially fruit and vegetables) under various conditions to ensure year-round supply and save energy on transport and storage;- wind energy for electricity and pumping ground water (for the Phase II building only);- photovoltaic cells for electricity, especially in summer;- wood-burning co-generation of heat and electricity in winter using wood grown on site in order to save energy and non-renewable sources;- shading and ventilation devices to prevent overheating in sunspaces and interior rooms and thus save energy for cooling in summer.

The Phase 1 Building

The original three-storey cottages are illustrated in Figure 5. A model of the renovated building showing the south-facing greenhouse is illustrated in Figure 6. The project as it looked on site in December 1988 is shown in Figure 7. The ground floor and first floor plans are given in Figures 8 and 9 and sections through the greenhouses in Figures 10 and 11. The two top storeys are within the 44 ° pitched roof. There is a basement under part of the building. The building contains four offices, a small seminar room containing information resources, one apartment and two single guest rooms. (One of these guest rooms will be turned into an office once the apartments in the Phase 2 Building have been completed.) The south-facing greenhouse is used as a seminar room for small group meetings from early spring to late autumn. Different climatic environments are produced in the different parts of the greenhouses, depending on whether they are oriented south, west or east. By using the various parts separately or in combination, many different types of plant can be grown throughout the year. The original cottages were built with a traditional brick and wood construction and a tile roof. The external walls on the ground floor are 360 mm thick. The roof slope has been insulated inside with 200 mm recycled paper impregnated with borax plus wood panelling. The outside of the south-facing gable has 120 mm insulation plus wood weatherboarding. The greenhouses are double glazed. The exterior walls up to cill level and the foundations to below frost level have 800 mm insulation so that heat is retained in the planting beds. Depending on the weather, roof and wall vents open automatically. Shading prevents overheating in summer. To ensure year-round growth, auxiliary underfloor heating (preheated by solar energy) has been installed to protect against temperatures below 5 ° C.

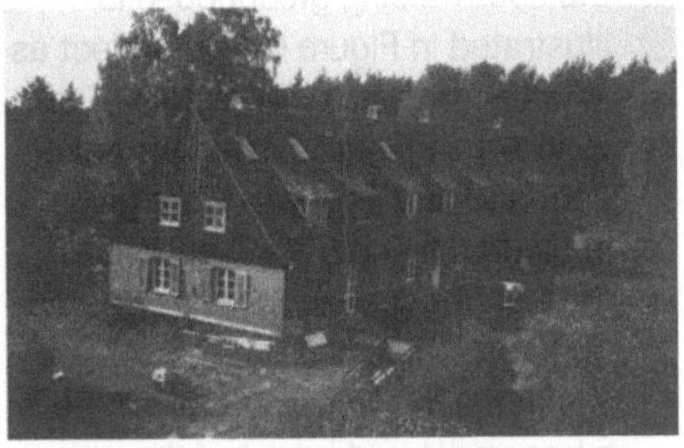

Figure 5. The original cottages in 1987

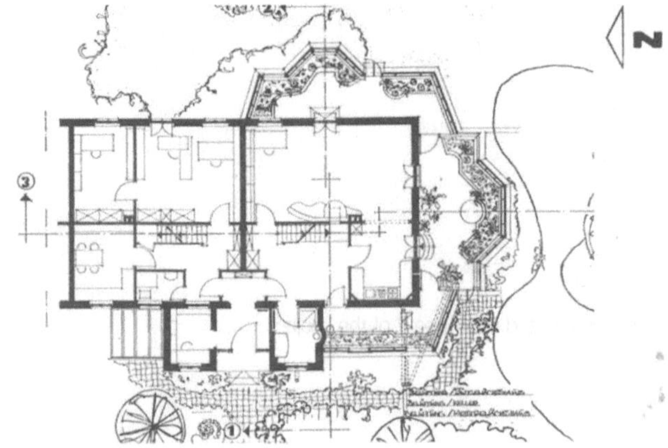

Figure 8. Ground floor plan of Phase 1 Building

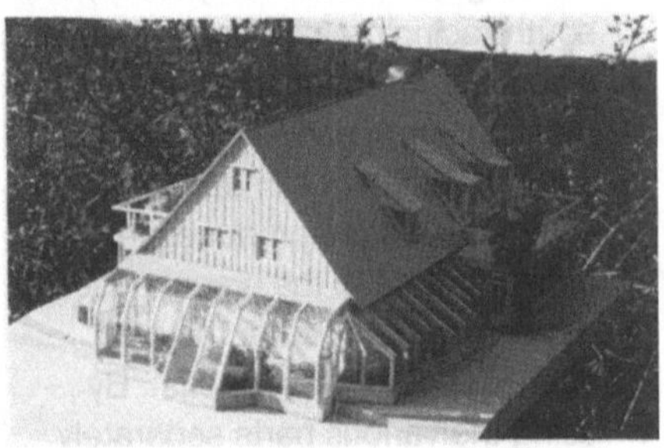

Figure 6. Model of the renovated Phase 1 Building

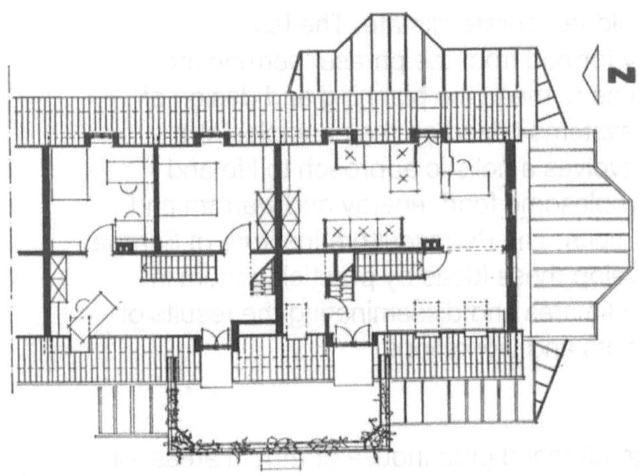

Figure 9. First floor plan of Phase 1 Building

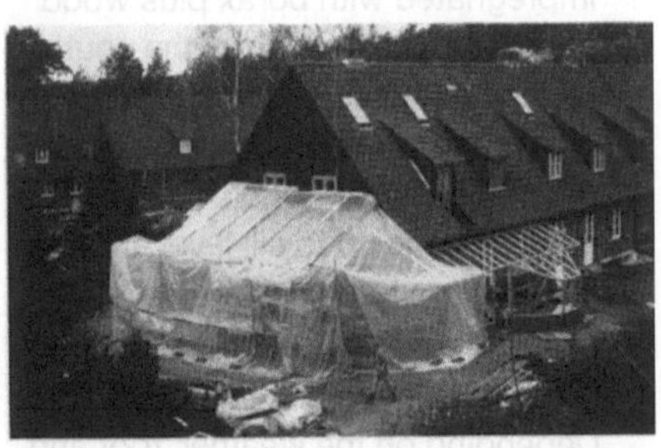

Figure 7. Renovated Phase 1 Building during construction in December 1988.

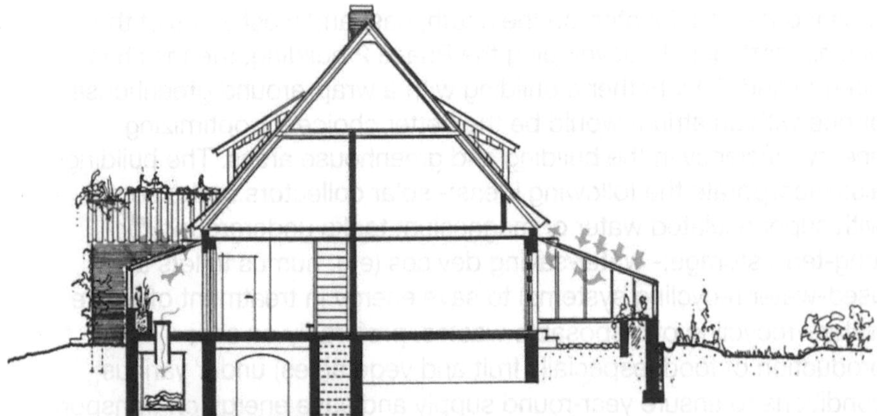

Figure 10. Cross-section through east and west greenhouses of Phase 1 Building

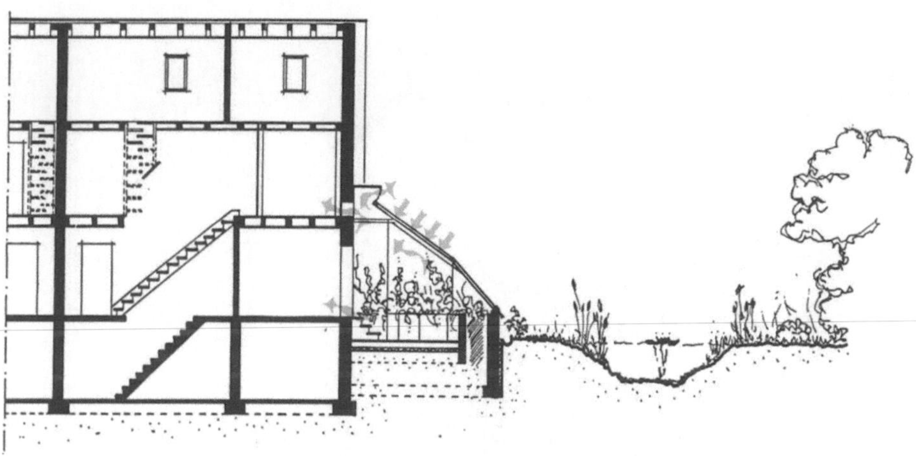

Figure 11. Section through south greenhouse and reflecting pond of Phase 1 Building

The Phase 2 Building

The new building will contain apartments (flats), offices, a large seminar room, kitchen and eating spaces and a barn. Two basic designs have been prepared. In Design I, greenhouses with 120 m^2 usable space are wrapped around the south, south-east and south-west sides of the main building. In Design II, the greenhouses have been pulled into the building to form atria with 130 m^3 usable space.

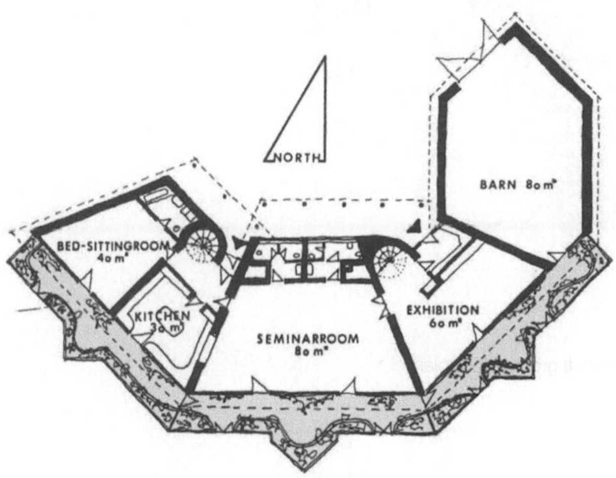

Figure 12. Ground floor plan of Phase 2 Building Design I

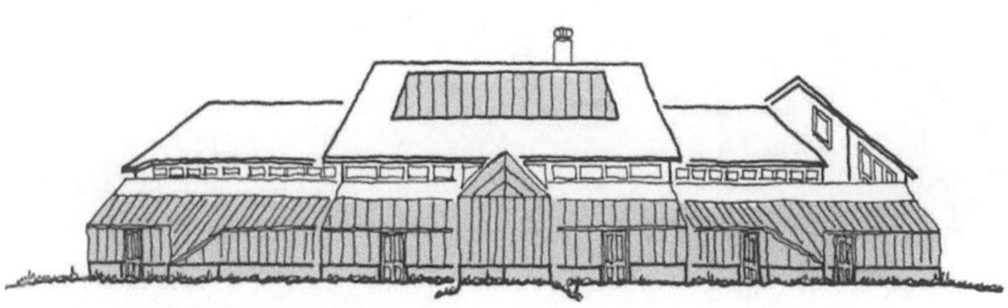

Figure 13. South elevation of Phase 2 Building Design I

Phase 2 Building - Design I

Design I is mostly two storeys high although there is an additional roof space in the middle section. There is a basement under part of the building. The ground and first floor plans, south elevation and north-south sections are shown in Figures 12-15. The main rooms face south, south- east or south-west and there is a narrow (2.5 m) greenhouse along the front. On the ground floor there is an exhibition room at the entrance which also serves as a foyer for the large seminar room. Eating takes place in the seminar room. The adjacent kitchen, with its basement and two flats, has a separate entrance from the outside. All the bathrooms, toilets and staircases are on the north side of the building so that they form buffer spaces for the main rooms. This buffer principle is also used on the first floor.Windows have been placed almost exclusively on the south side. This has been found to provide adequate daylighting for all rooms except the upstairs library. Here, to ensure good daylighting, the greenhouse (which is in general one-storey high) has been raised to two storeys and pulled back to the roof. A spiral staircase provides access from the library to an upper floor containing workspaces for apprentices and visiting researchers.The interior walls are mud-brick to add to the heat storage capacity of the building. All the outside walls of the greenhouse are highly insulated from below frost level to sill height. The glass wall between the greenhouse and the main part of the building will allow a high degree of heat transfer between the sunspace and the interior. All the glass roofs have shading devices which are also air collectors to provide hot water (up to 85 ° C) through an air-to-water heat exchanger. The surplus hot water is stored in tanks two storeys high and used to preheat water for the auxiliary heating system.

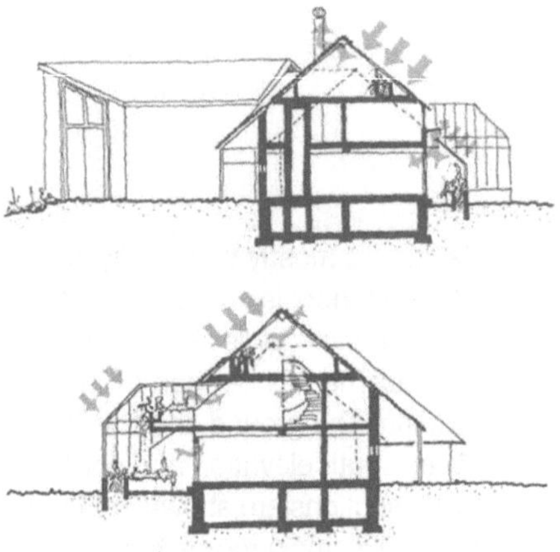

Figure 14. North-south and south- north cross sections of Phase 2 Building Design I

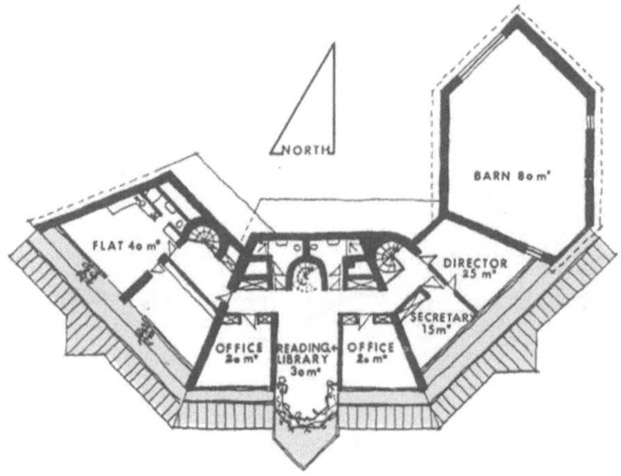

Figure 15. Phase 2 Building Design I first floor plan

Phase 2 Building - Design II

The ground floor, first floor and basement plans, south elevation and north-south sections of Design II are shown in Figures 16-20. The building has three atria, facing south, south-east and south-west. They have much the same area as the wrap-around greenhouses of Design I but seem more spacious and give occupants direct contact with plants and the outside world. The atrium facing south is for potted plants. Those facing south-east and south-west have direct contact with the earth so fruit trees from southern climates can be grown there. The atrium roofs are higher than the roof of the main building to ensure optimum ventilation. The main entrance is from the north, on a split level between the basement and ground floor. The entrance hall forms an exhibition area. The foyer for seminars is one floor below. The south-facing atrium is partly used as a dining area. The large seminar room is situated under it in a half-basement and obtains its daylight through the atrium facade. A second entrance from the north provides access to the kitchen, its basement store and two staff flats. The first floor contains the library and offices which are connected by galleries overlooking the atria. All the atria are fitted with shading devices/air collectors similar to those in Design I.

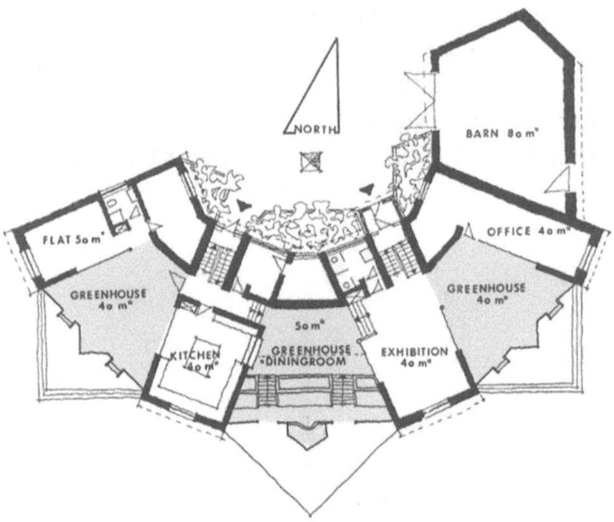

Figure 16. Phase 2 Building Design II ground floor plan

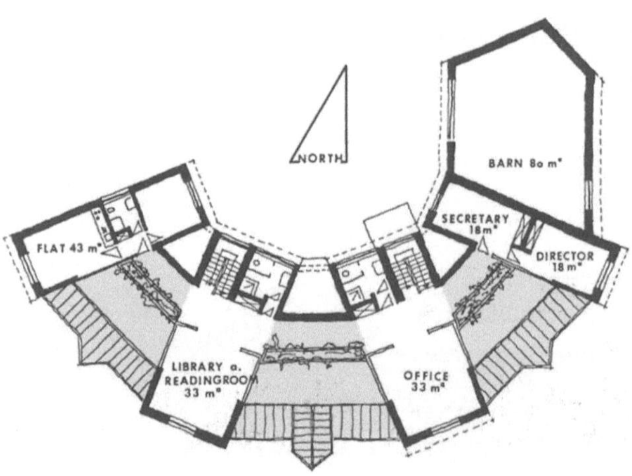

Figure 17. Phase 2 Building Design II first floor plan

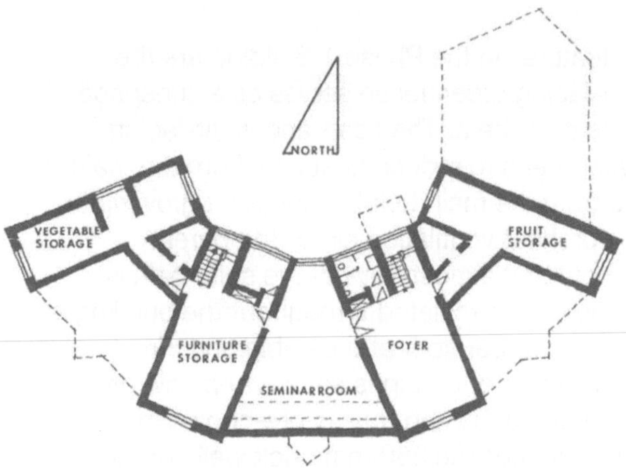

Figure 18. Phase 2 Building Design II basement plan

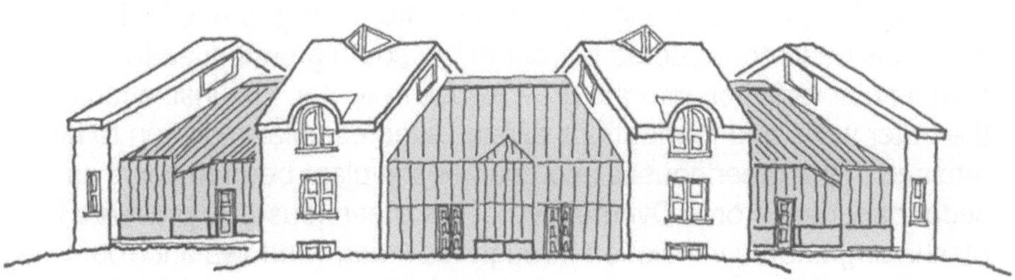

Figure 19. South elevation of Phase 2 Building Design II

	m^2	greenhouse/ atria	m^3
Design I	510	120	1400
Design II	584	130	1459

Table 2. Comparison of usable floor area and volume of the two Phase 2 Building designs

	volume m^3	costs/m^3 (DM)	total costs (DM)
Main building	1,400	800	1,120,000
Barn	560	500	280,000
Greenhouse	48	500	24,000
Glazing	305	480	146,400
Total			1,570,000

Table 3. Estimated construction costs of Phase 2 Building Design I

	volume m^3	costs/m^3 (DM)	total costs (DM)
Main building	1,459	800	1,167,200
Barn	560	500	280,000
Atria	52	500	26,000
Glazing	258	480	123,840
Total			1,597,040

Table 4. Estimated construction costs of Phase 2 Building Design II

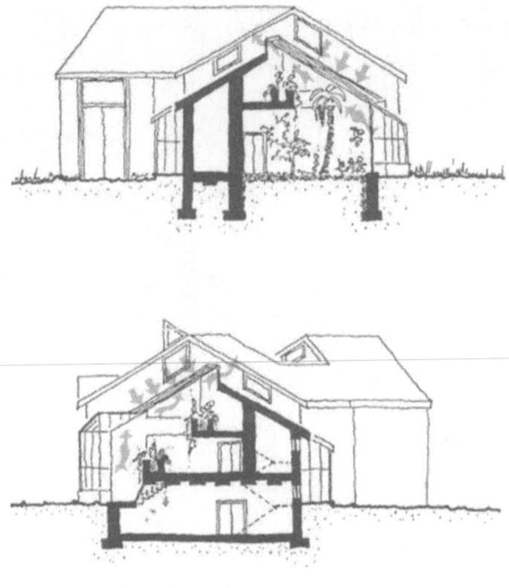

Figure 20. North-south and south- north cross sections of Phase 2 Building Design II

Comparison of Area and Construction Costs of the Two Phase 2 Building Designs

The floor area and volume of the two designs are compared in Table 2. Design II has more usable floor area because it has greater circulation space. In Design II the roof office space is omitted. It could, however, be added without altering the results of the costs and performance evaluations described below.At first thought, it seemed likely that the atrium design (Design II) would be much more costly to build. However, when the costs were estimated from the actual costs of the Phase 1 Building it was found that Design II would be less than 2% more expensive to construct (see Tables 3 and 4). Design II has the advantage of providing glasshouses where the height, width and orientation are more suited to growing plants in accordance with the principles of permaculture.

DESCRIPTION OF PASSIVE SOLAR FEATURES/COMPONENTS

The Phase 1 Building

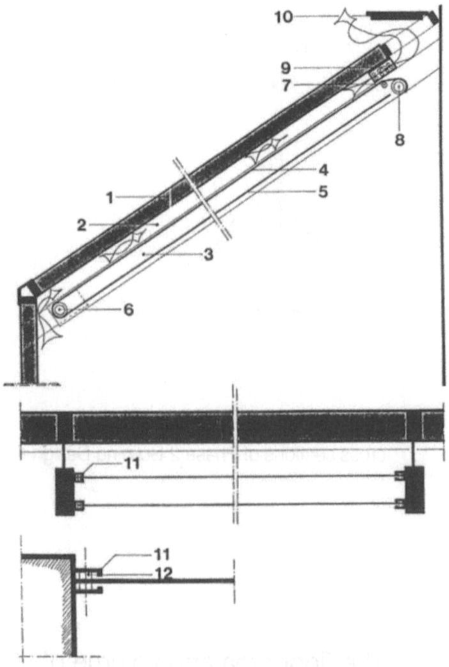

Figure 21. The greenhouse shades/solar collectors

Auxiliary Heating

Auxiliary space heating is provided by a natural gas-fired furnace/low temperature radiator system with thermostats on the individual radiators and an earthen wood- burning stove installed on the ground floor.

The Greenhouses

The main passive solar features in the Phase 1 Building are the greenhouses. The south-facing greenhouse serves as a sunspace which collects and stores solar heat. The east- and west-facing greenhouses act as buffer areas to reduce heat loss from the main building and utilize heat from the main building for plant growing.In winter, the greenhouses preheat ventilation air for the parent building. Calculations show that almost 60% of the solar gain is used for this purpose. The air is circulated throughout the building by means of vents, entering the upper floor at the gable end. Fresh air is brought into the greenhouses through pipes set 2.5 m below ground. The temperature of this air is 10 ° C and thus is pre-cooled in summer and pre-heated in winter.The 360 mm brick walls of the original building act as short-term primary heat stores. These help warm the building and assist the growth of plants by retaining some of the heat collected during the day well into the night. A reflecting pond has been constructed in front of the south greenhouse to reflect solar radiation into the greenhouse in winter. The insulation on the outer walls and foundations allows reflected solar radiation to be retained in the greenhouses and enables the plant beds to serve as secondary heat stores.Overheating of the greenhouses is prevented by shading and by ventilation through top-hung windows and roof-top flaps which automatically open at 35 ° C and close at 24 ° C. The top flaps also close automatically in case of rain.The performance of the greenhouses is being monitored over the period 1989 to 1991, prior to the construction of the Phase 2 Building.Greenhouse

Shades/Solar Collectors for Domestic Hot Water

Shading in the south-facing greenhouse is provided by a movable sheet membrane installed under the double-glazed roof. This prevents overheating. It also allows the layer of hot air between the membrane and glass to act as an air collector (see Figure 21) which is transformed by a heat exchanger into domestic hot water. The electricity for this and for the compressor for the automatic vents is run on a 12 volt circuit provided by photovoltaic cells on the south facade.

Other Energy-Saving Features

For the humus toilet in the bathroom on the west side, a large tank (Clivius Multrum) has been built underneath the west greenhouse, the floor of which has been constructed so that conductive heat is let through to allow composting to take place at 18 ° C.At the most northern part of the annex, a glass-covered open air wood stove has been constructed to dry new wood for the earthen stove and store it under cover.Under this structure is a new basement with a packed earth floor to provide an ideal storage temperature for vegetables throughout the winter.

The Phase 2 Building

Greenhouses/Atria and Other Energy-Saving Features

The main passive solar features in the Phase 2 Building are the greenhouses or atria. These and other energy-saving features will be constructed and operated in a similar way to those in the Phase 1 Building. However, there is in a new building the advantage of being able to place all the different components in the right place from the start. For instance, heat storage in interior walls and in large one-storey high water tanks can be better integrated. The hot air flow from the greenhouses to the upper floors can be placed all along the south walls to provide a better distribution than in Phase 1. To obtain the best energy-saving from the greenhouses, natural air flows from them to the heated occupied spaces can be encouraged by extracting air from heavily occupied spaces using the stack effect and taking it to the highest part of the building. The north side of the building should be airtight and highly insulated. In winter in northern Germany solar gains from the global reflection of the sky are higher than those from the direct sun. As a result, the angle of the glass roof should be as flat as possible - but not so flat that snow cannot slide off. In addition, the Phase 2 Building will take into account points learned in Phase 1. For instance:

- buffer zones and longer roof overhangs on the north side can reduce heat loss and serve for drying firewood;
- orientation and conductivity and storage capacity of materials all have to be taken into account in the overall building design;
- the walls of the main building need not all be built of the same materials or have the same thickness. Depending on orientation, they can be built of materials found on or near the site, thus reducing transport costs;
- roof ridge dormer windows can improve daylighting and reduce the necessity of north windows, thereby cutting heat loss;
- rather than having a grass sod covering, the roof should be covered with tiles made in a nearby brickyard to enable rainwater to be collected and stored in cisterns underground.

Auxiliary Heating

In winter, auxiliary space and domestic hot water heating will be provided by a co-generation system using a Stirling motor. This will run on wood fuel found on site. The motor runs an electric generator. The heat from the exhaust will be channelled through ducts in the interior walls, creating radiant wall panels. The air is circulated back to the motor's cooler without entering the rooms. The co- generator will be supplemented by woodburning stoves as in the Phase 1 Building.

ENERGY CALCULATIONS PERFORMED AND DESIGN TOOLS USED

Design Tools

The performance of Designs I and II for the Phase 2 Building were evaluated using a number of design tools. A preliminary energy analysis was conducted using a simplified thermal network method developed in the UK by Dr Nick Baker. A more detailed comparison of energy performance characteristics was carried out using the ESP computer model developed at the University of Strathclyde in Scotland. An idea of the daylighting performance was obtained using a 1:50 scale model. The results of the energy analyses are given below.

	Ratio		
	glass :	walls :	roof
Design I	4 :	3 :	3
Design II	3 :	4 :	3

Table 5. Ratio of glass to opaque surface areas in Designs I and II

Results of Preliminary Energy Analyses

Thermal Performance

The results of the preliminary calculations (which were based on rough U- values for the proposed materials and simplified sections for the building) are given in Table 6. Designs I and II require approximately the same amount of energy to run. There would, however, be a lower incidence of frost in the atria of Design II than in the wrap-around greenhouses of Design I. Furthermore, some parts of the atria would be virtually frost- free, permitting growth of perennial shrubs and trees from the Mediterranean region such as citrus fruit, melon, fig, etc. Conditions for such plants would be improved if a relative humidity of 70% and temperature of 5 ° C were maintained in the atria. This would require auxiliary heating at certain times - when, say, the external temperature was -20 ° C.

	Design I	Design II
Average January temperature in greenhouses or atria (° C)	4.0	7.2
Annual energy consumption (kWh)		
with no solar preheating of ventilation air	25,052	24,903
with preheating of ventilation air	21,225	20,902

Table 6. Preliminary comparison of the energy performance of Designs I and II for Phase 2 Building

Ratio of Areas of Glazed and Opaque Surfaces

It was found that Design II has not only a better energy performance but also a smaller proportion of glass (see Table 5). In traditional houses (including those which have been subject to modernization and renovation) in Lower Saxony, the ratio of openings:walls:roof is 2:4:4. In these buildings, only about half of the openings are filled with glass; the remainder consist of wooden doors, etc.

Additional Considerations

In addition to the above points, Design II would fit better into the local landscape because the glazing in the outside walls is less apparent. In addition, the opaque surfaces use local building materials and can be constructed with local labour.

Results of More Detailed Energy Analyses

Thermal Performance

The more detailed analyses of the Phase 2 Building were carried out for three different designs: one with wrap-around greenhouses where there is 72 m^2 single glazing between the greenhouses and the main part of the building (Design I); one with atria integrated into the building where there is no glazing between the atria and the rest of the building (a modification of Design II); and one where 10 m^2 glazed window has been introduced between each atrium and the parent building (a second modification of Design II).

It was found that the atrium designs consume less heating energy than Design I (see Figures 22 and 23). The total annual heating energy consumption for Design I is 43.1 kWh/m^3 building, that for the first modification of Design II (atrium type with no internal windows) 41.6 kWh/m^3 and for the second modification of Design II (atrium type with a total of 30 m^2 internal windows) 43.5 kWh/m^3.

The atrium designs will, however, require auxiliary heating for frost protection whereas the wrap-around greenhouses will not need any additional heating. This is because the 72 m^2 single glazing between the greenhouses and building in Design I will allow heat loss from the parent building to the greenhouses. The required peak heating power for frost protection in the south atrium is 1.7 kW, for the west atrium 0.5 kW and for the east atrium 1.5 kW. For the complete building, the peak heating power is 23.6 kW for Design I and 25 kW for Design II.

The annual cumulative distribution of air temperature in the west and east atria is given in Figures 24 and 25. The west atrium generates a warmer climate so that it will have a temperature over 20 ° C for 35% of the year. In the east atrium, the temperature will be above 20 ° C for only 20% of the time. The yearly mean air temperature of the greenhouses/atria are given in Table 7.

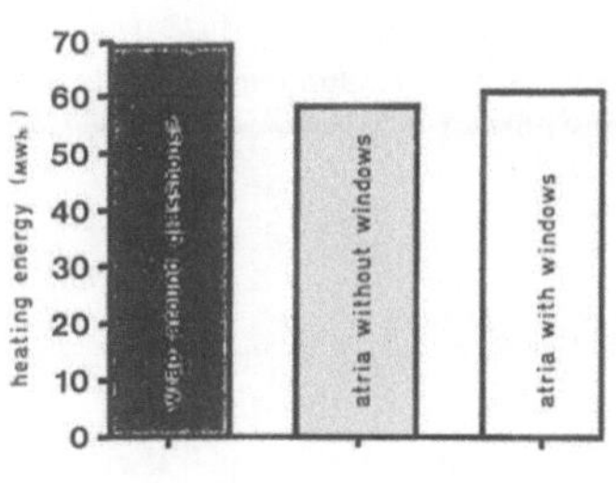

Figure 22. Annual heating energy consumption

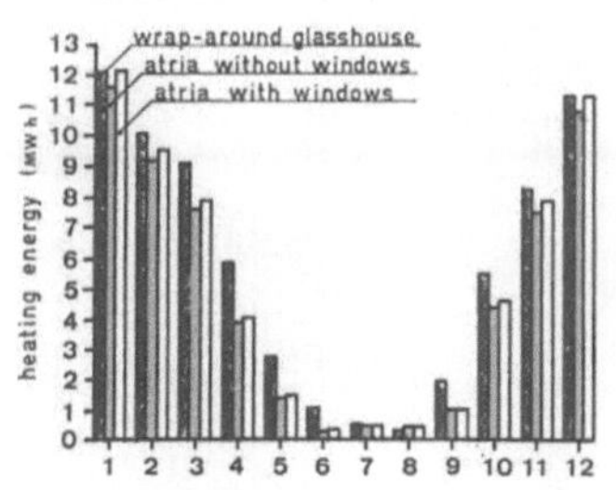

Figure 23. Monthly heating energy consumption

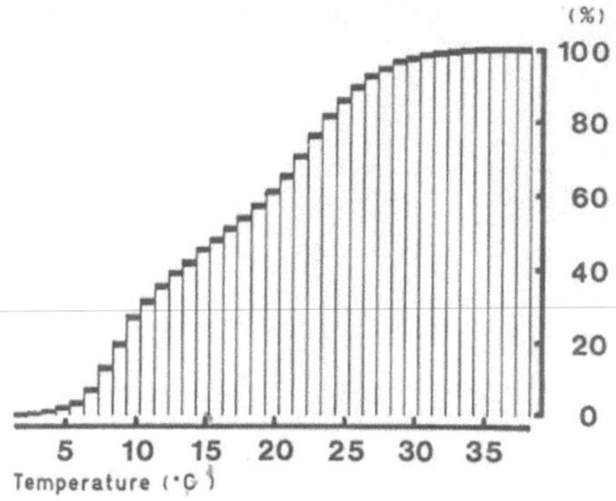

Figure 24. Annual cumulative distribution of air temperature in the west atrium

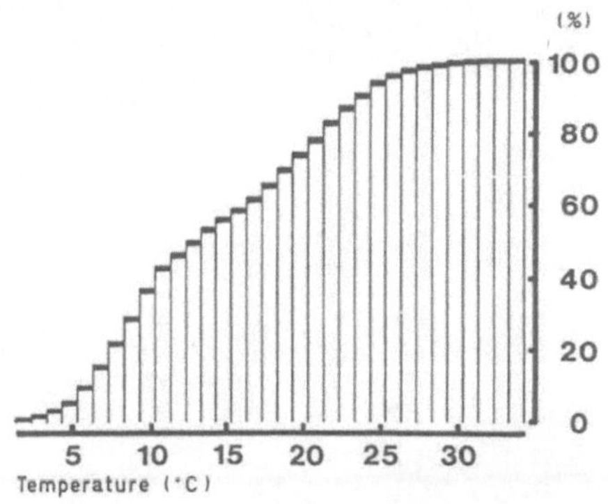

Figure 25. Annual cumulative distribution of air temperature in the east atrium

	°C
Wrap-around greenhouses	16.0
East atrium	14.6
South atrium	15.7
West atrium	17.0

Table 7. Yearly mean air temperature of greenhouses or atria

Final Modifications to the Design

From the above results, it was concluded that the atrium design without internal glazing in the connecting wall is to be preferred to the other two solutions. The possibility was considered of pulling the atria out two metres beyond the facade of the building and joining them up using narrow buffer greenhouses. It was felt that this might provide a combination of the best attributes of all the designs considered earlier. Simulation studies showed, however, that the utilizability of solar gain diminished as the area of glazing increased.The Building 2000 experts, therefore, recommended that the final design combine just one atrium (built on the earth and south-facing with a larger area than the south-facing atrium in Design II) with east- and west-facing wrap-around greenhouses. This solution would generate all the different climate possibilities but at the same time minimize heat loss. The atrium and wrap-around greenhouses would need shading and top and bottom ventilation to prevent summer-time overheating. Shading collectors would also be included to provide insulation on winter nights.

BUILDING 2000

Building 2000 brochures are published by Directorate General XII of the Commission of the European Communities to show how design studies can help architects and other building designers use passive solar principles to the best effect to produce attractive energy-efficient buildings. Each brochure describes studies carried out with the support of the Commission during the design phase of one of thirty-six non-domestic buildings in the EC Member States. The studies were on such topics as daylighting, heating, cooling, ventilation, comfort, control systems and urban design. They were carried out with the help of acknowledged European experts in these fields and drew heavily on lessons learned and techniques developed through the Commision's research and development programme on solar energy applications to buildings.

Commission of the European Communities/Directorate-General for Science, Research and Development

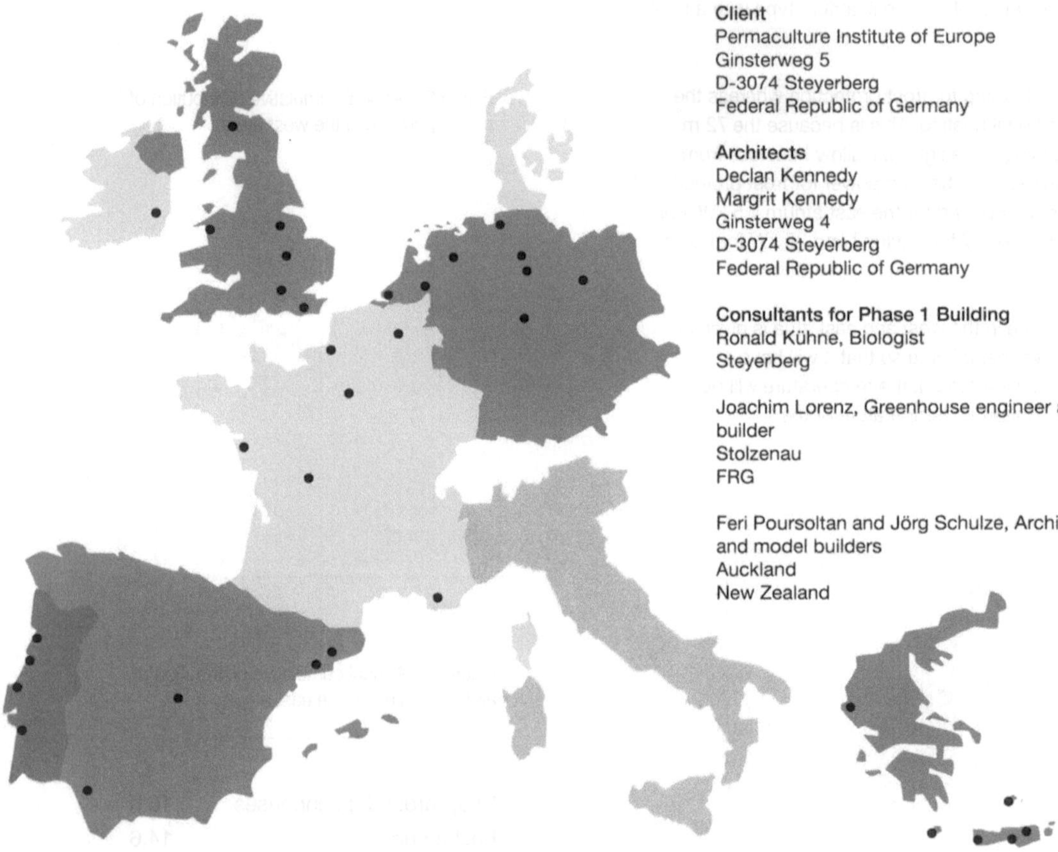

List of Design Team Participants and Advisers

Client
Permaculture Institute of Europe
Ginsterweg 5
D-3074 Steyerberg
Federal Republic of Germany

Architects
Declan Kennedy
Margrit Kennedy
Ginsterweg 4
D-3074 Steyerberg
Federal Republic of Germany

Consultants for Phase 1 Building
Ronald Kühne, Biologist
Steyerberg

Joachim Lorenz, Greenhouse engineer and builder
Stolzenau
FRG

Feri Poursoltan and Jörg Schulze, Architects and model builders
Auckland
New Zealand

Peter Rabien, Heating and stirling motor specialist
Nuremberg
FRG

Bengt Warne, Architect and builder
Stockholm
Sweden

Consultants for Phase 2 Building
Nick Baker
Cambridge
United Kingdom

Jörn Behnsen, Architect and electronic engineer
Oldenburg and Hannover
FRG

Herbert Schulz, Plant expert and landsca designer
Bad Zwischenahn

Ekkehard Wagner, Glass specialist and engineer
Westerstede
FRG

Peter Voit, Heating and computer engine
Stuttgart
FRG

This set of Building 2000 brochures illustrates how architects and other building designers can successfully apply passive solar principles to produce energy-efficient buildings.

BUILDING 2000
Participants

Project Director
Theo C. Steemers

Coordinator
Cees den Ouden

Technical Steering Committee
Dean Hawkes
Nick Baker
Alex Lohr
Jean P. Lepoivre

Regional Liaison Agents
(D) Jörn Behnsen
(E) Vicente Sifre
(F) Michel Raoust
(GB) Alan Hildon
(GR) Matheus Santamouris
(P) Eduardo Maldonado

Further information or copies of the brochures can be obtained from prof. ir. Cees den Ouden, EGM Engineering BV, P.O. Box 1042, 3300 BA Dordrecht, The Netherlands.

BUILDING 2000

Commission of the European Communities

- Renovation of existing building to form residential and teaching facilities for ecological academy.
- Demonstrates use of passive solar features, daylighting and traditional adobe building techniques to academy visitors.
- Glazing in lecture room roof improves daylighting.
- Sunspace and glazed walkway improve daylighting, create buffer spaces whice reduce heat losses and provide solar gains.

EDUCATIONAL AND VOCATIONAL TRAINING CENTRE

HUNTLOSEN/ FEDERAL REPUBLIC OF GERMANY

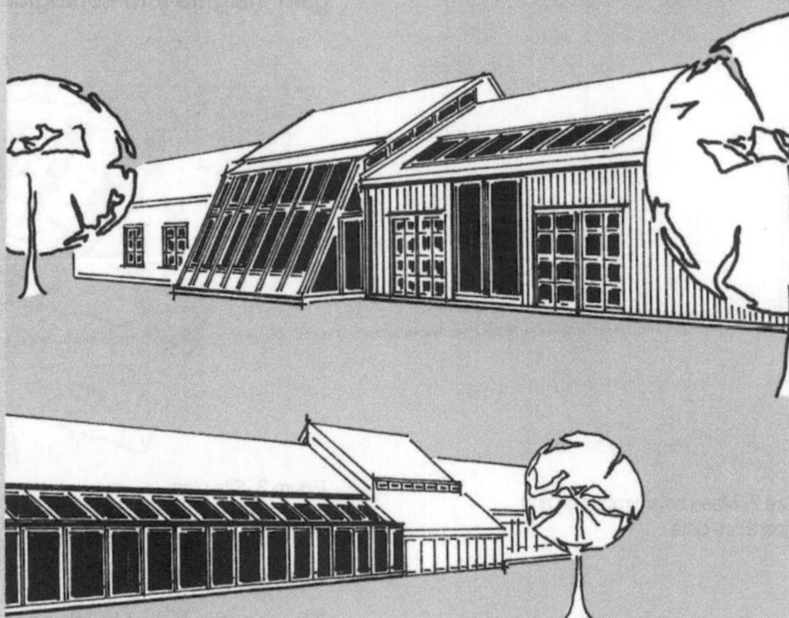

Building 2000 is a series of design studies illustrating passive solar architecture in buildings in the European Community.

FEB 1991

PROJECT DESCRIPTION

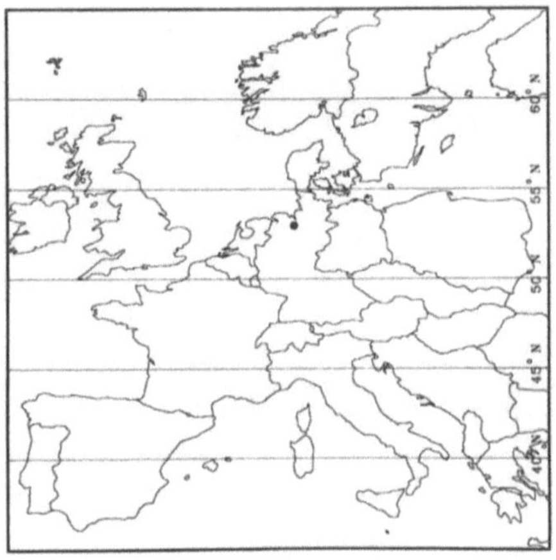

Figure 1. Location

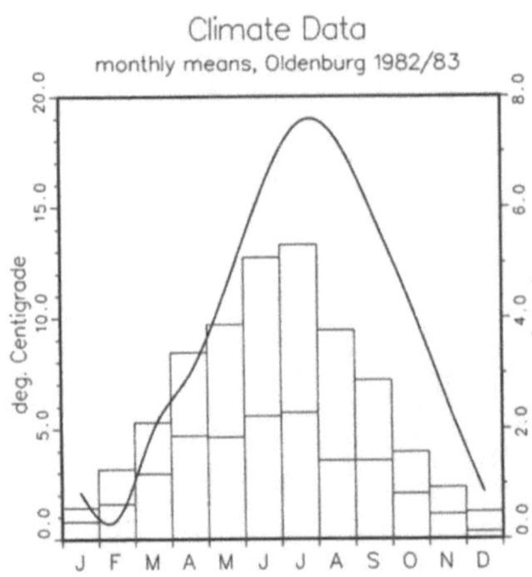

Key: — mean ambient temperature
direct irradiation
diffuse irradiation

Figure 2. Mean solar irradiation and temperature data

Building Type

This project concerns the conversion of former tuberculosis clinic buildings into an educational and vocational training centre for the Okologische Akademie Hosüne, an academy specializing in ecologically- oriented sciences, technologies, lifestyles and health care. The development, which involves refurbishment and partial rebuilding, is being carried out in stages. The first phase, which is described here, concerns refurbishment of part of one building (marked with a circle on the site plan in Figure 3) into residential and teaching accommodation for twenty people most of whom will be attending weekend or week-long courses. The refurbished part of the building contains bedrooms, a lecture room, kitchen and breakfast room. Other phases of the development are concerned with preparation of a central kitchen/dining area, an exhibition area, a library and canteen/theatre complex, offices for Hosüne eV and storage facilities for a number of enterprises operating in the areas of ecological building, renewable energy and marketing of organically-grown produce. Most of the academy staff live on site so that visitors gain insights into ecological lifestyles.

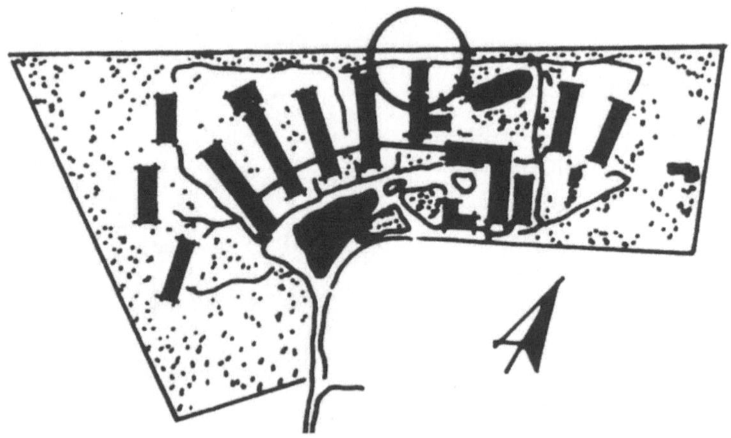

Figure 3. Site plan

Location

The centre is in Huntlosen, 20 km south of Oldenburg in north west Germany (see Figure 1). The 100,000 m^2 site is slightly hilly and sparsely wooded. A large forest stretches to the north, west and south west.

Site Microclimate

Mean solar irradiation and temperature data for Oldenburg are given in Figure 2. The climate in this area is influenced by the North Sea. Winter temperatures are higher and summer temperatures lower than those found in central Germany. The microclimate on the Huntlosen site is affected by the presence of tall trees on site and the surrounding forest. This reduces the amount of solar radiation falling on the buildings. The obstruction is, however, less in the heating season than in the summer because most of the trees shed their leaves or needles in winter.

Design and Construction Details

Introduction

In their work on the Huntlosen site, the academy's aim is to refurbish the original buildings (which date from 1943/44 and have a total floor area of 10,000 m^2) to a high degree of comfort using traditional building techniques involving materials such as wood and adobe which have been adapted to the requirements of today's building sites.

The academy regards adobe as an ecological building material par excellence because it does not have to be manufactured but is available practically everywhere. The energy required to transport it is negligible. Its physical properties make it excellent for use in ecological buildings. Its vapour diffusion, boundary layer and semi-permeability characteristics, for instance, are good. Its production costs are low, although its use is labour-intensive. The reverse stability of adobe, however, makes it a building material which requires unconventional processing using techniques which are generally forgotten today.

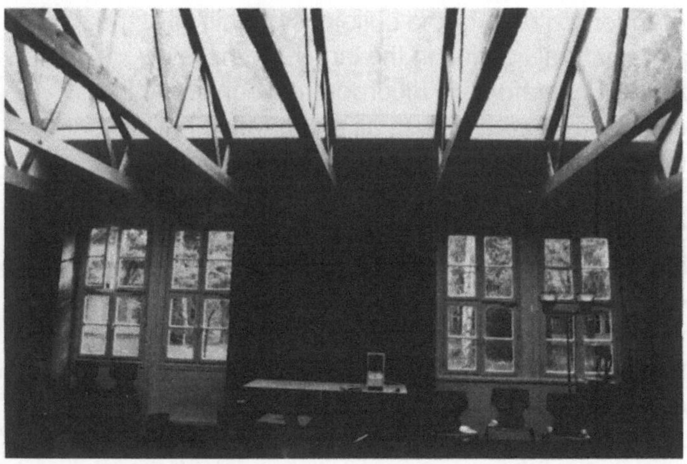

Figure 4. The Phase 1 building

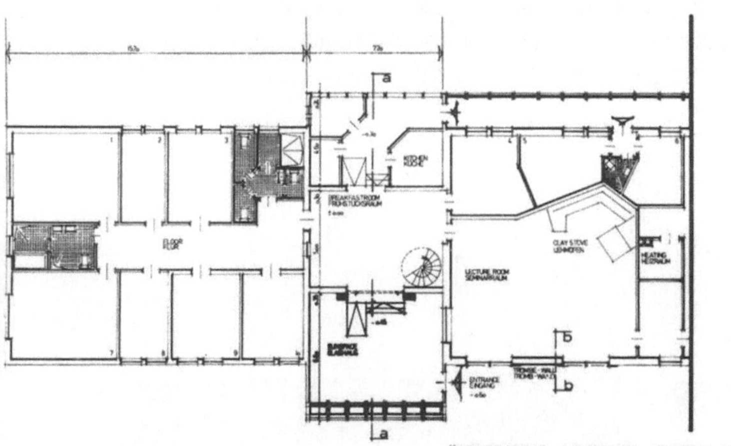

Figure 5. Ground floor plan

Details of the Phase 1 Refurbishment

The refurbishment described in this brochure has a total floor area of 300 m^2. The ground floor plan and section are shown in Figures 5 and 7. The walls are formed from an inner layer of 300 mm hollow bricks plus an outer layer of the original external wall, made of 270 mm massive bricks. The ceilings are insulated with 200 mm cellulose fibre/flakes of recycled paper. U- values of some building elements are given in Table 1. Heating is provided by a 62 kW central heating boiler which also serves the 400 m^2 of the building still to be refurbished. The lecture room can also be heated with a 11 kW wood stove.

Walls	0.41
Ceiling	0.33
Roof	0.33
Windows	3.0
Glazed roof	1.5
Floors	0.33

Table 1. U-values (in W/m^2 K) of some building elements

DESCRIPTION OF PASSIVE SOLAR FEATURES/COMPONENTS

Introduction

The main problem to overcome in incorporating passive solar features into this refurbishment was the orientation of the original building. The prime objective, therefore, has been to demonstrate passive solar features so that visitors to the site learn what they look like and how they operate and are constructed. The strategy has been to improve the daylighting where it is needed, reduce heat losses by using special glazing and creating buffer spaces, provide usable living spaces and, where possible, increase solar gains. The passive solar features used fall into two categories, those which aim to provide light and solar gains for the lecture room (i.e. the roof glazing and Trombe wall) and those which create daylit and sheltered buffer spaces for general use (i.e. the sunspace and glazed walkway).

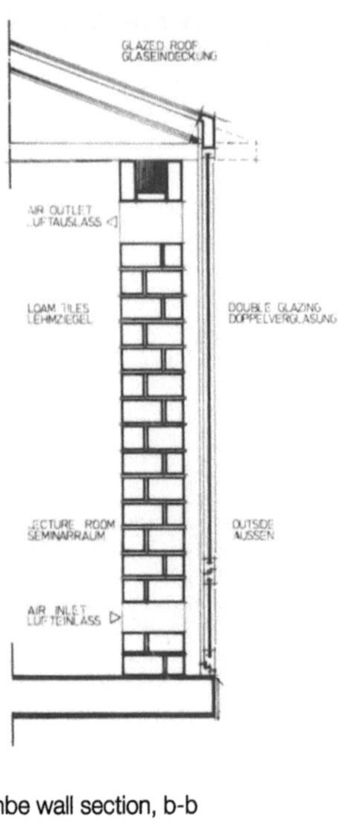

Figure 6. Trombe wall section, b-b

Lecture Room Roof Glazing

The lecture room, which has a floor area of approximately 74 m^2 and a maximum depth of 9 m, was insufficiently lit by the original windows. To improve the daylighting it was decided to increase the window area and glaze part of the roof. Because the orientation of the room is west-south- west and the roof inclination 30 , this did not help the heating load. Heat losses were, however, reduced by the use of a special glazing with U-value 1.4 W/m^2 K. It was found that the glazed area of 15 m^2 lit the room well, considerably reducing the need for artificial lighting.

Lecture Room Trombe Wall

Although it was appreciated that the Trombe wall could not contribute much to the heating of the lecture room because of its orientation, it was however constructed using adobe bricks produced during one of the academy's seminars in order to demonstrate an adobe-glass method of construction (see Figure 6).

Sunspace and Glazed Walkway

Because of the depth/length ratio of the buildings on this site, the provision of adequate natural lighting in the larger rooms is a problem. It was decided, therefore, to incorporate a sunspace into the building to create a daylit area for use as an entrance hall and dining area. Two bedrooms were put into the upper storey. The sunspace is not heated but serves as a buffer space between the main part of the building and the external environment. Most of the energy savings stemming from the presence of the sunspace are due to this buffer effect. To maximize solar gains, however, the sunspace was oriented as much to the south as possible by creating oblique angles on its main facade (see Figure 8). The sunspace is used as an entrance hall all year round and as a dining area when it is comfortable enough. The period of the year when this is possible was examined in some detail (see section on energy calculations below).

The walkway along the north side of the building acts as a convenient passage between the accommodation area and kitchen. It is fully glazed so that it provides daylight and acts as a thermal buffer space.

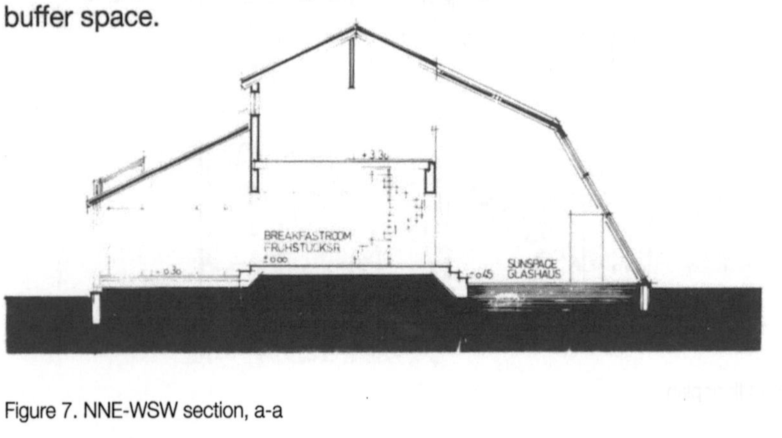

Figure 7. NNE-WSW section, a-a

ENERGY CALCULATIONS PERFORMED AND DESIGN TOOLS USED

Design Tools

A number of design tools were used in the course of the design. The annual heating load of the building was calculated according to the German standard procedure outlined in DIN 4701. The annual heating energy demand was determined from these results using the standard procedure in VDI 2067. More detailed analyses of the heating load and energy demand and identification of useful insulation measures were obtained using the STATBIL software package. This contains calculation schemes based on DIN 4701 and VDI 2067 using a thermal network model.

The solar energy collected by the sunspace and Trombe wall was obtained using the Method 5000 procedure developed by Dialogic in France. The detailed computer simulation program ESP developed by the University of Strathclyde in Scotland was used to determine the temperature distribution and comfort conditions reached in the sunspace at different times. These calculations were carried out in co-operation with the Institute of Thermodynamics at the University of Stuttgart.

The effect of shading on the solar thermal performance of the building was evaluated using a sun graph. This consisted of a Plexiglas sight on which was marked the position of the sun each month of the year. The outline of the obstructing trees was sketched onto an overlay and the resulting shaded area calculated.

Solar Gains and Shading Factor

The contribution of solar energy to the refurbished building's total energy requirements was found to be fairly low because of the poor orientation and the relatively small amount of glazing compared to the overall building size.

The shading factor due to trees was calculated by the sun graph method to be 70%. The trees on the site, however, are tall birches and larches which shed their leaves and needles in winter. It was assumed, therefore, that the overall shading was 50% at times of high solar altitude and somewhat less when the sun is low in the sky.

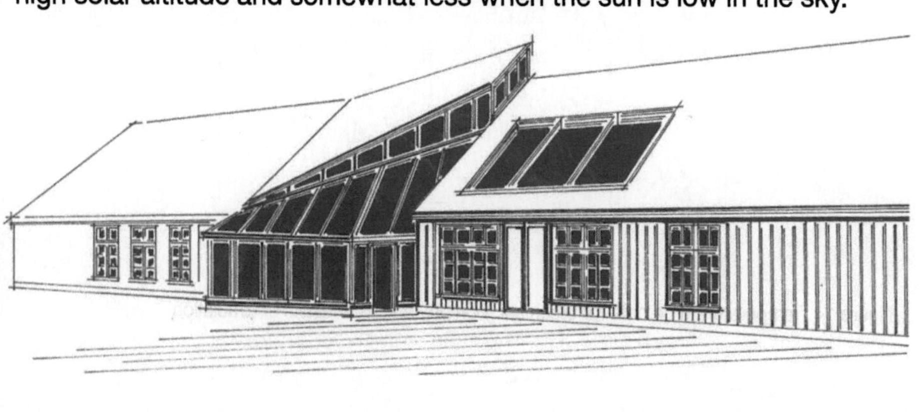

Figure 8. The oblique angles on the sunspace facade aim to collect as much solar energy as possible

Heating Energy Consumption

The annual heating load of the original building was calculated to be 56 MWh. The predicted heating energy requirements of the refurbished building are given in Table 2. The refurbished building requires 21.4% less energy to run than the original building. The auxiliary heating requirements of the refurbished building are only half the original heating load (see Figure 9). They amount to less than 100 kWh/m of floor area of the heated part of the building.

	MWh/year
Heating load of original building	56.0
Heating load of refurbished building	44.0
solar contribution	10.6
internal gains	3.6
gas heating	29.8

Table 2. Heating energy requirements of original and refurbished buildings

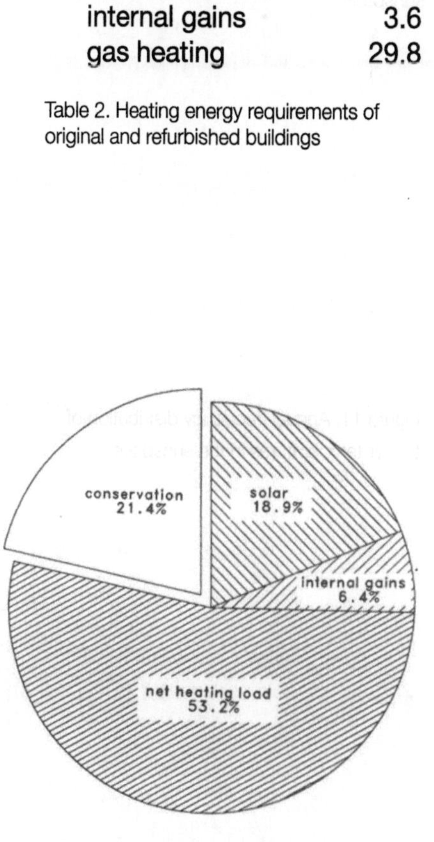

Figure 9. Heating requirements of refurbished building compared with original

Comfort Conditions in Sunspace and Breakfast Room

The conditions in the sunspace during cold January and hot July periods were evaluated in order to establish whether the sunspace can be used as a dining area. Some of the resulting temperature-time data are shown in Figures 10-14.

It was found that the sunspace would never suffer from overheating but the bedrooms in the upper storey might become too hot in summer. Figures 10 and 11 show that the sunspace will only be thermally comfortable for 30% of the year. However, the sunspace will tend to be used more at meal times and, because the sunspace is oriented in a westerly direction, the higher temperatures will probably be achieved during the afternoon and dinner hours. The temperature-time charts indicate that in January there will indeed be a rise in temperature in the afternoon when the sun shines. This suggests that a more detailed time analysis could reveal a utilizability factor greater than 30%. It should also be noted that building occupants may choose to have their meals in the sunspace whenever it looks inviting - for instance, in cold but sunny weather - even if the temperature is below 18 ° C.

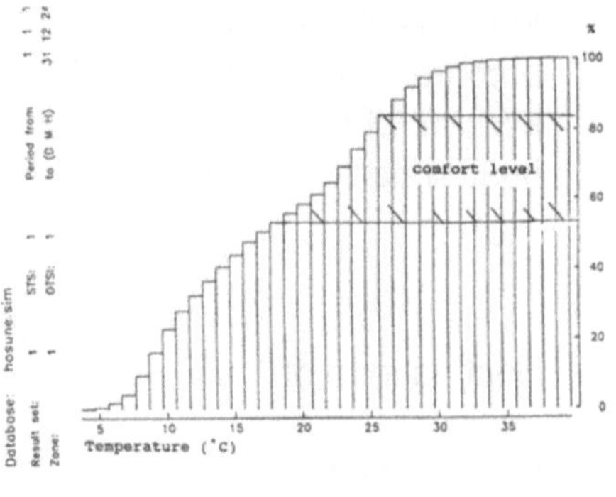

Figure 10. Annual cumulative distribution of the hourly average air temperatures in the sunspace

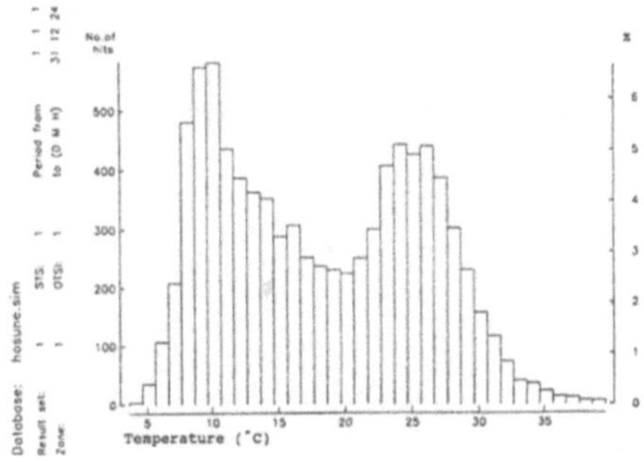

Figure 11. Annual frequency distribution of the air temperatures in the sunspace

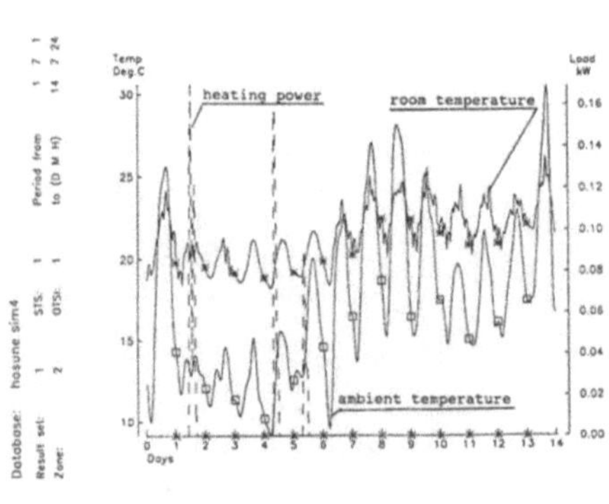

Figure 12. Temperature-time data for breakfast room in July - results of simulation

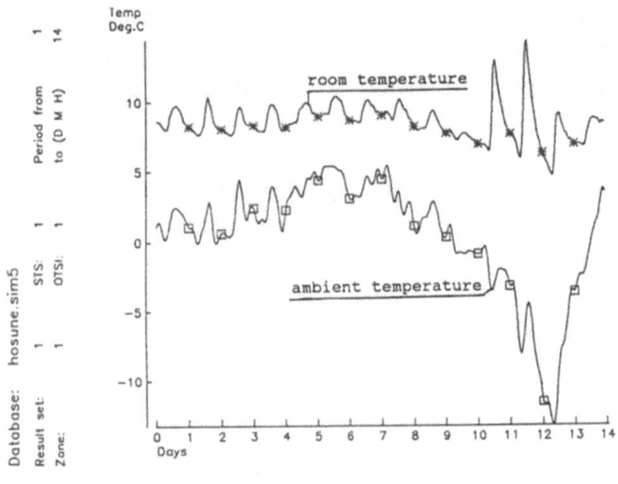

Figure 13. Temperature-time data for sunspace in January - results of simulation

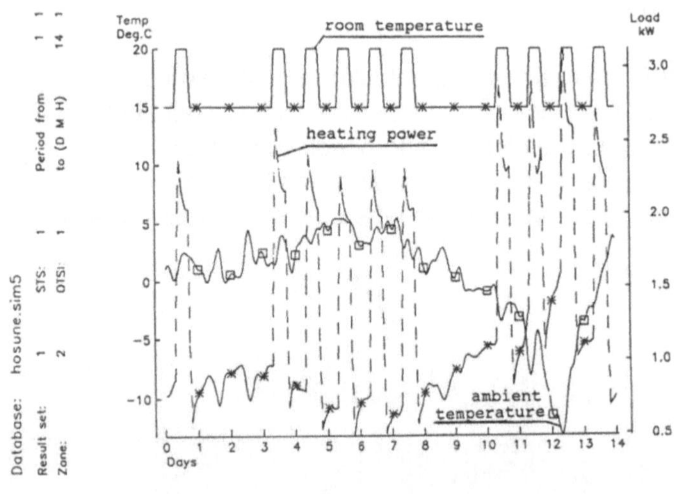

Figure 14. Temperature-time data for breakfast room in January - results of simulation

GENERAL DESIGN GUIDELINES/POINTS OF INTEREST RESULTING FROM THIS PROJECT

Reconsidering the Sunspace

An interesting point regarding unheated sunspaces is that they provide a sheltered environment halfway between the outside and inside worlds. The provision of a dining area in such a space will invite students to look at the outdoor world more than they would if they remained in the main part of the building. Thus the sunpace makes them more aware of the outside environment as well as providing amenable surroundings for various activities.

Adding Solar Features to Existing Buildings

Retrofit of passive solar features to existing buildings frequently involves compromises. The west south-west orientation of the buildings in this project discouraged incorporation of a sunspace or other passive solar features. The designers felt, however, that although they are not the most cost-effective energy saving measures, sunspaces do offer living spaces which, when carefully planned, display a certain charm and invite use even when comfort conditions are not optimal. They therefore decided to include a sunspace in this building and their original design is shown in Figure 15. Following discussions with the Building 2000 consultants, however, they changed the design so that more of the glazing was oriented to the south. This was done by creating oblique angles for the main sunspace facade (see Figure 8).

When considering whether to use passive solar features in future refurbishments it is recommended that the architect looks with great care at any glazing which does not face south. Setting aside for the moment the possibility of using the sunspace as a living area, it could be more effective to provide a high level of insulation in a building which is oriented away from south rather than incorporate solar features. In the Hosune complex, however, sunspaces and roof glazing will always be a favoured option because they improve daylighting - a vitally important factor in these particular buildings.

Figure 15. The original sunspace design

BUILDING 2000

Building 2000 brochures are published by Directorate General XII of the Commission of the European Communities to show how design studies can help architects and other building designers use passive solar principles to the best effect to produce attractive energy-efficient buildings. Each brochure describes studies carried out with the support of the Commission during the design phase of one of thirty-six non-domestic buildings in the EC Member States. The studies were on such topics as daylighting, heating, cooling, ventilation, comfort, control systems and urban design. They were carried out with the help of acknowledged European experts in these fields and drew heavily on lessons learned and techniques developed through the Commision's research and development programme on solar energy applications to buildings.

Commission of the European Communities/Directorate-General for Science, Research and Development

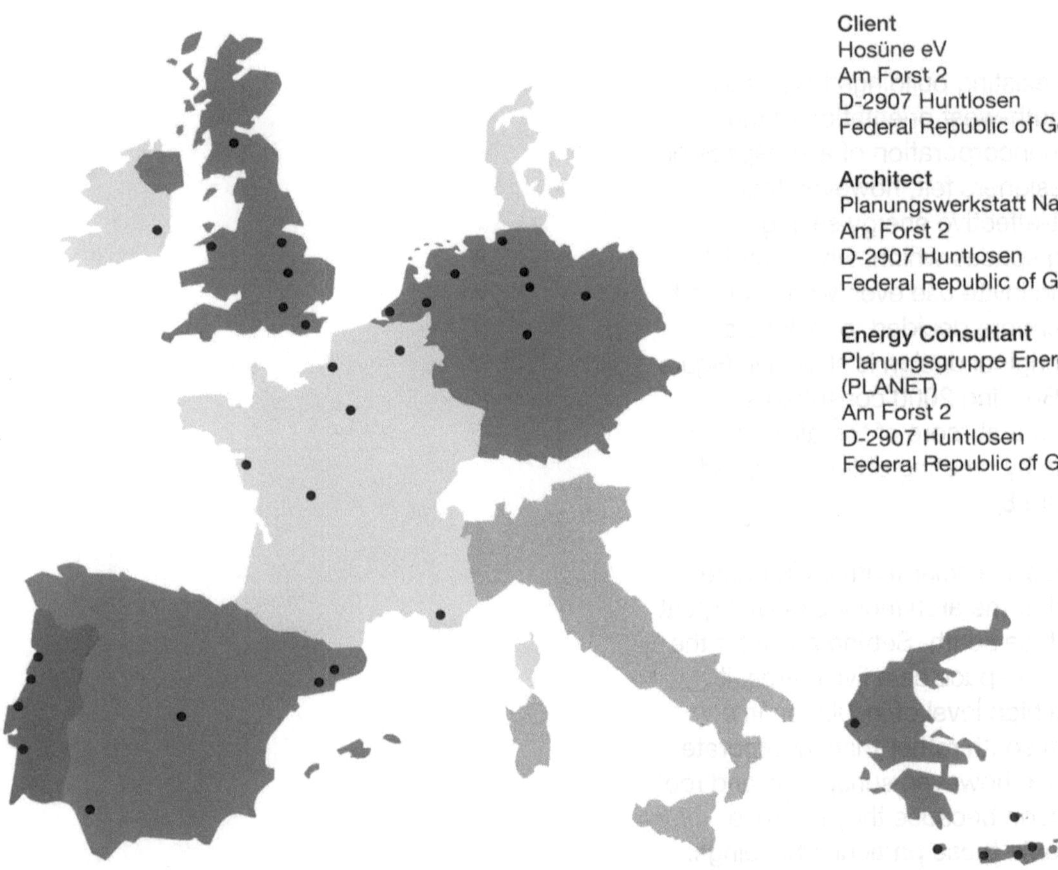

List of Design Team Participants

Client
Hosüne eV
Am Forst 2
D-2907 Huntlosen
Federal Republic of Germany

Architect
Planungswerkstatt Naturbauweisen
Am Forst 2
D-2907 Huntlosen
Federal Republic of Germany

Energy Consultant
Planungsgruppe Energie und Technik (PLANET)
Am Forst 2
D-2907 Huntlosen
Federal Republic of Germany

Building 2000 Consultan
Nick Baker
Cambridge Architectural Limited
Cambridge
United Kingdom

Jörn Behnsen
Universität Oldenburg
D-2900 Oldenburg
Federal Republic of Gern

Ekkehard Wagner
Glaszentrum Westersted
D-2910 Westerstede
Federal Republic of Gern

ESP Calculations
Peter Voit
Institut für Thermodynam
Universität Stuttgart
D-7000 Stuttgart 80
Federal Republic of Gern

This set of **Building 2000** brochures illustrates how architects and other building designers can successfully apply passive solar principles to produce energy-efficient buildings.

BUILDING 2000 Participants

Project Director
Theo C. Steemers

Coordinator
Cees den Ouden

Technical Steering Committee
Dean Hawkes
Nick Baker
Alex Lohr
Jean P. Lepoivre

Regional Liaison Agents
(D) Jörn Behnsen
(E) Vicente Sifre
(F) Michel Raoust
(GB) Alan Hildon
(GR) Matheus Santamouris
(P) Eduardo Maldonado

Further information or copies of the brochures can be obtained from prof. ir. Cees den Ouden, EGM Engineering BV, P.O. Box 1042, 3300 BA Dordrecht, The Netherlands.

Introduction in Design Support Activities and summaries of Design Support Reports.

C. den Ouden

During the period between the start of Building 2000 (officially 1-1-1988, but effectively since May 1988) and June 1990, a wide range of Design Support Activities has been carried out for almost all Building 2000 projects (only a few projects were delayed and, therefore, design support was not useful).

Three different types of design (advice) support activities have been offered to the Building 2000 design teams:

a. The consultation of an individual European Passive Solar Expert or a National Design Support Expert on Passive Solar and related topics.

Note:
Since several of the participating projects were already in the detailed design phase, this seemed to be the only effective way to improve the design on certain component detailing and/or minor changes in the passive solar and daylighting aspects already integrated in the design concept.

b. The carrying out of a small case-study by a recognised Design Support Unit having experience with certain passive solar/ daylighting design tools. In this case-study a sensitivity analysis for certain design parameters to be subject to change could be made so as to build up to the selection of the final size of these parameters.

Note:
This seemed to be of value for those projects in which the preliminary design phase had not yet been concluded.

c. The conducting of a more detailed modelling study. In several Building 2000 projects detailed simulation models might have to be used to provide answers on specific design aspects, such as air movements in atria, approaches to passive cooling etc.

Note:
When these requirements were identified to the coordinator, he contacted the right modelling expert to conduct a more detailed study on a subject preferably of use to a group of similar Building 2000 projects.

Moreover, several Design Support Workshops were organised during the Building 2000 programme during which Building 2000 design teams could directly exchange design ideas with the various design advice experts represented at these workshops.

In the second part of the Building 2000 final report a summary of a selection of many reports is presented, as produced by Design Support experts. Most of the design support activities resulted in changes of the various designs, as have been reported by the design teams in the brochures presented in the first part of the book.

The willingness to improve the building concepts by collaborating with R & D experts was encouraging for the Commission's action in this field. It is to be expected that design aids and simulation tools for passive solar options, daylighting concepts, comfort criteria etc. under development in the Commission's R & D-programme will be utilised more frequently in the future.

This will result in a better exchange of information between the actual design practitioners and the European R & D community, and most importang that this technology transfer, as initiated by this Building 2000 programme, will result in buildings with a higher quality with respect to energy and environmental issues.

LEARNING FROM BUILDING 2000

J Owen Lewis and Loughlin Kealy
School of Architecture, University College Dublin
Richview, Clonskeagh, Dublin 14 Ireland

1. INTRODUCTION

A number of expert studies were undertaken as part of the Building 2000 project. The purpose of this particular study was to distil and analyse the operation of Building 2000 in order to obtain feedback which might be of value in refining future CEC programmes in this area.

Originally the study was to have three principal stages:
- preliminary analysis of the reporting formats of the Regional Liaison Agents;
- the carrying out of structured interviews;
- analysis of these interviews, and the integration of the outcome with observations of the reporting process of Building 2000.

As the study unfolded, it became apparent that the reporting formats would not be available before the other stages were initiated, and that accordingly, structured interviews would be based on a "proforma questionnaire" which had been prepared in the absence of preliminary analysis.

Three "proformas" were prepared as the basis for the structured interviews of the selected groups: members of design teams, experts involved in the programme and Regional Liaison Agents.

This paper is based primarily on responses to the questions put during these structured interviews. Use was made of opportunities of contact generated in the course of Building 2000 or other CEC meetings. Thus, informal comments of participants have also been taken into account. In the selection of participants to be interviewed, preference was accorded to those involved in projects featuring daylighting and atria mainly because of the importance of such systems in contemporary architecture and their implications for energy use in non-domestic buildings.

This study is necessarily incomplete: it is the result of a limited investigation and should ideally be augmented by a wider sampling of opinion from the various groups who participated in the programme.

The paper is in two principal sections: the first summarising the experience of the design teams as communicated to us; the second summarising the experience of the supporting groups, the experts and the Regional Liaison Agents.

2. EXPERIENCE OF THE DESIGN TEAMS

The experience of the design teams can be summarised under two headings:

- how the design support they received related to their needs;
- what lessons do they draw from the experience of the project.

In describing design team experience under these general headings, we are not concerned to describe the particular circumstances which underlay the experience. This report can only be synoptic: each project presented an individual situation, which merits description in its own right. However, since such a task was beyond the scope of available resources, our endeavour was to extract general observations.

2.1 How Design Support Matched Needs

The basic reaction of design teams was favourable: the Building 2000 experience was positive, projects had benefited significantly as a result. Within this overall picture there was considerable variation in emphasis. The principal elements of the picture are as follows:

- The integration of members of the design team with the external assistance provided by Building 2000 experts was a major factor. Where a high level of integration was achieved, design teams reported that the impact on the design was pronounced. This did not always involve a transformation of a design project, since in some cases design teams had already considerable experience in the area. The impact which teams identified was often in the stimulus provided by the collaboration and the sometimes novel and challenging relationships engendered by the interplay of experts and the design team. The Building 2000 experience has lead some participants to re-assess their work practices, and may therefore have generated longer-term impacts. While, on the other side of the coin, none of the design teams in our sample reported negative impacts from the intervention of experts, it should be pointed out that not all design teams reported a high level of integration.

One particular factor emerged from the reports. While respondents pointed out that within the context of the Building 2000 programme the timing of intervention could not be faulted, nevertheless, from the perspective of the individual project it sometimes came too late to affect strategic decisions. In such cases all that could be achieved was a degree of modification, and there is evident frustration in some of the reporting, since the full potential of the collaboration became apparent without the capacity to benefit fully from it. In this respect also, it appears that not all respondents were sufficiently attuned to the purpose and potential of Building 2000 when first becoming involved with the programme, so that there is an awareness of potential impact expressed in terms of reflections on the composition of design teams, which had been established before these aspects became clear.

- The element of timing emerges also with respect to the informational material made available through Building 2000. For example, the design guidelines for the "Working in the City" competition are singled out for praise by several respondents, while regretting that these were not available to them from the outset.

- There is wide variation in reports of the use and impact of computer simulations. Some reports simply record the use of SERI-RES for example; others record that they used programmes of their own or acquired programmes from sources other than Building 2000. There is also some evaluative comment, both on the usefulness of simulation work and on the relative allocation of resources within Building 2000 in this regard, with the opinion expressed that more concentration on "training" of design teams would have yielded better results. Without a full analysis of the particular circumstances we can only mark the comment, and relate it to the reported experience of the actors in the programme. Some were accustomed to computer simulation work and design teams had this fully integrated into their design method. Here again, closer analysis would be necessary in order to provide an insight into the specific contribution of Building 2000 in this respect.

- In comments on the effectiveness of workshops, there is a number of salient points. Most notable are those which single out workshops as forums through which design teams learned from each other's experiences. It was observed that in some cases, it was possible for teams to present their work and to communicate at a high level with other participants. One comment speculated on whether it would be possible to generate this kind of communication early in the programme rather than late, while another favourably contrasted this type of experience with the frustrations of the usual conference reporting formats. Not everyone was completely happy. The sentiment was expressed that presentations (at all stages) tended to be theoretical rather than practical and thus of limited value after a point. One may infer that workshops are a powerful learning vehicle if well timed and structured, and that they need to cater for a range of approach in order to realise their potential.

2.2 Lessons for the Future

The experience of the project teams reviewed in this study was diverse, and the lessons drawn from those experiences reflect this diversity. Initially therefore, it may be useful to list those future-related comments which emerge most clearly. It should be emphasised that this listing reflects the opinions of a particular range of project personnel, and has to be read in conjunction with other opinions cited in this paper, and with opinions recorded elsewhere.

1. The need for defined goals for each project within the design programme, articulated at the outset.

2 The need for comprehensive good quality theoretical and practical information at the commencement of the project and throughout. There was a strongly expressed need for support in the form of additional and better quality design and evaluation tools.

3 Involvement of design support initiatives at early stages of a project, before strategic decisions are taken.

4 The need for a training element which would maximise the benefits derived from expert input and simulation exercises.

5 The provision of measures to facilitate communication between projects, and to facilitate fuller reporting and analysis of project experiences.

In addition, there were two issues raised in forceful comment, which require consideration. These are:

1. The creation of integrated design teams which incorporate the expert input as an alternative to design support in the form of "consultancy";and

2 The appraisal of the effectiveness of the computer simulations carried out within the Building 2000 programme.

From the reviewers' perspective, two matters emerge from contact with members of the design teams. First, that, when comments are made with clarity and by several respondents, we must present them as such, but without fully knowing how representative they are. Second, it seems to us that design teams would greatly value a thorough feedback mechanism, both as an objective evaluation of their own projects, and also as a way of relating their experiences to those of others. These matters will recur in the final remarks in this paper.

We note that it had been hoped that Building 2000 would provide a valuable testing ground for CEC-generated research results. This hope appears not to have been directly realised, largely because relevant results and materials were not available in time.

3. EXPERIENCE OF EXPERTS AND LIAISON AGENTS

Reporting on the experience of experts and liaison agents in summary form presents rather different problems from those of reporting the experience of design teams. While each design team is concerned with its own project, and there are variations between projects, there is a certain congruity of experience apparent. Other actors in the design support initiative of Building 2000 all have dealt with a range of projects, and a multiplicity of situations. The comments these people make therefore, tend to be concerned either with especially notable experiences with particular projects, or else with aspects of the design support process which seemed to them to be significant in terms of Building 2000 as a whole. This summary account concentrates upon the latter.

Comments are grouped under three headings: the design team/expert relationship; the impact of design support on the projects and the need for comprehensive evaluation.

3.1 The Design Team/Expert Relationship

There is far greater variation in the accounts given by these groups than is the case in those given by the design teams. The design team accounts are positive by and large, while the other groups report experiences ranging from excellent to highly problematical. A number of themes are discernible in the range:

- there were differences between what some design teams expected and what could be provided. In some instances teams expected a "service", whereas experts expected to offer advice. In others, what was expected was financial assistance.
- the reaction of design teams to advice varied considerably. Some accepted advice reluctantly, others did not act on advice given. There were, in contrast, instances where the relationship was highly fruitful and satisfactory to both parties. This appears to have occurred both in situations where experts integrated to a marked degree with the teams, and also where advice was accepted in a consultancy format and schemes modified as a result. Variables in experiences, attributed to participants' personal sensitivities, and the influence of external factors such as institutional relationships for example, form a significant part of observations under this heading.
- there was considerable variation in the prior experience of design teams, some being new to the practice of energy-conscious design while others were already highly experienced in the field. It is not possible however, to make a clear correlation between prior experience and the success of the collaboration, beyond noting both instances in which lack of prior experience is cited as a factor in problematical collaboration, and instances in which extensive experience accompanied obviously successful collaboration.
- in some instances it is clear that good working relationships were facilitated by the fact that participants knew each other beforehand, and had already considerable knowledge and appreciation of each other's work.

3.2 The Impact of Design Support

Comment under this heading falls into three categories: workshops, simulation exercises and project design.

There were few direct comments on the subject of the effectiveness of workshops. But some points of significance arise from a comparison between examples cited by experts. It seems clear that when participating design teams are relatively inexperienced, and in a situation where they are working with strangers, an inordinate amount of time can be spent in lengthy exposition of schemes which in any event are destined to change, once a critical energy perspective is applied. Experience here would indicate that it is in this type of situation that a great deal of resistance to change is encountered. Alternatively, experienced teams working with known advisors can benefit quickly from the exchange of experience which the workshop provides. The implication would appear to be that inexperienced designers would benefit from one-to-one consultation before being asked to participate in group occasions.

While the sample base for our observations under the 'simulation' heading is also small, there is evident a sentiment that considerable time and resources were, in some instances, devoted to simulation exercises with trivial outcomes. The implication here is not that such exercises were irrelevant, but rather that they require careful integration into the design process, so that appropriate questions are being asked. It is also evident that the exploration of alternatives through simulation contributed significantly to the successful modification of schemes within the programme.

It is apparent that the consultation process resulted in some notable modifications of designs. It is also clear that this was not always the case, even in situations where the need for modification was obvious to the experts concerned. Various contributory factors have been mentioned already in different sections of this report, the timing of intervention in relation to the stage of the individual project or over-defensive attitudes, to cite just two examples.

3.3 The Need for Comprehensive Evaluation

While there is no clear preference for a particular form for evaluation, both experts and regional liaison agents are agreed on the necessity for comprehensive and rigorous analysis of the Building 2000 experience. In some cases there is considerable strength of feeling on this subject, based on the view that the widest possible review process is needed in order to achieve an accurate picture of this complex programme.

Suggestions are made on various aspects of the review task:

- there is a need for thorough de-briefing of designers in order to establish the effectiveness of the design support initiatives;
- the outcomes should be published. This suggestion encompassed both the projects themselves, and also those informational materials which were produced in the context of the programme.

4. OBSERVATIONS

Since this paper is already given in summary form, in this section we attempt to extract significant threads from the comments received during the course of the study, rather than to reiterate what has gone before.

Two principal observations were made:

1. A further CEC design-support programme would be welcomed by participants, and would appear to be justified on the basis of the feedback monitored in this study.

2. Any subsequent programme would need to take into account observations regarding communication of intentions, project goals, timing of interventions, availability of information, training and inter-project communication outlined in earlier sections.

Within this framework, two further observations are made as follows:

1. A number of comments indicate that there have been projects in which the provision of appropriate support and the utilisation of support or its monitoring have been problematical. We do not have sufficient detail information to make useful comments other than to point out that such matters are sure to be known by considerable numbers of participants. It is important that the final reporting on such matters should be widely available so that confidence in future CEC initiative in this area is assured.

2. There is a case to be made for design support initiatives targetted more specifically towards problem areas identified in this initial programme. Four areas emerge as particularly salient:

a) simulation studies and other design and evaluation tools, and their appropriate integration in the design process;

b) the potential for integrated design teams incorporating appropriate expertise as distinct from a consultancy-type relationship;

c) projects incorporating a training element for beginners in this area;

d) targetting of related issues; for example region-related such as cooling, context-related such as existing urban settings and those related to building type.

5. ACKNOWLEDGEMENTS

In the preparation of this report, feedback was obtained on some fifteen projects, either through the structured interviews, through informal interview with Regional Liaison Agents and/or experts or through a combination of these routes. We are indepted to the following for their assistance: C.J. Ashford; Dr. N. Baker; J. Behnsen; C. Cortesão; Yvonne Dean; M. Fontoynont; G. Gandei; A.J. Howrie; A. Lohr; Prof. E. Maldonado; Prof. A. Gomes Martens; M. Santamouris; Prof. R. Serra; V. Sifre; T. de Silva.

We would emphasise that the conclusions and opinions expressed in this paper are those of the authors.

PASSIVE AND HYBRID COOLING OF BUILDINGS. RESULTS OF A STUDY.

M.Antinucci[1], B.Fleury[2], J.Lopez d Asiain[3], E.Maldonado[4], M.Santamouris[5], A.Tombazis[6], S.Yannas[7]
(In alphabetical order)

1.Conphoebus Scrl, Zona Industriale Passo Martino, Fermo Posta, 95030 Piano d Arci,Catania, Italy.
2.LASH/ENTPE, Rue M.Aduin 69120, Vaulx en Velin, France.
3.Seminario de Arqitectura Bioclimatica, Escuela Superior de Arqitectura,Avda Reina Mercedes S/N 41012 Sevilla Spain.
4.Universidade do Porto, Rua D.Manuel II, 4003 Porto Codex Portugal.
5.University of Athens, Physics Department, Ippokratous 33,Athens,Greece.Author to whom correspondence should be addressed.
6. Meletitiki Ltd, 75 Vassilissis Sofias str, 11521 Athens Greece.
7. Architectural Association of London, Energy Studies Programme,34-36 Bedford Str. London,WCIB 3ES UK.

1. Introduction

Proper design of buildings in hot climates requires a balance of heating,cooling and daylighting performance. Achievement of thermal comfort during summer and reduction of the cooling load by natural means, is considered now a first priority for electric utilities and consumers.The resulting peak electricity demand profiles and the very high cost of conventional cooling systems makes the strategic management of air conditioning growth an important exercise for industrialized countries planners.
The development, in the 30's, of inexpensive,reliable Carnot cycle refrigeration systems, using electrically driven compressors, has made cooling widely feasible. During the past decades, air conditioning became widespread in residential and commercial construction. Annual purchases of air conditioning equipment have exceeded the $20 billion mark by the end of the 80's, (1), while, the world trade in air conditioning and industrial refrigeration equipment has more than tripled in real terms during the decade 1976-1985, (2).
The increase of family income in Europe has made the use of air conditioning systems highly popular. Annual purchases of air conditioners in Greece have increased approximately 900 per cent during the last three years,while sales of air conditioners in Italy and Spain go side by side, (3).
The impact of air conditioners usage on electricity demand is a serious problem for many developed countries. Peak electric loads oblige utilities to build additional power plants in order to satisfy the demand, thus increasing the average cost of electricity.
This situation has become a serious problem in the United States, where the total megawatts of air conditioning induced electric peak load is estimated to be 175 MW or 38 percent of noncoincident peak demand, (4).
Important peak load problems during summer have been also experienced in Greece during the last few years (5).
Environmental problems associated with the use of ozone depleting CFC refrigerants used mainly by the conventional air conditioners are an additional impulse for the definition of an alternative cooling policy.
Finally, recent comparative studies on the indoor air quality in air conditioned and naturally ventilated office buildings have shown that illness indices are higher in air conditioned buildings (6-7).

Alternate cooling strategies based on an improved thermal protection of the buildings envelope, as well as on the dissipation of the buildings thermal load to a lower temperature heat sink, appear to be very effective.These strategies and techniques known as passive and hybrid cooling have already reached a certain level of architectural and industrial acceptance. It is reported that imports of evaporative coolers in Greece reach values close to 23 million dollars per year,(8). A survey of the European industry has shown that there are more than 200 industrial units producing more than 500 different products related to Passive and Hybrid cooling of buildings,(8).

2. Prevention and Modulation of Thermal Gains

2.1. The role of microclimate

Modifications of the microclimate around buildings can help to improve indoor comfort conditions and to reduce cooling loads,while also providing protected spaces for outdoor uses.
Siting and site layout as well as landscaping represent the two important groups of strategies to improve the microclimate around buildings.
Appropriate siting of a building can provide natural solar protection and help to take advantage of local winds and breezes. Nearby mountains, valleys, lakes and sea may enhance these effects. Important information on staggered layouts enhancing the potential for cooling load is given in (9).
Landscaping can play an important role in microclimate modifications in both overheated and underheated conditions. Useful information to modify microclimate using appropriate landscaping is given in (10-11).
Vegetation provides natural protection from the sun. Moreover vegetation is a natural form of an evaporative cooler. Trees are commonly used for shading and their characteristics are listed in (12-13). A detailed sample of landscape materials and the corresponding characteristics are given in (14).
Current knowledge on the role of vegetation results mainly from observations. Temperature reductions of about 2-3 C due to evapotranspiration are reported. Important information can also be gained from theoretical analyses. It is reported that evapotranspiration from one tree can save 1-2.4 MJ of electricity in air conditioning per year (15), while an average full sized tree evaporates 1460 kg of water per sunny day which is the equivalent of 870 MJ cooling capacity (16). It is also stated that the latent heat transfer from wet grass can result in temperature 6-8 C cooler than exposed soil, and that one acre of grass can transfer more than 50 GJ/day, (16).
Other landscape techniques include the use of pools or ponds, fountains or sprays and cascades for evaporation. According to (17), under mean conditions of wind,dry and wet bulb temperature, the energy released by a square meter of open water surface is close to 200 Watts.

2.2. The role of mass

Internal walls, partitions, and furniture, as well as floors, can provide thermal storage capacity. This is usually referred to under the generic term : thermal mass. Thermal mass can play the role of a regulator, smoothing temperature swings, delaying peak temperature, decreasing mean radiant temperature and providing better comfort conditions.
Optimisation of the thermal mass levels inside the building depends on various parameters. Climatic conditions, occupancy patterns, building orientation, and the use of auxiliary heating and cooling system define the more important of them.
Classification of buildings related to its thermal mass and the related thermal performance is given in (18-19). Various methods to optimize the thermal mass of a building have been proposed. An evaluating discussion on some of these methods is given in (20-21).
Repertories of many wall compositions has been developed and proposed in (22), and the attenuation and time delay of a 24-hour temperature wave is calculated.
Thermophysical characteristics of local materials used for building

constructions are also reported in (23).
Experiments using variable levels of thermal mass in similar buildings are reported in (24). It is found that increase of the internal mass from 21 Kg per square meter of floor to 201 kg, in close and in ventilated buildings, reduces the peak indoor temperature by about 1.0 and 2 C respectively.
Theoretical analysis based on intensive monitoring of office buildings, (24), have shown that important reductions in peak heating and cooling loads was observed for high thermal mass designs. Energy savings varied from 18 to 20 per cent compared to a base case.
However, optimization of thermal mass in buildings requires further research.

2.3. The role of solar control and shading

Solar control is one of the most important strategies for heat gain prevention. Its role is focused on :
a) The prevention of solar gains on buildings or landscape surfaces, occupant clothing or skin and
b) The regulation of daylight and the protection of occupants and room contents from glare and ultraviolet radiation.
Effective design of solar control in a building should aim to balance cooling, heating and daylighting in the building. Reduction of the cooling load during summer, maximum utilization of the solar radiation during winter, and achievement of optimal daylight levels are among the basic objectives of the solar control techniques.
A complete review of solar control techniques as well as design considerations and solar control objectives are given in (25).
Relative design procedures have been proposed by various teams. However, the work of Aladar and Victor Olgyay's, published some twenty five years ago, is still the most comprehensive and most widely quoted work on the subject.
Modelling of shading devices and their impact on ventilation, lighting and occupant comfort remains the major weakness of most building performance evaluation models. This weakness in knowledge is reflected in practice, leading to a number of not technically satisfactory applications.
Considerable developments have been achieved in glazing technology, both through improvements in thermal properties and through applications of optically controllable materials. Optical switching films, multilayer low-emissivity coatings and transparent insulation materials are some of the more important expected products.

3. Use of Environmental Heat Sinks

3.1. Ventilation

Ventilation provides cooling by using air to carry heat away from the building and/or from the human body itself. Air movement may be created either by natural forces, (wind and stack effect) or mechanical power (fan).
The distribution and shape of the openings is a key element for efficient natural ventilation. Quantitative and qualitative aspects of cross and single sided ventilation techniques as well as roof opening techniques are discussed in (26-28).
Wind tower techniques are used as architectural elements to generate air movement within the building. Various architectural integrations of the system are presented in (26). Design information and performance calculation procedures are given in (29-30).
Solar chimneys are completely integrated in the building architecture elements which use the sun to warm-up an internal surface of the chimney. Performance analysis of such systems are performed in (31-33).
The use of wingwalls, proposed in (34), can considerably enhance cross ventilation (35). It is found that for oblique winds, the average room airspeed in a room with wingwalls was about the 40 percent of the outside wind compared with about 15 percent for the room without the wingwall. Vegetation affects the air flow pattern in the same way as outside buildings or wingwalls (36). Moreover, it also offer air filtering, noise reduction and shading.

Selection of the appropriate window type and shape is necessary in order to optimize cross ventilation. However, a compromise has to be realised between daylighting, weatherproofing, structural strength, solar gain, operation, maintenance, cost and air movement control. Research has shown that horizontally shaped windows create greater wind speeds than vertical windows. A typology of various window styles and their impact to ventilation is given in (26).

Nighttime ventilation offers a high potential for building cooling because of the lower outside temperature. Performance data reported in (37-39), show that this is a powerful technique. A method to estimate the night ventilation potential is proposed in (40). Nighttime ventilation techniques, when natural, could face problems of building security and occupant privacy.

A box fan, an oscillating fan or a ceiling fan could supplement the effectiveness of natural ventilation by increasing the interior air velocities and convection exchange (41-42). Recent research, (43-44), has demonstrated the capability of fans to extend the summer comfort within the building outside the typical comfort zones. Oscillating fans present high potential at the local level to reduce the cooling load of the building. More research is needed to present guides on the use and potential of fans in hot climates.

3.2. Evaporative Cooling

Evaporation occurs whenever the vapour pressure of water, in the form of droplets or a wetted surface, is higher than the partial pressure of the water vapour in the adjacent atmosphere. Evaporative cooling techniques are classified also as passive or hybrid.
and direct or indirect.

Passive direct systems and techniques include the use of the vegetation for evapotranspiration, the use of fountains, sprays and pools, previouslydiscussed as well as the use of the volume and tower cooling techniques.

Volume cooling techniques are known from traditional architecture. A contemporary version of this technique was presented in (45).

Passive indirect evaporative techniques include mainly the roof spray and the open water pond technique. In the roof spray technique, the exterior surface of the roof is kept wet using sprinklers. Cooling of the building is achieved due to the conversion of the sensible heat of the roof surface into latent heat of vaporization. Experience acquired through many commercial applications in United States, (46) show that reduction in cooling load close to 25 per cent is possible . However, important operational problems, are associated with this system which has been abolished from any further developments.

In an open roof water pond, evaporation of the water to the dry atmosphere occurs during night time. Thus, indoor and radiant temperatures can be lowered without elevating indoor humidity levels. Knowledge on this technique is developed through experiments described in (47-49), and simulations given in (49). A decrease of the ceiling temperature between 2 to 13 C has been reported. However, important limitations associated with this technique has not permitted its wide application.

Hybrid direct evaporative systems involve the use of fibrous pads where the air is drawn and cooled by evaporation of flowing water. Thus, dry bulb temperature is reduced while its moisture content is increased. In order to increase the efficiency of the system various type of technologies have been proposed.

Hybrid direct evaporative coolers are in use for more than 50 years. Important industrial experience is achieved mainly in Australia and in United States, while last figures show an important market in Europe (8). The efficiency of the system is seriously affected by local climatic conditions. The coolers are mainly suitable for places characterized by low wet bulb temperature. Performance data for fourteen Mediterranean locations are given in (3).

Hybrid indirect evaporative cooling systems are based on the use of a heat exchanger where the indoor ventilated air passes through the primary circuit where evaporation occurs while the outdoor air passes through the secondary circuit. Thus the temperature of the air is decreased without any moisture

increase.
Important industrial development of the system has been achieved during the past few years. Three main type of indirect coolers are available : Plate, tubular and rotary type coolers.
Performance data on plate coolers developed at CSIRO, Australia (50), show that energy savings of up to 60 per cent can be achieved compared to compression refrigeration systems. However, the efficiency of the system is strongly influenced by the wet bulb temperature of the ambient air. Performance evaluations of the system for various European locations are given in (8).
Where the ambient air is too high, a two stage evaporative system can be used. This consists of an indirect cooler coupled with a direct one. There are many applications of these systems, especially in California. Energy savings for such a system have been reported at close to 50 percent compared to an equivalent A/C system. Performance data for 14 European locations are given in (3).

3.3. Ground Cooling

Ground or earth cooling is based on the temperature difference between soil and the atmospheric environment. During the summer, the soil temperature at certain depth is considerably lower than the ambient temperature and therefore offers an important source for the dissipation of the building excess heat.
There are two main strategies for the dissipation of the heat to the ground. The direct ground contact cooling and the indirect or buried pipes technique.
Direct ground contact cooling involves partial or total placing of the buildings envelope in direct contact with the soil. This technique based on the transfer of heat from the building to the ground through the walls, the floor and possibly the ceiling is used at different times in history and in different parts of the globe. A classification of the earth sheltered buildings is given in (51).
Important advantages and disantvantages of earth contact buildings are reported in the literature (51). However during the last years this technique has become popular. Estimates of the number of earth sheltered houses in USA in 1982 range from 4000 to 8000, (51).
Knowledge on the performance of earth sheltered buildings comes mainly from measurements of traditional buildings, (52), from monitoring of new constructions, (53-54), as well as from theoretical analysis, (51). Energy gains of about 50 to 90 percent are reported for various monitored earth sheltered buildings (53-54).
Underground cooling tunnels is a concept that can be traced back several centuries. The concept involves the use of a metallic, PVC, or concrete pipe buried at 1 to 3m depth. Ambient or indoor air is delivered inside the tubes where it is precooled and then delivered to the building.
Knowledge on the topic is coming mainly from single and multi pipe experiments as well as from theoretical analysis. Various techniques to estimate the efficiency of such systems have been developed and are evaluated in (55).
Results reported on the performance of buried pipes , show that for a 30 m long PVC pipe buried under shaded soil at 1.5 m depth, the maximum temperature drop was 18 C. However the temperature reduction is time dependent.
Special problems related to this technique is the limited potential for dehumidification, reducing thus the possibility for latent heat cooling. Condensation inside the tubes can occur only with low air flow and high ambient dew point temperatures. Also in the event of damaged tubes, water, could enter into the tubes. Moisture accumulation can lead to biological growth and cause odour problems. However, no such problems have been reported.

3.4. Radiative Cooling

Radiative cooling is based on the effect of heat loss by a body due to its long wavelength radiation to the night sky.
Existing radiative cooling techniques are classified as Passive or Hybrid. Passive systems, mainly involve the use of the building roof as the radiative

component,while hybrid systems involve the use of special metallic surfaces characterized by high emissivity in the longwave range.

Painting the roof white increases the reflectivity of the roof to the solar radiation thus reducing its temperature. Measurements of the performance of this technique reported in (56) give a cooling potential of 0.014 KWH per square meter per day. Similar results are also reported in (57). However, the effect of color is more important with light structures than with structures of hight thermal capacity.

Exposing the cold storey mass to the sky during the night while protecting it during the day by a movable insulation is an other passive radiative cooling technique which tries to optimise the potential of radiative cooling. Measurements of the performance of this technique reported in (56) give a cooling potential of 0.266 KWH per square meter per day.

The Skytherm system which is proposed by H.Hay involves the use of water bags which are placed on the roof. A movable insulation is placed above the bags during day to keep the solar heat away from the water and is removed during night where when the water looses heat by convection, radiation and evaporation and thus the water cools the living spaces. Knowledge on this system comes from monitoring of the Hay's house at Atascadero in California,(58). It was observed that on a typical summer day, the variation in outdoor temperature was from 13 to 34 C, while the corresponding indoor temperature ranged from 21 to 23 C only.

The main problem associated with passive radiative techniques is that are effective only in single storey buildings or to the upper floor of multistorey buildings. Specific problems with the above presented techniques are discussed in (3).

Hybrid radiative cooling systems involve the use of specialized metallic longwave radiators which can reach temperatures below those achieved when the roof is cooled directly. The operation of the system involves the flow of a transfer medium, water or air, flowing above, under or within the radiator which is cooled and then is used directly or it is stored in order to cool the building.

The cooling potential of the metallic radiators for some European locations is discussed in (3,59). It is found that under stagnation and for mean climatic conditions, the expected temperature depression rarely exceeds 3 C. Under optimum climatic conditions, the calculated maximum depression temperature range from 5 to 7 C.

Use of selective surfaces as well as windscreens transparent to infrared radiation can, under certain conditions, improve the performance of the metallic radiators.

Experiments with selective surfaces report temperature's depression from 5 to 17 C. Comparisons with ordinary surfaces have shown however that the difference between them is negligible. Also when the temperature depression is important, condensation occurs on the surface and thus the surface loses its selective radiative properties.

Experiments using windscreens, described in (60), report that a radiator covered by a windscreen was 3-4 C lower than the corresponding temperature of an ordinary radiator. However when the temperature of the polyethylene film drops below the ambient dew point, dew is formatted over the windscreen, and thus the transmissivity of the polyethylene to the infrared radiation is reduced significantly.

4. Conclusions

Demand for indoor cooling has increased rapidly over the past few years rapidly, creating important environmental, health, capital and mainly peak electricity load problems.

Passive and Hybrid cooling techniques and components can minimize the use of mechanical energy and its impact on the environment and on occupant health, while they can decrease significantly peak electric demand.

Important knowledge on the topic has been produced during the last years and demonstration buildings designed using passive and hybrid cooling systems, present a significant energy performance and excellent indoor comfort

conditions. However the impact on professional building design practice has been modest.
Important future research actions are necessary in order to increase our knowledge base on the topic. Also an important effort should be made in order to increase the accessibility of the existing and new information.

5. References

1.Predicasts Inc : Predicasts Forecasts, Issue 108,4th Quarter, Refrigeration and Air Conditioning Equipment Data, 1987.
2.United Nations : International Trade Statistics Yearbook,VolII,Trade by Commodity, 1980-1985.
3.M.Santamouris : Natural Cooling Techniques. Proc.Workshop on Passive Cooling.E.Aranovich, E.Oliveira Fernandes and T.C.Steemers(eds).Ispra,Italy 2-4 April 1990.
4.C.J.Andrews , The Energy Journal 4,1,79, 1987.
5.M.Santamouris et al : Correlations of peak electricity profiles with some climatological indices. Greek Productivity Center,Athens,1991.
6.A.Hedge, E.M.Sterling and T.D.Sterling : Building Indices based on questionnaire responses. Proc IAQ 86, ASHRAE,p.p.31-43,Atlanta,USA, 1986.
7.M.Santamouris, E.Daskalaki, D.Kitari, K.Kostopoulos and V.Papadimitriou:A comparative study on the air quality of office buildings. Greek Productivity Center, Athens, 1991.
8.PROTECHNA Ltd : Passive and Hybrid Cooling in Greece. Study carried out for the Center for Renewable Energy Sources, Greece. Athens, 1991.
9.Z.Warsi and I.Chand : Indian Journal of Technology,6,1,23-25,1968.
10.M.Girgis : Landscaping for energy conservation .Florida Solar Energy Center,EN,12,1983.
11.H.Taha, H.Akbari and A.Rosenfeld : Vegetation canopy microclimate. A field project in Davis. California LBL Report 24593,1988.
12.D.A.Montgomery : Passive Solar Journal, 4,1,79,1987.
13.S.Risvi and K.Talib : Landscape as energy and environmental conservation in the arid regions Saudi Arabi. Proc. International Conf. Passive and Hybrid Cooling. Miami Beach, 1981.
14. R.Stirling, J.Carmody and G.Elnicky : Earth sheltered community design, Energy efficient residential development. N.Y. Van Nostrand Reinheld Company,1981.
15.H.Taha, H.Akbari, A.Rosenfeld and J.Huang : Residential cooling loads and the urban heat island. The effect of albedo. LBL Report 24008,1988.
16.A.Moffat and M.Schiller : Landscape design hot save energy,N.Y.William Norrowand Company,1981.
17. J.Y.Yellot : Passive and Hybrid Cooling Research. Advances in Solar Energy,1983.
18. N. Billington : Building physics : heat. Pergamon Press. London,1967.
19. J. Anquez, M. Croiset and J. Le Quere : Calcul simplifie des economies de chauffage apportees par l energie solaire. CSTB, Cahier 1704, Paris 1981.
20. G. Paparsenos : Heat attenuation and the thermal mass effect. Ibid 4, 111-119.
21. M. Antinucci : Heat attenuation as a mean to limit cooling load and improve comfort. Ibid 4, 101-110.
22. C. N. R. : Repertorio delle caratteristiche termofisiche dei componenti edilizi opachi e transparenti. Edizioni CNR-PFE, 1982.
23. E. Athanassakos and E. Gratsia : Suitability of building materials for passive solar applications. Report I. Institute for Technological Applications, Greek Productivity Center. 1988.
24. M.J. Brown : J. of Solar Energy Engineering, 112, 273-279, 1990.
25. S. Yannas : Solar Control Techniques. Ibid 3, 75-97, 1990.
26. B. Fleury : Ventilative Cooling. State of the art. Ibid 3, 1990.
27. T. Boutet : Controling air movement. Mc Graw Hill Company, 1987.
28. S. Chandra, P. Fairey and M. Houston : Cooling with ventilation. Florida Solar Energy Cooling, SERI Report Sp 273-2966, Dec. 1986.
29. M. Bahadori : A Passive cooling heating system for hot arid regions. Proc. 13th National Passive Solar Conference, Cambridge, USA, 364-367, June 1988.

30. C. Karakatsanis, M.Bahadori and B. Vickery : Solar Energy 37, 5, 363-374, 1986.
31. A. Bouchair, D. Fitzgerald and J. Tinker : Moving air, using stored solar energy. Proc. American Solar Energy Conference, Cambridge, 1988.
32. P. Koronakis : Submitted to Int. J. Solar Energy, 1991.
33. A. Dimoudi : An evaluation of the use of solar chimneys to promote the Natural ventilation of buildings. MSc. Thesis, Cranfield Institute of Technology, Sept. 1989.
34. B. Givoni : Ventilation Problems in hot countries. Research report to Ford Foundation. Haifa, Israel, Building Research Station, Israel Institute of Technology, 1968.
35. S. Chandra, P. Fairey, M. Houston and A. Kerestecioglu : Wingwalls to improve natural ventilation : Full scale results and design strategies. Florida Solar Energy Center, PF-47-83.1983.
36. I. Chand, V. Sharma and N. Krishak : Indian Journal of Technology 15, 369-374, 1977.
37. M. Santamouris and E. Daskalaki : Performance of nighttime ventilation of buildings in Greece. Report for the Greek Productivity Center, 1991.
38. S. Baer : Cooling with night air. Albuquerque, NM, Zomeworks. 1984.
39. M. E. Hoffman and M. Gideon : Window design practical directions for passive heating and cooling in heavy and light buildings. Proc. Windows in building design and maintenance, 277-288, Goteborg, Sweden, 1984.
40. N.V. Baker : Passive Cooling evaluation method - PACE. Proc. PLEA 86. Pecs, Hungary, 1986.
41. I. Chand : Research and Industry, 18, 3,50-53, June, 1973.
42. S. Chandra : Fans to reduce cooling costs in the Southeast. Florida Solar Energy Center, EN-13-85, 1985.
43. H. Wu : The use of oscillating fans to extend the summer comfort envelope in hot, arid climates. ASHRAE Far East Conference on air Conditioning in hot climates. Malaysia, Oct. 1989.
44. D. Scheatzle, H. Wu and J. Yellot : Extending the summer comfort envelope with ceiling fans in hot arid climates. ASHRAE Transactions, V. 95, 1, 1989.
45. W.A. Cunningham and T.L. Thompson : Passive Cooling with Natural draft cooling towers in combination with Solar Chimneys. Ibid 61, 1986.
46. H.R. Hay and I. J. Yellot : ASHRAE Trans. 75, 1969.
47. S.P. Jain and K.R. Rao : Building Science 9,9, 1974.
48. A.L. Pittinger, W.R. White and K.I. Yellot : Proc. Sec. Nat. Passive Solar Conference AS/ ISES, 773, Philadelphia, 1978.
49. M. S. Sodha, A.K. Khatry and M. A. S. Malik : Solar Energy 20, 189, 1978.
50. D. Pescod and R.K. Prudhoe : Telecommunications Journal Australia, 30, 2, 1980.
51. C. Carmody, G.D. Meixel, K. Labs and L.S. Shen : Advances in Solar Energy, 2, 297, 1985.
52. E. Chronaki : Ph. D. Thesis, University of Thessaloniki, 1983.
53. D. Carter : Underground Space 1, 317-323.
54. W.J. Rivers, B.Helm, W.D. Warde and W. Grondzik : Ibid 14, 126, 1981.
55. A. Tzaferis and D. Liparakis : A comparative study of the algorithms predicting the performance of earth to air heat exchangers. Interim Report, TEI Pirea, Greece.
56. B. Givoni. Ibid 14, 279, 1981.
57. B. Givoni and M. Hoffman : Build. International 6,525.
58. H. Hay : Proc. Conference Passive Solar Heating and Cooling, Albuquerque, N.M. 1976.
59. A.Argyriou and M. Santamouris : Possibilities of radiative cooling in Greece. Proc. Conf. Energy and Environment. A. Sayigh (ed), Reading, 1990, Pergamon Press.
60. M.Mostrel and B. Givoni : Passive Solar Journal, 1,4, 229, 1982.

BUILDING 2000

VENTILATION STUDIES DISCUSSION MEETING

Held at the Department of Applied Mathematics and Theoretical Physics
University of Cambridge
31 Oct 89

Introduction

Considerable progress has been made in the thermal analysis of buildings, using mathematical models,and also large scale test cells. Much less progress has been made in airflow modelling - indeed it is the modelling of convective heat flows in thermal models which remain their major weakness.

Mathematical models tend to fall into two broad categories - those describing inter-zonal mass flows, and those describing three dimensional flow within a zone often referred to as computational fluid dynamics or CFD codes. Neither of these are familiar to building designers; the former kind require unfamiliar input data such as air leakage distribution and pressure coefficients - the latter requires specialist input and expert interpretation of output, and heavy computing resources. Furthermore, unlike certain other environmental parameters such as lighting and acoustics, air movement is not readily perceptible, and thus designers have not developed an intuitive understanding of it.

The result of this is a lack of familiarity in the building design profession of air movement behaviour. This has lead to several unfortunate trends in building design, for example, the lowering of ceilings in offices, schools and hospitals, and problems in heating and ventilating large interconnected spaces, and inappropriate positioning of inlets in naturally ventilated spaces.

One area where this lack of knowledge is particularly evident is in the design of atria. Sadly, the demand for this particular building feature has outstripped the development of our understanding of the environmental physics of these spaces. In particular, we are short of knowledge on air movement, both between the atrium and the parent building, and the air flow within in the atrium and how this leads to thermal stratification, and over- and under heating. The response of the designer, and the consultant, has been to rely heavily on mechanical servicing of these spaces, where there is more certainty of their performance. This leads to unnecessary investment in plant and running costs, and even then does not always guarantee success.

In the Building 2000, where environmental analysis expertise is provided for designers of innovative low energy buildings, air-movement and ventilation have been identified as the major uncertainties in almost every consultancy.

With the growing concern for environmental issues, in particular the association of electrical energy use and CO_2 production, and the connection between refrigeration plant and CFC release, there is renewed interest in natural ventilation as an alternative to air-conditioning. The size and complexity of modern buildings, and the expectations for comfort of the occupants, will place heavy demands on designers who adopt natural ventilation as an option.

A parallel topic, that of daylighting, is already much better served. Daylight levels are relatively easy to calculate, image synthesizing models are being developed, and physical daylight modelling is a simple and well validated technique. Physical scale modelling of daylight is of particular relevance for comparison here. Its value lies in that it links a visual simulation which can be judged subjectively with accurate objective quantitative measurement.

Preliminary experience with the physical airflow modelling technique, one of the subjects of this meeting, shows the same virtue. In this technique, developed at DAMTP, Cambridge, buoyancy driven air flows are modelled in reduced scale models. Salt water solutions are used to represent the changes in density due to thermal effects in the real building. The application of the method for a number of real building projects has lead to both useful quantitative "design proving" results, and "design leading" proposals (1) (2).

Within the B2000 Programme, a physical modelling study by this technique has been commissioned for the University of Seville Dept of Humanities building on the 1992 World Exhibition site. It is a four storey atrium building where the atrium is used to drive buoyancy ventilation of the surrounding rooms, at the same time as retaining a cool pool of air at the ground level, as a refuge from the intense summer heat.

This meeting was held to discuss the theoretical and practical background to the technique and to present preliminary results.

A secondary purpose of the meeting was to present a CFD model developed at the TNO Institute of Applied Physics (TPD) in Delft by A D Lemaire. This permitted some comparison to be made between the physical and mathematical modelling approaches.

The Meeting

The programme for the meeting was as follows -

1) Visit it to laboratory and demonstration of flow model

2) Theoretical background presented by Dr Paul Linden (DAMTP)

3) Video of flow simulation, G Lane-Serff (DAMTP)

4) Results and future work

5) Presentation of CFD model WISH 3D by A D Lemaire (TNO/TPD)

6) Discussion on comparison between physical and mathematical models

The following notes follow the structure of the meeting and refer to accompanying papers Annex 1 to 3.

1) The Laboratory Technique

The basic experimental arrangement is shown in fig 1. A perspex model built to a scale of between 1:20 and 1:100 is suspended upside down in a tank of water. A parallel beam of light traverses the model and forms a shadowgraph on a translucent screen, enabling variations in fluid density to be visualized. This image is filmed by a video camera. Normally the direction of the beam lies in the plane of the floors of the building model and thus the image represents a kind of integrated section in two dimensions.

Air hotter than ambient is modelled using salt solutions which are of greater density (rather than lower density) than the ambient fluid. This is why the modelled is inverted. By inverting the video camera the image can be viewed right way up. Temperatures lower than ambient are simulated using alcohol solutions. Solutions are dyed to aid visibility.

Heat sources are modelled by delivering salt solution to the point where the heat source is located. This means that the technique can only represent a convective input. The power of the source is controlled by two parameters, the concentration of the salt solution and the flow rate. Further discussion on the scaling of dimensions, temperature and time are given in Annex 1.

The Theoretical Background

This is reviewed briefly in Annex 2 and in more detail in Annex 1.

The Dept of Humanities Building, University of Seville.

The proposed building, fig 2, is to be naturally ventilated and daylit. The critical situation is in summer, when daytime temperatures are typically over 30°C. The proposition is that high rates of ventilation required in the densely occupied lecture rooms (and offices) will be driven by thermal buoyancy pressures. This flow is provided by allowing the upper part of the atrium to become heated by solar gains made above the shading level. Air from the lecture rooms will be elevated to about 3°C above ambient and will be relatively buoyant when entering the middle zone of the atrium, which will be at about ambient temperature. This will minimise disturbance to the cool pool of air, maintained by the structure, cooled by night ventilation

Video of Flow Simulation

A video of the experiments on VHS format is available from ...... Photographs from the video are shown in plates 1-4 inclusive.

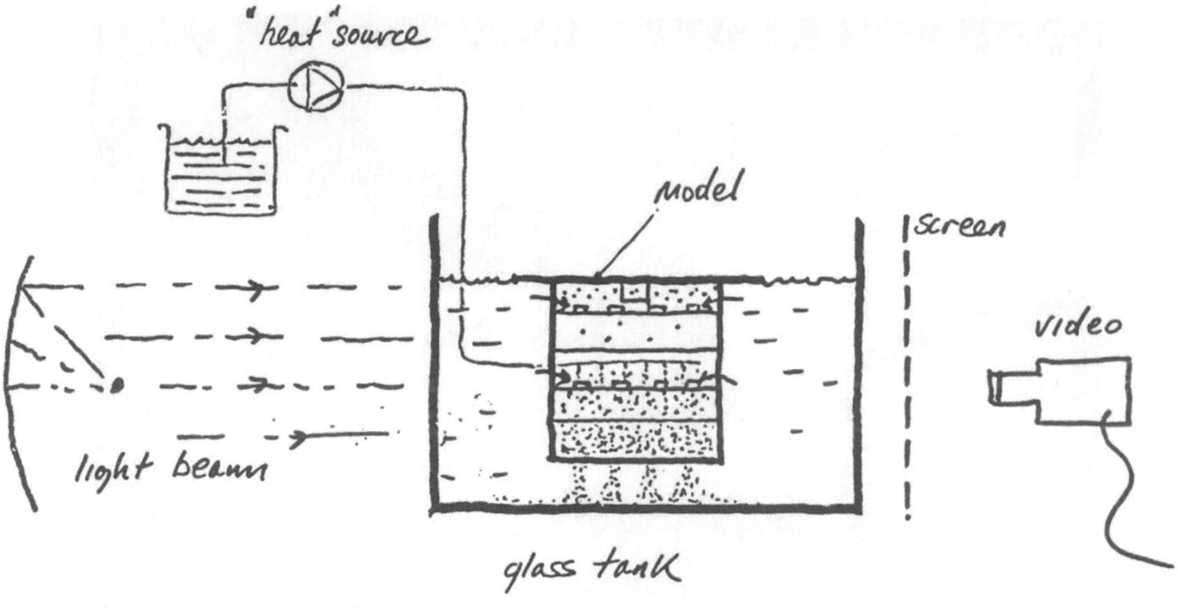

Fig 1 Experimental arrangement for laboratory flow modelling

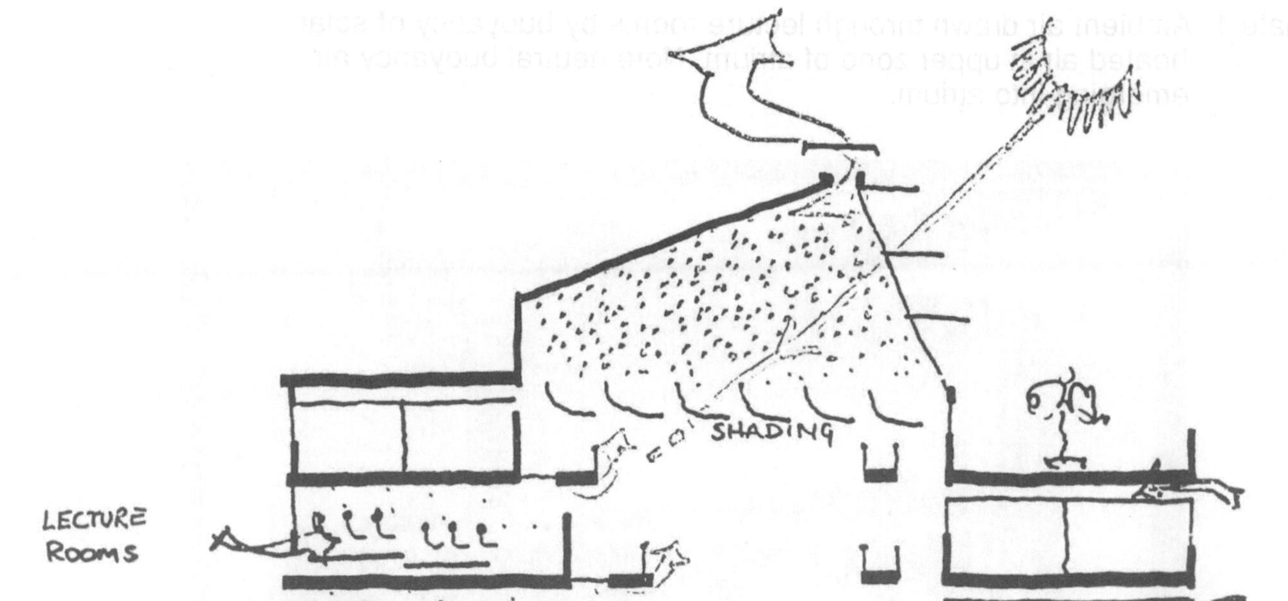

Fig 2 Proposed Department of Humanities, University of Seville

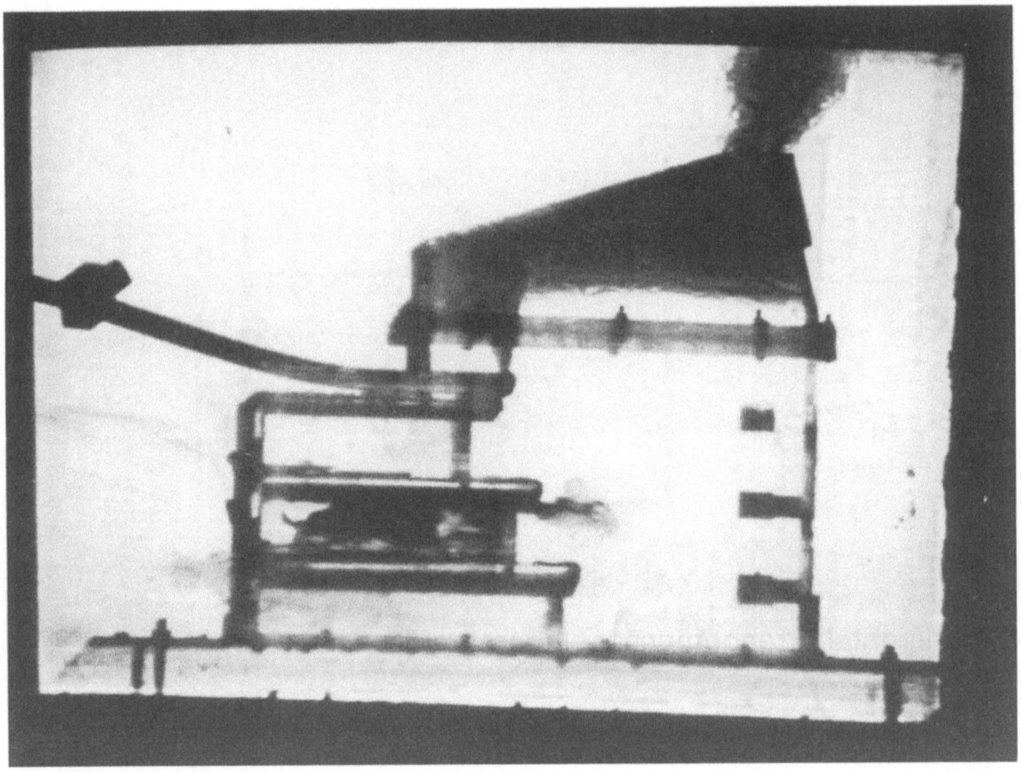

Plate 1 Ambient air drawn through lecture rooms by buoyancy of solar heated air in upper zone of atrium. Note neutral buoyancy air emerging into atrium.

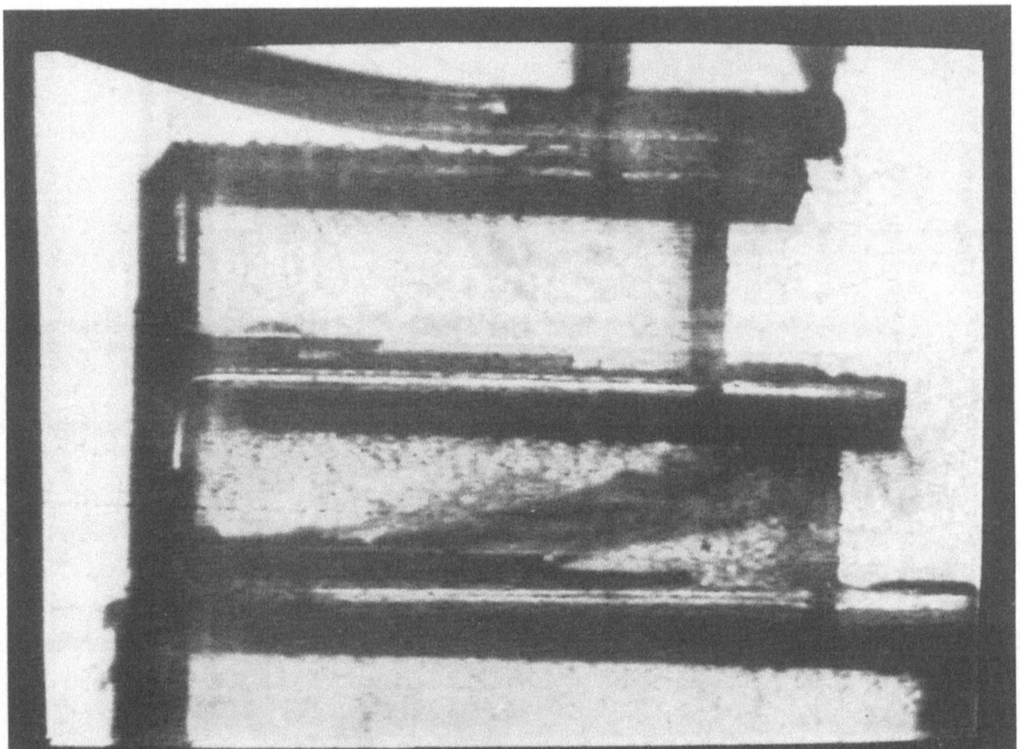

Plate 2 Close up of fresh air delivery via underfloor ducts.

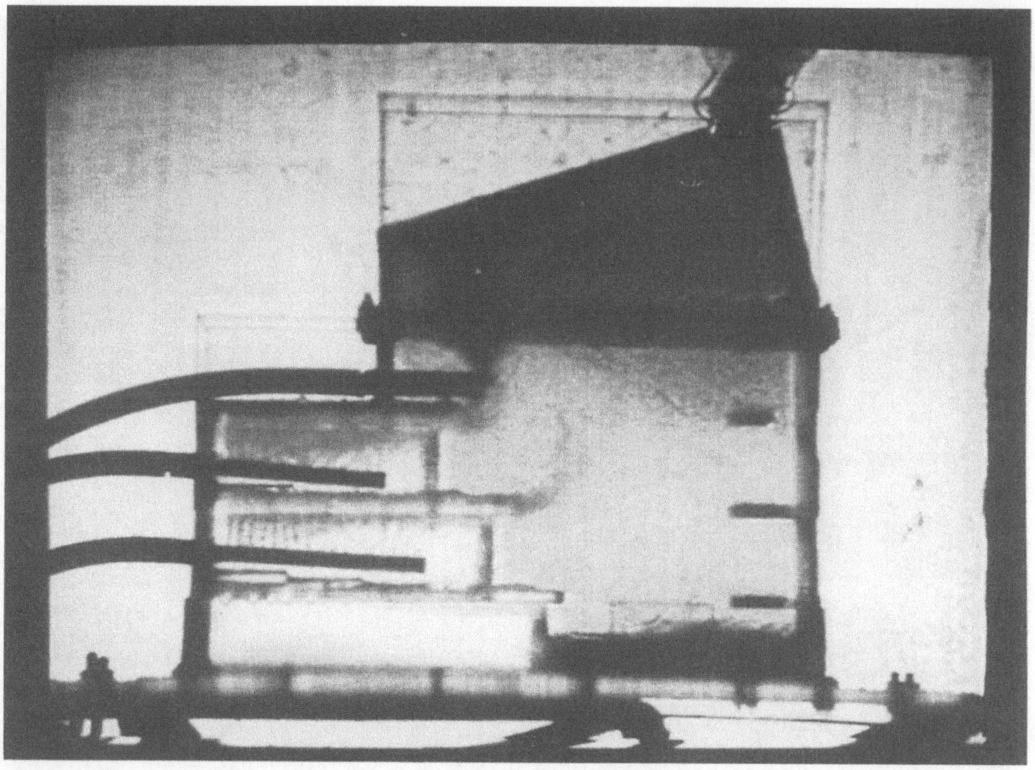

Plate 3 Heat gains from occupants in 1st and 2nd floor lecture rooms. Warm air plumes emerging rise immediately. Cool pool at base of atrium showing slight erosion due to ambient air being drawn through ground floor library.

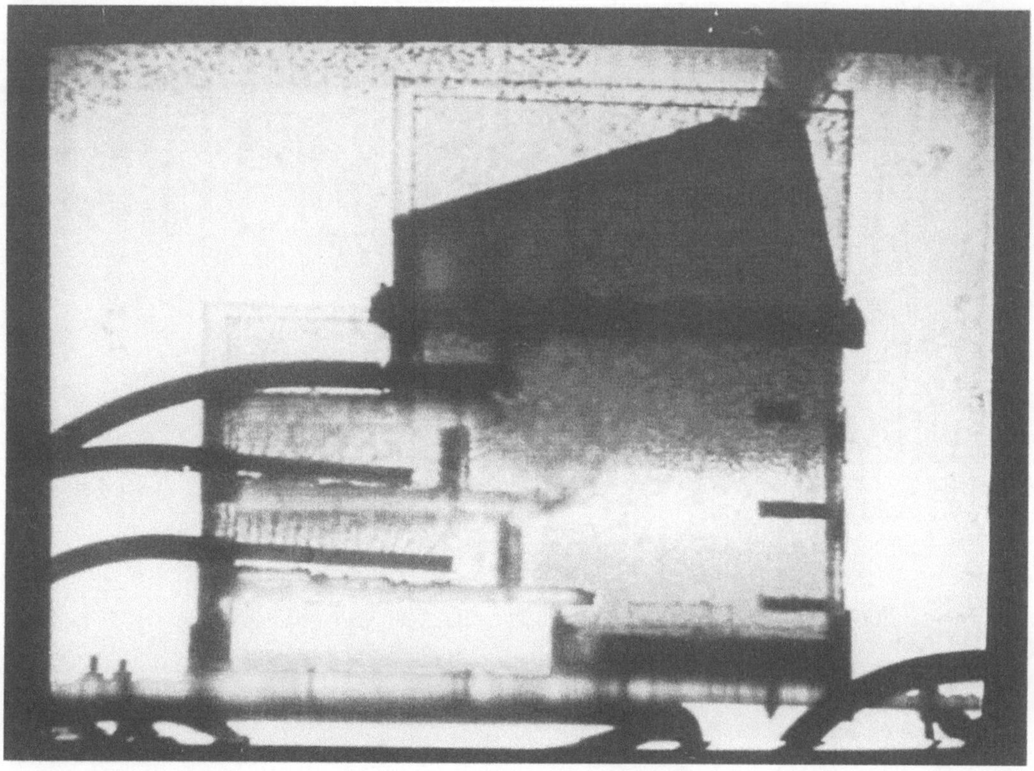

Plate 4 Reduced openings lower stratification layers. Cool pool stable due to zero vent through library.

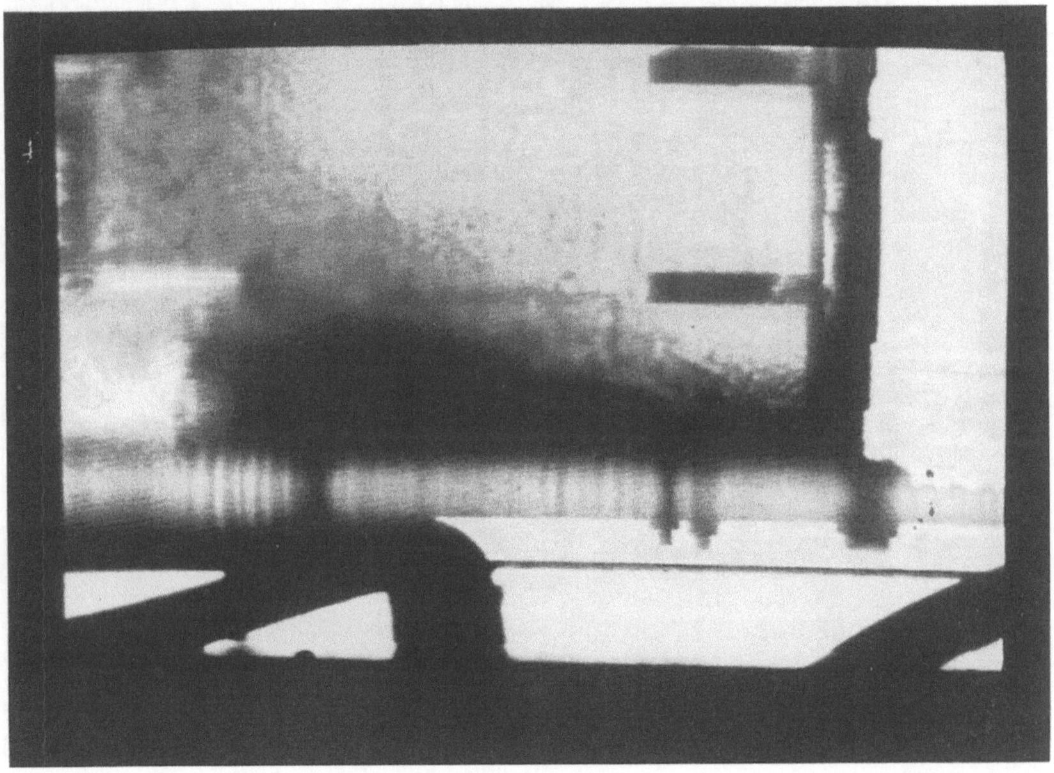

Plate 5 Erosion of cool pool due to ingress of ambient air from ground floor doorway

Results and future work

1) Solar heated upper atrium will drive ventilation through the lecture rooms.

2) Position of separation layer between hot air and air at near ambient temperature, can be controlled vertically by relative sizes of top and bottom openings and magnitude of heat input. In the real case the latter could be controlled by the use of shading devices.

3) Generally the position of the separation layer is above the occupied zones of the atrium.

4) The underfloor ducts into the lecture rooms show good distribution of fresh air.

5) Heated (and humidified) air from the occupants appears to move vertically into the upper zone of the room creating a stratified layer close to the ceiling, which is high. The hot air exhausting into the atrium is replaced by fresh air via the underfloor ducts.

6) The pool of cool air (5.4^{o}C below ambient) at the bottom of the atrium is only slightly disturbed by ventilation through the lecture rooms.

7) If air is drawn through the lecture rooms which are unoccupied i.e. no heat input, the emerging jet of air at neutral temperature is much more destructive to stratification than the warm air from the occupied rooms.

8) Suction from the heated upper part of the atrium can maintain the cool pool even when large openings exist at ground floor.

9) Aerodynamic design of the balconies may be important, to deflect warm air into the centre of the atrium, away from the balconies and rooms.

The following future work was identified -

a) Quantify temperatures and flow rates in lecture rooms in relation to opening sizes and temperature differences.

b) Model occupants in library

c) Test different balcony and opening designs.

d) Test the effect of vegetation on stabilising cool pool.

e) Investigate flow patterns within lecture rooms using larger scale model of one room.

Presentation of CFD Model WISH 3D

This is described in Annex 3.

Comparison between physical and mathematical models

The main advantages of the physical model are -

1) Flow is fully three-dimensional and turbulent.

2) Accurate representation and measurement of full scale stratification is achieved.

3) The air flow patterns are readily visualised in an easily disseminated manner (by video). This is of particular value in demonstrating design implications to a non-specialist.

4) Complex building geometries can be represented and changed relatively easily.

The main disadvantages are -

1) Radiation and heat transfer from surfaces can not be modelled.

2) Diffusion of heat within the flow is poorly represented.

The main advantages of CFD are -

1) Good diagnostics of flow fields.

2) Detailed quantitative output of velocities and temperatures

3) Radiative and heat transfers from surfaces can be included.

The main disadvantages are -

1) Require large amounts of input data

2) Complex geometries are difficult to model

3) Accurate representation of stratification and effects due to turbulent flow hard to achieve.

General observations

Both techniques make considerable demands on hardware. In the first case the laboratory requirements whilst not highly sophisticated, do require specialised equipment, i.e. tank, illumination system and pumps, and instrumentation for quantitative evaluation, i.e. conductivity meter or micro -

hygrometer. Furthermore, the physical model requires machine shop facilities beyond those normally required for architectural working models.

In the case of CFD modelling, the hardware demands are of course in computing power. The model WISH 3D typically requires 18 hours on a super mini (Alliant FX 40) for one flow and temperature mapping.

In terms of expert's time, the physical modelling, once the technical facility is set up, could actually be carried out by a technician, with expert advice required only for the specification of the model (simplification) and the experimental runs. In the case of the mathematical modelling, CFD specialist time is typically 5 - 20 hours. Differences in the kind of work prevent a direct comparison.

Both approaches seem to suffer from a common limitation, for different reasons. This is the modelling of heat flows into surfaces. In the case of the physical model conductive flows cannot be represented at all, thus, for example, the flow of heat from convective gains into thermally massive elements cannot be modelled. In the case CFD modelling, boundary conditions at surfaces are modelled by an independent thermal model which has to assume values for convection coefficients. However these are dependent on the surface flow conditions, and in the case of glazing where the major part of the thermal resistance is due to the boundary layer, this must lead to errors.

Whilst both techniques could (and have) been used to assess specific design proposals, their real value probably lies in the development of generic solutions and understanding of airflow problems. The generation of highly visual output by the physical modelling, lends itself to interactive video techniques for the retrieval and management of generic solutions.

Development of computing power/cost ratio, and speed, could increase the accessibility of the CFD approach to the non-specialist, provided unambiguous interpretation of the output could be provided. It seems unlikely that technical developments will significantly change the situation with the physical modelling technique

List of Annexes

Annex 1 Laboratory and Mathematical Modelling: The Design of a Naturally Ventilated Building.
G F Lane-Serff, P F Linden, M Galtry, N V Baker
Proc Vent 90, Lisbon

Annex 2 Notes on the Natural Ventilation of Atria
N V Baker

Annex 3 Numerical Simulation of the Air Movements and Temperatures in Large Glazed Spaces
A D Lemaire, TNO Institute of Physics, Delft.

ABOUT VENTILATION IN BUILDING 2000 PROJECTS

Bernard FLEURY*
ENTPE LASH, Rue Audin
69518 Vaulx en Velin Cedex
France

1 INTRODUCTION

As expert, I have been involved in three projects : an educational building, a holiday complex and a hotel. The buildings were at different stages of design : preliminary, intermediate and advanced.They were not located in the same geographical regions. Weather conditions, occupancy, lay-out, typology differed significantly for one building to another. However, ventilation principles apply in any case. In this paper, a synthesis of my experience with the Building 2000 projects is presented.

2 A NEED FOR A MINIMUM KNOWLEDGE IN VENTILATION

First, I should say that I have been surprised by the diversity in awareness and motivations of the architects.

Repeating dull mistakes might signify that the project leader was still at the exercise level and had never faced the construction of a real building. In this type of EEC competition, the reference to previous realized buildings should be included in the submission form and used as a selection criteria. The preliminary design of a passive solar feature is simple in comparison to the work to have the system built.

Surprisingly, strange blunders with regard to ventilation and physical principles were next to some complex concepts well understood and integrated. Architecture is a complex, multidisciplinary approach made of compromise and difficult choices. However, the missing of a design criteria (acoustic, environment, ventilation,interior organization, ...) at any step in the building design could lead to the project failure. Focusing on the use of passive solar technologies could push insidiously the architect to ignore other key constraints. The building includes advanced technologies and products (solar or wind chimney, lightwell, atria, advanced glazing,...) but may present functional deficiencies, harmful for a comfortable interior environment.
Fortunately, some buildings had integrated the ventilation constraints and advantages at every stage of the design. Architects did not go further in their approach because of the lack of user-friendly simulation tools and previous recorded experience.

---

* Dr B. FLEURY is now in charge of the R&D department at Aéroplast Vim , Azay le Brulé, 79400 St Maixent l'Ecole, France.

3 EXPERTISE AND PRACTICE

The level of knowledge is so diverse in the design teams that the expert can level the database from his experience, present the state of the art, illustrate solutions, stimulate the dialogue among the design group and solve some critical issues. Therefore, offering design assistance and support of consultant is beneficial. However, expertise is necessary anf successful if an intense back and forth is initiated. The expert should not appear as the providential teacher who will cure for ever a problem. He can only improve the design approach and the system performance, diagnostic and prevent doubtful options and quantify the impact of various parameters. He should always stay within the frame of the constraints of the project.The expert should be considered as a full partner and not as an afterward person that will only take care of his specific problems. Add-on techniques and afterward expertise should be excluded in the design of a new building.

On the other hand, experts are limited by the state of knowledge in every specific field. As far as ventilation is concerned, we should mention that summer ventilation is not well apprehended today and a thorough research program should be lauched.

4 DESIGN APPROACH

Through this paragraph, we review the key issues, part of an optimum design as far as ventilation is concerned. It is always difficult to present recommandations outside a specific project. From the remarks during the expertise, a checklist can be drawn. It may serve as a preliminary design tool for winter and summer ventilation. Some questions may appear too simple or too obvious to the reader but I have been so astonished by elementary issues not well tackled that I do not elude them.

Climatic conditions, air quality, fire safety, occupant's comfort, air flow pattern, load modulation, mechanical ventilation enhancement and side problems such as acoustic, safety, privacy, are the main parameters. The list is not exhaustive and can be considered only as the minimum requirements of a ventilation conscious design. It could be completed, refined and formatted in a more comprehensive way. In order to reach a final stage, a confrontation between experts should be organized. This list represents the questions which arise during our discussions.

This list and the building plans should be used in parallel and the checkpoints should be modulated with the case study. Projets are so diverse that some questions are more relevant to a specific design than others.

5 WINTER VENTILATION

5.1 Building approach

What are the direction and intensity of the winter prevailing winds?

Can you represent the site by a few design days ?

Identify the various sources of pollutants, their location and occurence (water vapor, smoke, CO2, ..)

Mark the building zone where supply air is required. Living spaces should benefit of fresh and clean air.

Notice all the air inlet and outlet in every aeraulic zone with an arrow. Check if every zone has at least one inlet and one outlet.

According to the pressure field around the building and the flow resistance, joint the arrows to identify the air flow pattern within the building and verify the physics of fluid dynamics.

Notice the spreading of pollutants with your ventilation system. Could you implement an exhaust next to the source?

Check local ventilation rates with air quality requirements

Does your exhaust network cross any cold zones? If so, is there any condensation risk?

Is your attic ventilated?

Does your polluted air exhaust in a cold zone (attic, garage)?

Could you use any buffer space to preheat the air?

Can your ventilation system react to critical situations during winter such high internal or solar gains or intense temporary production of contaminants?

In case of fire, what is the strategy ? Do air inlets and outlets have a similar size?

Could they satisfy a high exchange rate and exhaust the deadful fumes generated by the fyre?

Do you prevent spreading of fire over the space and through the HVAC system?

5.2 Within the room

Identify potential draft due to low position of air inlet? Could you move them in a higher position?

Does your interior door allow air circulation between the rooms (grilles, air paths,...)?

Could you have any shortcircuiting of air and dead zones in your room?

Do you have any large glazed area that could generate convective currents and induce air draft within the occupied space? If so, can you deviate them or reduce air velocity?

In case of zone connexions through ducts, did you consider the problem of acoustic transmission?

5.3 Systems

In presence of combustion system such as fireplaces, is combustion air specially driven to the system or do you suck room air?

When the fireplace is not running, is the trap closed if any?

In case of hybrid systems, did you consider the parasitic power consumption of the mechanical devices?

Does the ventilation system require maintenance? How easy is it?

Do you have any previous experience with the designed system? If not, complete your search.

6 SUMMER VENTILATION

6.1 Cooling strategy

Is natural ventilation used as a cooling strategy?

Do you plan to use night time and/or daytime ventilation?

If your answer is "Yes", pursue through the questionnaire

Do you know the prevailing winds (speed, direction, intensity, frequency) and the temperature variations?

Are local weather conditions compatible with natural ventilation?

Do you first consider to reduce to the minimum the cooling loads of your building?

Check possible cooling load reduction strategy: window shading, landscaping, radiant barriers, minimum infiltration during hot summer days, reflective surfaces, natural lighting instead of artificial.

6.2 Ventilative cooling

Do you try to enhance wind resources next to the building? What is your strategy ? wingwall, trees and landscaping, channeling the winds through building lay-out.

Do you favor natural ventilation within the building?

Do you consider horizontally shaped windows, optimum distribution of openings for doubled sided ventilation, movable shades to direct the air to a specific zone, similar inlet and outlet area?

Do you have zones in series? Can you modify your plans to get parallel cross ventilation?

Does natural ventilation work whatever the wind direction is?

Consider any strategy to enhance the potential of natural ventilation especially in the absence of wind? Solar or wind chimneys, ceiling fans are some answers.

Can you exhaust high intemal gains through boyancy effect?

Does natural ventilation respect privacy and intimity? Large openings favor noise transmission.

Does your window have any screen (sand, insects, ..)?

Do you try to precool the air before entering the building (evaporation,trees,ponds)?

Do you take in account latent loads in your energy calculations?

Does night ventilation compatible with safety?

Are the night air streams directed toward the massive walls for optimum storage of coolth?

If summer ventilation compatible with winter considerations?

Evalate roughly the air change rate and air velocity within each space. Do you achieve your design criteria?

7 BENEFITS

Experts were not isolated and devoted to a specific restricted project. They confronted their experience from diverse fields and horizons, resulting in a more general debate on multidisciplinary approaches. A strong desire of collaboration arose, leading to the creation of working groups. For example, it appeared that passive cooling represented a fundamental issues not well apprehended. An horizontal study was initiated and the status of the art has been reviewed, illustrating the potentials of passive cooling techniques and pointing out the needs for fututre research. These working groups are still active even after the conclusion of Building 2000. These Europeen networks are the illustration of the success of this program.

ATRIUM DESIGN GUIDELINES

Nick Baker
The Martin Centre for Architectural and Urban Studies
University of Cambridge, Department of Architecture
6 Chaucer Rd, Cambridge, UK

The atrium is one of most popular features of the contemporary non-domestic building. However, its promotion as an architectural feature does not ensure a clear functional and environmental brief. It is often unclear whether the atrium is an unserviced 'intermediate' space between inside and outside, or just a rather extravagant and large top-lit room. Furthermore, the unfamiliarity of handling such spaces, with large heat losses and solar gains via the glazing, and the unknown effectsof stratification over the large floor to ceiling height, often promts designers to opt for a highly serviced solution.

Ironiocally, atria are often regarded as passive features and associated with energy saving. However the energy consumption of a parent building may be increased if compared with a similar shaped building without an atrium. This is due to the increased need for artificial lighting, mechanical ventilation, and possibly cooling. Though the atrium may save heating energy, this is usually the smallest component of Primary energy consumption. The atrium itself can consume energy if it is artificially lit and conditioned, - often for the health of the planting rather than the comfort of the occupants.

These notes offer guidance in the design of atria which are passive and energy saving, and whilst these atria cannot guarantee the environmental standards of a fully conditioned space, they may actually benefit from the free running "atrium climate"

1. EVOLUTION OF ATRIUM FROM OPEN COURT

It is helpful to consider the evolution of an atrium or sunspace from an open space or courtyard, which would provide the parent building with daylight, natural ventilation, and if south-facing, useful passive solar gains. All of these functions are potentially energy saving.

If the open space is now glazed over to form an atrium, (figure 1) we must ensure that these original functions are not jeopardised.

In figure 1, the original functions of the open space are:-
1) cross ventilation, 2) single sided ventilation, 3) daylight;

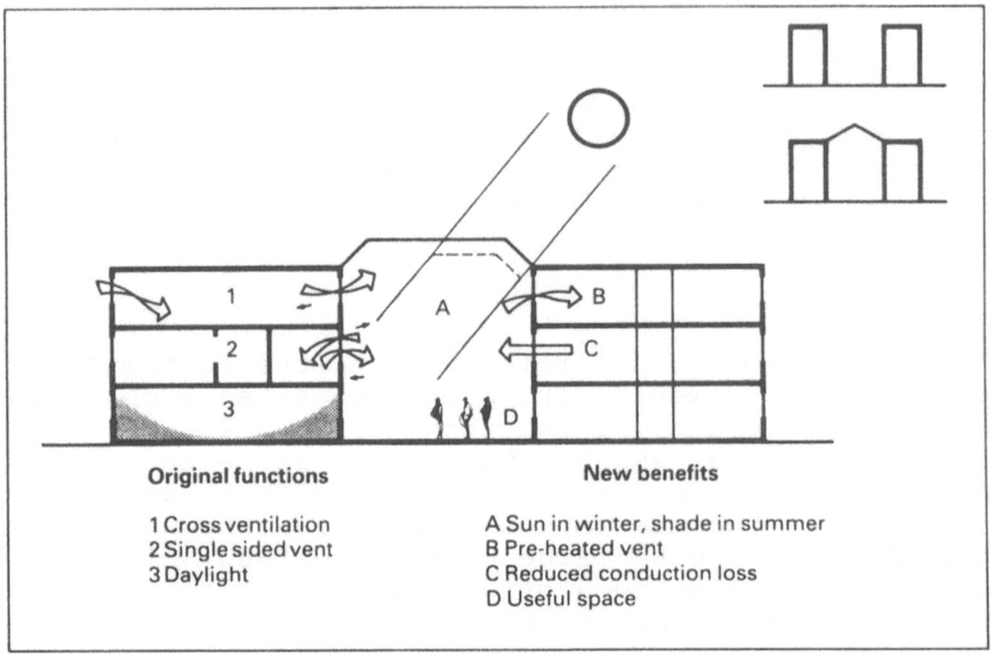

Figure 1: Evolution of atrium from open court.

and the new functions of the atrium space are:-
A) sun in winter, shade in summer, B) pre-heated ventilation, C) reduced conduction loss and D) useful space.

To ensure that the presence of an atrium space does not increase the energy consumption of the parent building the following points must be observed:-

1) The daylighting levels in the glazed space must be maximised by the use of reflective finishes and clear glazing, figure 2. This will allow the surrounding rooms to be daylit and avoid the need for daytime artificial lighting in the atrium.

2) The atrium must have a supply of fresh air which will permit natural ventilation of the surrounding rooms, (if local fire regulations permit).

3) Shading and high ventilation rates must be provided in the summer to prevent overheating.

4) Shading must be movable to allow maximum daylight into the atrium in winter and other times of low sky luminance.

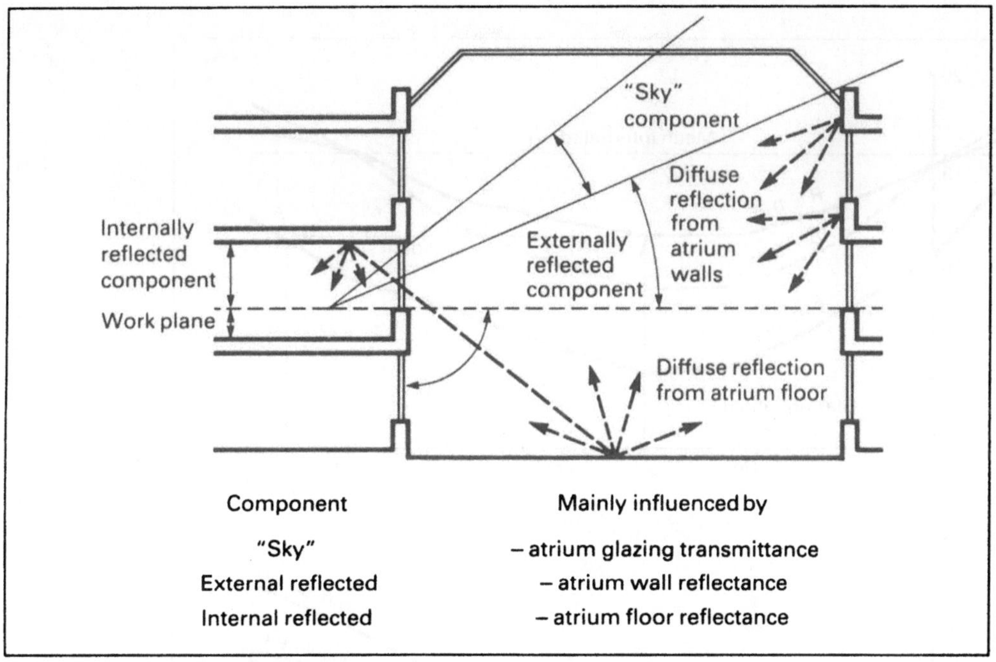

Figure 2: The atrium as a source of daylight.

2. WINTER PERFORMANCE

Even if the atrium space is unheated the temperature in the atrium will be above the ambient temperature. The increase will depend on a number of factors-

1) The ratio of external glazing area of the atrium to the wall area of the parent building which is protected by the atrium, figure 3.

2) The thermal transmittance of the separating wall between atrium and parent building. This will normally be determined by the amount of glazing in the wall.

3) The orientation, inclination and transmittance of the external glazing of the atrium or sunspace.

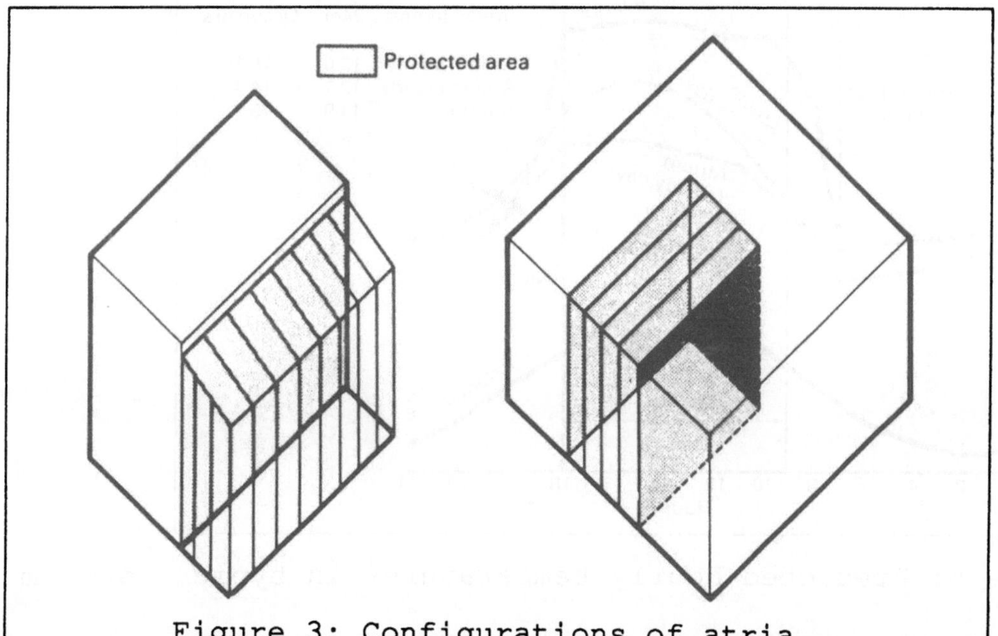

Figure 3: Configurations of atria.

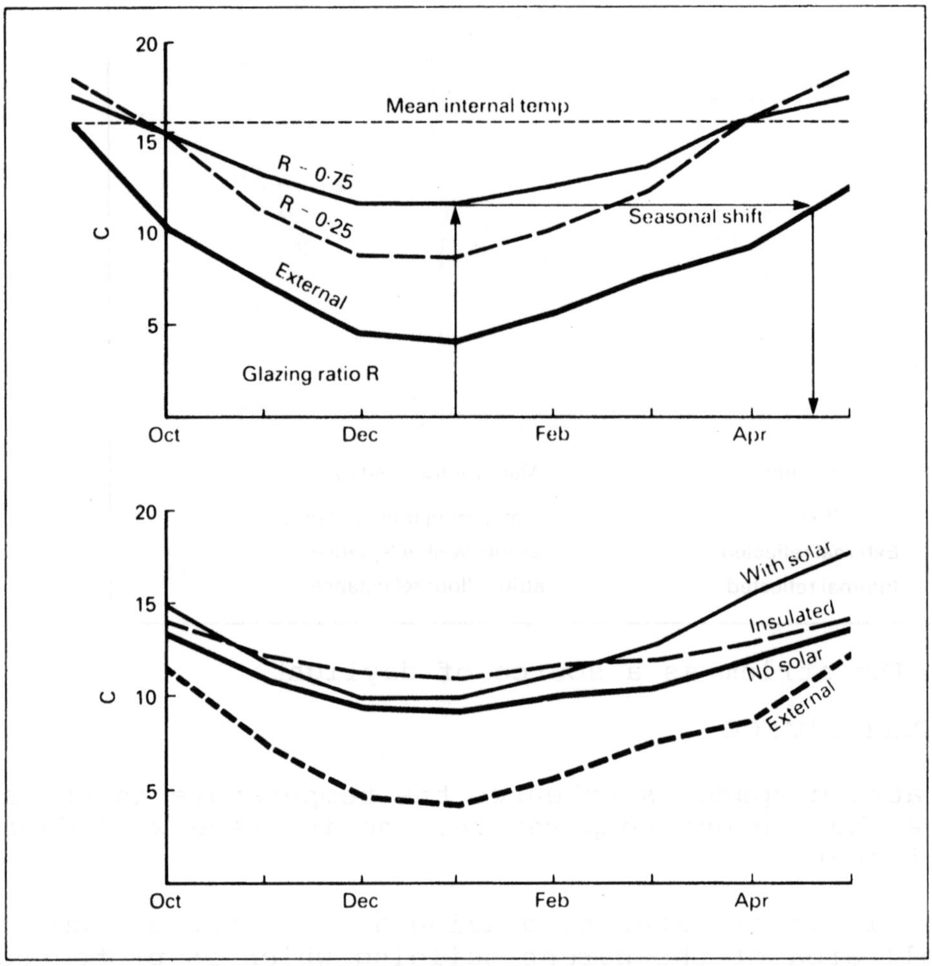

Figure 4: Predicted monthly temperatures in typical atrium.

The temperatures in figure 4 are predicted for a typical atrium in the UK. Note the "seasonal shift" from January to late April. Figure 5 shows typical hourly temperatures for a moderate winter day - note how the daytime atrium temperature is much higher than the 24-hour average.

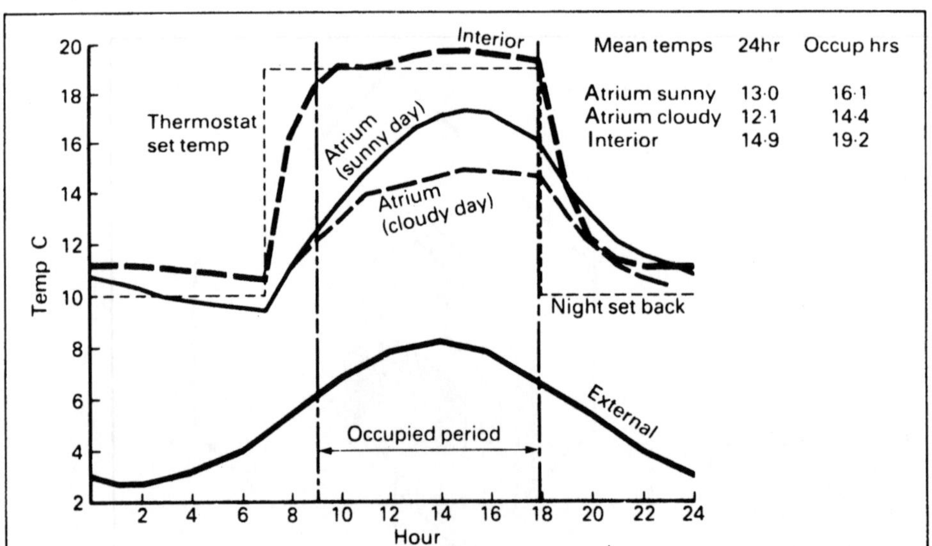

Figure 5: Predicted hourly temperatures in typical atrium.

Stratification will not be great, since there are no high temperature air sources in an unheated atrium, and the cool air plumes falling down from the cool glazing at high level will entrain warmer air from the heated building surfaces. There may be problems however, where there are large areas of vertical glazing down to floor level.

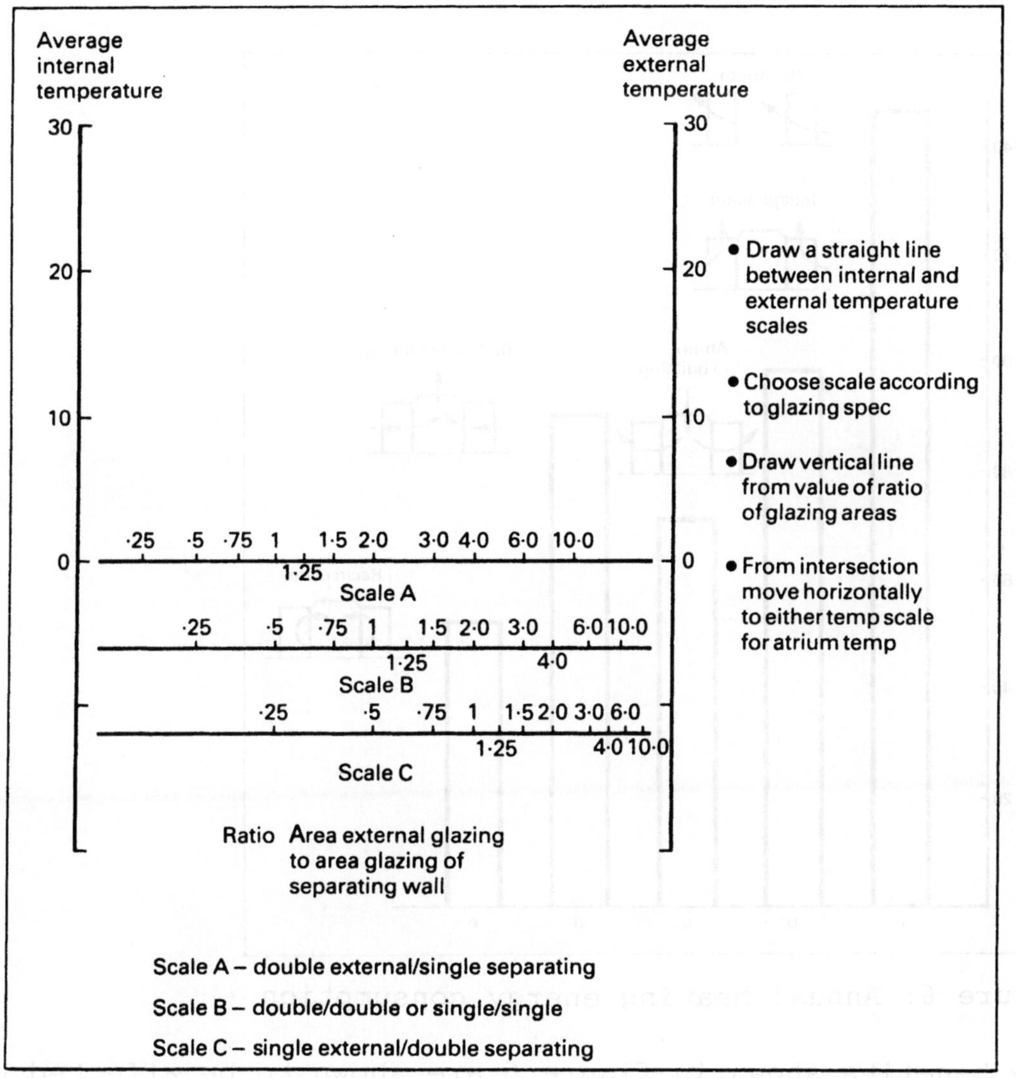

Table 1: Nomogram to calculate atrium average temperature.

This nomogram may be used to estimate the temperature conditions in an unheated atrium in winter. It takes account of heat loss into the atrium from the heated building, but not the beneficial effect of solar gains. If the average outdoor daytime temperature is used, and the internal comfort temperature for the building, the nomogram will give a good indication of conditions on days of poor radiation.

3. THERMAL ENERGY

The temperature increase in an atrium space reduces the heat losses from the parent building and can provide tempered fresh air to reduce the ventilation heating load. However, local fire safety regulations may prohibit some ventilation options. The temperature increments together with the protection from rain and wind will greatly increase the amenity value of the atrium space when compared with an open space.

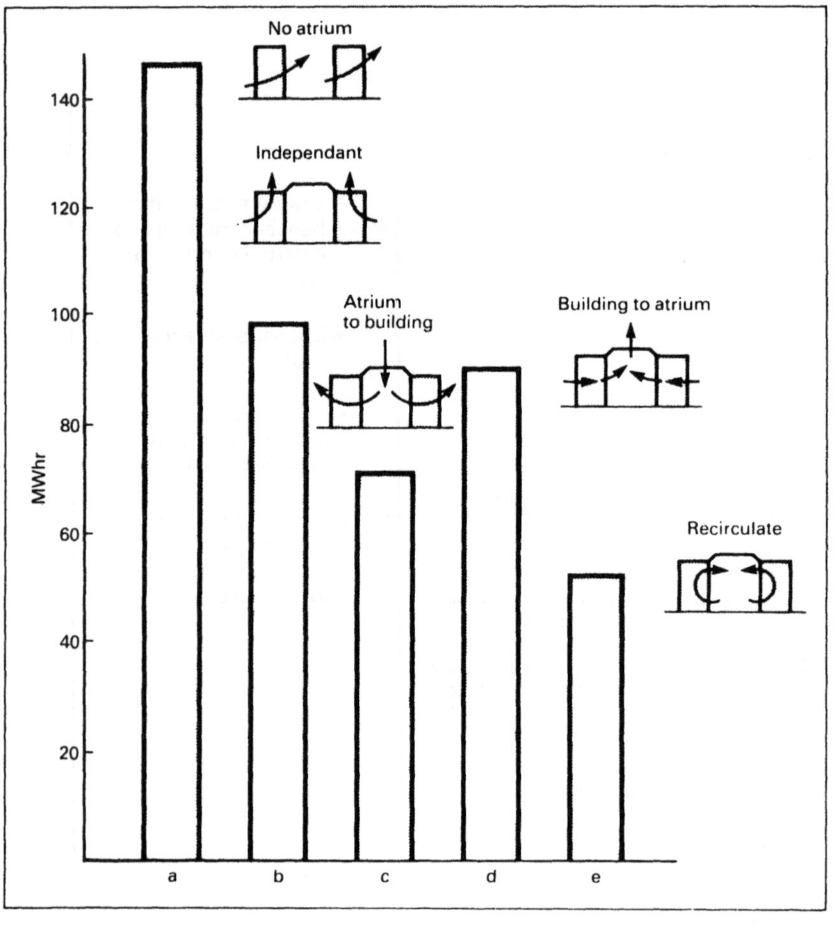

Figure 6: Annual heating energy consumption

The energy benefits shown in figure 6 are shown to be affected by the different ventilation options. This is just an example and they are also effected by the configuration of the atrium as described in figure 3. The main benefit occurs due to the reduction of conductive losses - 'independent' refers to the lack of ventilation or convective exchange between the atrium and the parent building. Further benefits occur with ventilation pre-heating and recirculation. Dumping air from the heated building mainly benefits the temperature conditions in the atrium. In practice, the optimum mode would probably be a combination of these under intelligent control.

However, it must be noted that heating energy is a small part of the total energy consumed by a large non-domestic building. The thermal benefit of an atrium can easily be compensated by the loss of daylighting and the resulting increase in artificial lighting energy.

4. SUMMER PERFORMANCES

The main concern is to prevent overheating. Three passive measures may be employed to attain this - shading, ventilation, and the incorporation of thermal mass. Figure 7 illustrates an atrium roof incorporating shading and ventilation.

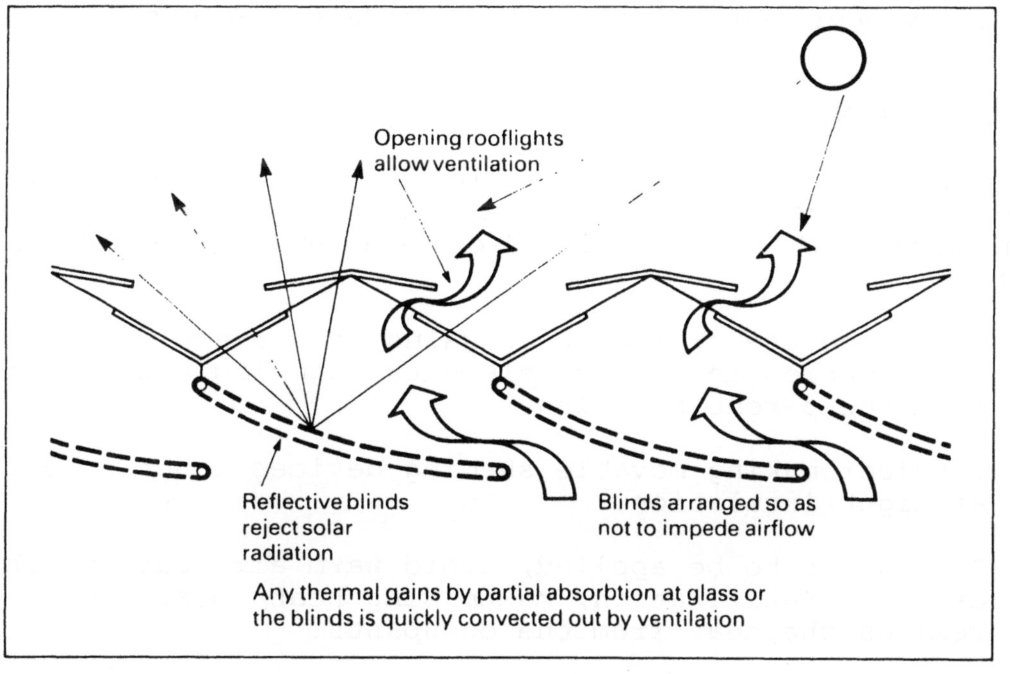

Figure 7: Shading and ventilating to prevent overheating.

Shading is the most effective, reducing the ingress of solar gain. Shade can also reduce the effective temperature as experienced by an occupant, by up to 8 degrees centrigade. The shading should be movable.

Fixed shading, or tinted glazing should be avoided. This will reduce daylighting levels all year round, creating a need for artificial lighting for the atrium, and the adjacent spaces, in winter.It is quite pointless to put expensive glazing in an atrium envelope and then to permenantly obstruct it. Fixed geometric shading, i.e. that which transmitts according to solar angles, is only appropriate for climates with very bright skys and major overheating risk. In predominantly diffuse skys it will reduce annual daylight yield significantly.

Ventilation can be used to reject heat, when the outside temperature is below the atrium temperature. The ventilation of the atrium and the parent building may be provided by stack effect; but openings will be required at the top and bottom of the stack of between 5% and 10% of the roof glazing area.With this arrangement, stratification will result in the temperature at the floor of the atrium being close to ambient, or even slightly below due to the effect of thermal mass.

It must be emphasised that the stack effect only works when the average temperature of the air in the stack is greater than the ambient temperature. In summer conditions this may be too hot for comfort. At night however, heat retained in massive elements in the atrium will drive the stack effect to provide useful night cooling.

One set of openings can often serve the function of summer ventilation and smoke extract vents. Both shading and vents can be mechanically operated and controlled by heat sensors.

5. HEATING IN ATRIA

Atria are often heated. This is to improve comfort conditions or to provide protection for planting. Heating an atrium can be very energy consuming. It can be avoided or reduced by observing the following:-

1) Make good use of free heat gains, for example, solar gains stored in massive elements, dumping exhaust air from the parent building.

2) Consider making movable shading devices function also as night insulation.

3) If heat is to be applied, avoid warm air heating. This causes stratification, maximising heat loss, and removes the heat from the occupants.

4) Radiant systems apply heat to the occupants, keeping air temperatures down, permitting local heating and minimising heat loss.

5) Choose plant species which can tolerate low temperatures. Remember that the plants occupy the building 24 hours per day, 7 days a week, 4 times longer than human occupants.

6. SUBJECTIVE FACTORS

The perception of comfort is complex and responds to many stimuli and expectations outside the purely thermal environment. For example, if the atrium provides the image of a "plushy" interior, with carpeted floors and soft furnishings, and artificial lighting, the occupants will expect "interior type" comfort standards. On the other hand, the atrium may offer outdoor wall finishes, and a paved floor. Bright and sunny in winter, with sky and clouds clearly visible, and shady in summer. Temperate plants allowed to loose their leaves in winter, bursting into new growth when the increased warmth and daylight indicates the approach of the "atrium spring" When (and if) heating is provided, it is a radiant source evocative of winter sunshine rather than the warmth of a cosy living room. People will accept these conditions as an "atrium climate" rather than an engineered environmental standard. And it may be a much nicer building for it.

DAYLIGHTING ADVICE

Marc Fontoynont, LASH - ENTPE, Rue M. Audin
F 69120 - Vaulx en Velin

1. INTRODUCTION

This report presents the conclusions of the advice and assistance in daylighting provided by the author to design teams of the CEC-sponsored BUILDING 2000 program. Projects were to be built in Spain, Greece, Portugal and France. Most of the advices were given during meetings with the design teams as well as engineering firms involved in the project. Detailed analysis was also conducted using computer programs to test performances of daylighting techniques. Teams appeared to be very motivated by achieving good daylit buildings. However, the difficulty stood in establishing proper sizing of apertures with respect to visual and thermal constraints.

2. APPROACH

The major interest toward daylighting is not among professionnals, but among the general public, able to appreciate pleasant indoor spaces where it feels good to stay. Architects are also interested, mainly because their architecture is a game of light in itself and they can put their art in it. Engineers are concerned because of the various problems involved with the presence of glazed areas. The major difficulty than one encounters in talking about daylighting is to put together the subjective appraisal and the logic of light propagation. The first action was therefore to include basic laws of daylighting in early discussions: typical optimum window sizes according to room dimensions, typical maximum depth of light wells, efficiency of solar protections.

3. SIMPLE SOLUTIONS

Recently, architectural publications have often presented projects where daylighting was achieved with rather complex systems, as if a good solution would require an important construction of glass, louvers, ..etc. Surprisingly, there is a large gap between this design direction and general building pratice leading to office spaces with little access to daylight.

Daylighting systems may however be very simple, if a global strategy is defined early enough. For instance, building pratice and energy conservation standard lead to glazed areas for offices which tend to converge toward average values: around 40% of glazed areas in facades for Northern Europe, versus 25 % for Southern Europe.

Good Daylighting will not be achieved by increasing these values, since thermal risks would become higher: more difficulty to reduce solar gains in summer. With clear glass, glare problem from diffuse light may become critical.

Improvement of building performance with respect to daylight will be achieved through the control of building thickness. Typically, the distance from facade to facade should remain lower than 5 times the ceiling height. Maximum use of daylight will be obtained if a good glare reduction device is employed on windows. Best results are found with venitian blinds, due to their low luminances under sunlight. With these standard approach, hallways can be lit with borrowed light only if the partition between offices is largely glazed (more than 40 % of the partition area).

With such standard techniques, it is possible to bring "good" amounts of daylight to more than 70 % of the building surface.

4. INCREASED LIGHT PENETRATION

Often building projects do not allow such reduction of building thickness. Also, concerns on energy conservation tend to aim at rather compact distribution of volume. To prevent a large fraction of the building to be deprived from daylight, new accesses need to be created. Since it it difficult to increase daylight penetration from windows unless large ceiling heights are employed, it appears attractive to collect light from the roof.

Linear light wells are more efficient than square or cylindrical ones: they absorb less light due to a lower number of reflexion in the longer axis. They also can bring light to larger areas.

Typically, it is difficult to conduct sufficient amounts of daylight if light well depth exceeds 4 times the section (square wells) or 6 times the section for linear ones. Highly reflective materials, such as anodized aluminium increase performances. Light wells are quickly innefficient if diffusing materials are employed.

Above circulations and other spaces of temporary use, higher ratios (depth to width) are acceptable since these spaces may operate with lower amounts of light.

5. ATRIA

Due to their limitations in depth, light wells width needs to be very large if the building height exceeds 3 or 4 floors. This bright space becomes an important architectural feature. Glazed atria are attractive in northern Europe, since they offer a compromize between access to daylight and reduction of thermal losses through the envelope.

The spaces around the daylit atrium could benefit from borrowed light. However, an atrium is darker than the external environment. Typically 4 to 10 times depending on the shading devices employed. If light transmission through its roof is too low, then adjacent spaces could not benefit from enough light fluxes.

In Northern Europe, atrium depth should not exceeds 1.5 times the width for efficient borrowed light penetration in spaces around the atrium. In southern Europe, the ratio can be increased to 3 for the sunniest climates.

Facades of lower floors should be largely glazed (100 % of facade if possible) since illuminances on them will not be high.

Facades of upperfloors should be smaller and special considerations should be brought to glare control. Thermal confort is a difficult issue there, due to the concentration of hot air under the glazed roof, even if shades are drawn. Ventilation is a key issue for atrium agreeableness.

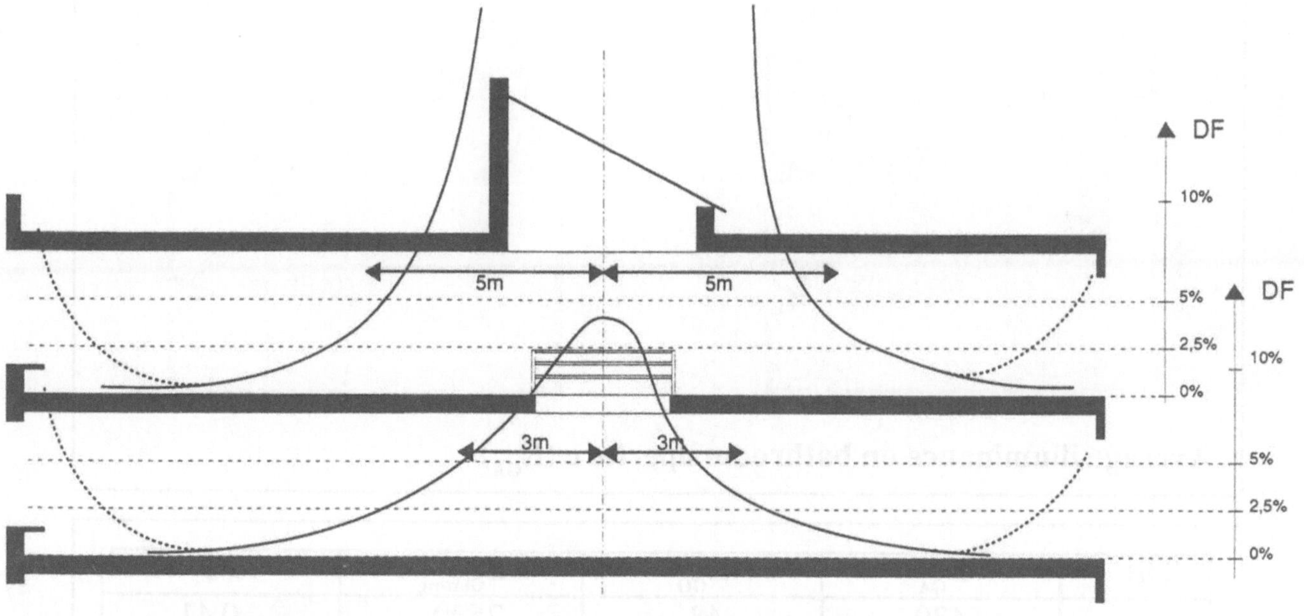

CONDITIONS OF SIMULATION = UNIFORM SKY (L= CONSTANT)
MATERIALS = DOUBLE GLAZING, CEILING REFLECTANCE = 60%, FLOOR = 30%, WALLS =40%

Figure 1. Daylight distribution under a light well, from computer simulations (GENELUX, ENTPE). Illuminances on the floor decrease sharply when one moves away from the center of the well. This is due to the direction of incoming light, falling like water from a shower.

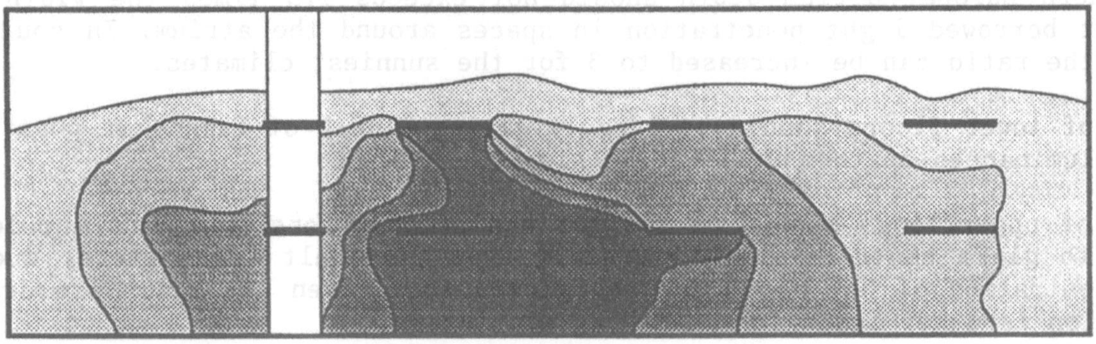

Illuminance distribution on the atrium wall (DF,%)

Simulation assumptions :
- Uniform Overcoast
- Outdoor horizontal illuminance (10000 Lux)

Reflections coefficients :
- Floors = 30%
- Walls = 40%
- Ceiling = 60%
- Glazing = Double Glazing

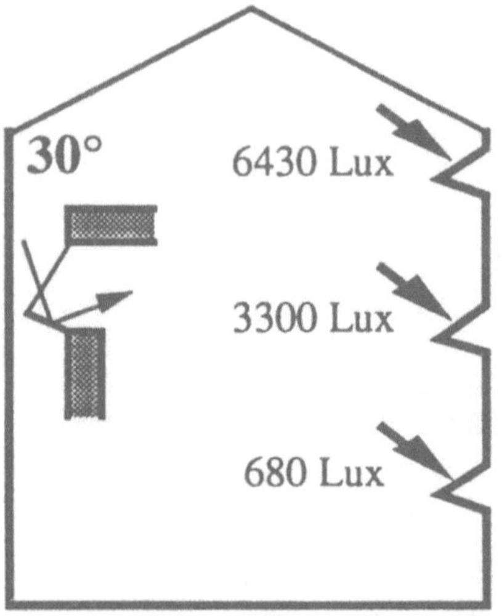

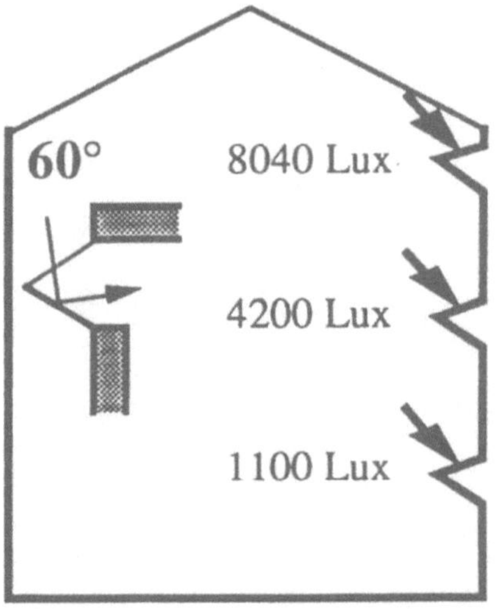

Average illuminance on bathroom aperture (E_{GA})

Tilt	E_{GA}	E_{GO}	$E_{Glazing}$	FJ
30°	6430	48	7520	0,41
	3300	50	7520	0,22
	680	40	7520	0,04
60°	8040	54	9370	0,46
	4200	56	9370	0,25
	1100	55	9370	0,06

Average Daylight factors on work plane in bathrooms by taking into account the effect of atrium

Figure 2. Example of detailed analysis of light collection techniques from an atrium. Illuminances decrease rapidly along the facades, and windows need to be tilted to collect more light. Simulations conducted with GENELUX, from ENTPE.

DAYLIGHT DESIGN FOR PASSIVE SOLAR BUILDINGS

A SUMMARY OF DAYLIGHT CONSULTANCY SERVICES IN THE BUILDING 2000 PROGRAMME

J. Lynes, Hull School of Architecture, Humberside Polytechnic, Strand Close, Hull HU2 9BT, England

1. NATURAL LIGHTING STUDIES

Hull School of Architecture has been involved in four projects for the Building 2000 Programme:

Rotherham Magistrates' Court

Newbury Central Library

Machynlleth Study Centre

Kent Schools Refurbishment

1.1 Rotherham Magistrates' Court

From the point of view of natural lighting this building marks a departure from customary practice in the design of magistrates' courts. For reasons of economy and security these have been conceived as deep-plan buildings.

In Rotherham orientation and massing were carefully assessed for their passive solar potential at the outset.

Advice was given on natural lighting for the two-storey public waiting areas and for the courtrooms.

1.2 Newbury Central Library

The dominant daylighting feature in this design is a central light well incorporating indoor reflectors and baffles. This will produce high daylight factors in a central space whose role may not be of sufficient significance to warrant such strong visual emphasis.

Advice was offered on bookstack lighting, on side windows, on study carrels, on solar protection and on the removal of overhead baffles.

1.3 Machynlleth Study Centre

The most generously daylit spaces in this project are the circulation spaces and the atrium/conservatory. Since people are generally attracted towards zones of high luminance, special provision is needed to avoid the impression of moving out of a bright circulation space into a relatively dark living or working space.

Advice was given on rooflight design, and on ways of encouraging occupants to switch lights off.

1.4 Kent Schools Refurbishment

The purpose of this project was to prepare guidelines, from a passive solar standpoint, for the refurbishment of system built schools.

Advice was given on how the quantity and the quality of natural lighting may be affected by outdoor obstructions including trees, by window geometry, and by outdoor and indoor surface textures and colours.

2. CONCLUSIONS

The notes which follow offer some tentative conclusions arising from experience on the above four projects. These observations may apply to many other passive solar buildings.

A first impression is that passive solar design is sometimes taken as a licence to use large windows. This seldom makes for energy conservation. Better by far to forget the passive solar label and concentrate on husbanding energy resources: the outcome, paradoxically, will be a passive solar building that really works!

A second impression is the need for simple, back-of-an-envelope, guidance on daylight design. Computer programs are legion, but intuition atrophies for lack of a launch-pad.

Examples of apt daylighting aphorisms might be:

1. The amount of daylight reaching a point indoors depends mainly on how much sky can be seen from that point.

2. The character of the natural lighting depends largely on position and sharpness of the no-sky line. (The no-sky line is the line, or to be more precise the surface in 3-dimensional space, which divides a room into two parts: one part is exposed to a view of the sky, the other part receives no direct daylight at all).

3. Areas beyond the no-sky line are likely to be inadequately daylit. The view from those areas is also impoverished.

One can use such aphorisms at successive stages in the design process:

1. To compare alternatives for the massing and orientation of passive solar buildings.

2. To determine whether a room is too deep to be satisfactorily daylit.

3. To optimise the shape and position of critical windows, bearing in mind the need to create meaningful patterns of natural light and shade, with significant items preferentially daylit.

2.1 Other Resources

We would commend the CIBSE Applications Manual on Window Design, published in 1987 by the Chartered Institution of Building Services Engineers, 222 Balham High Road, London SW12 9BS, England. This contains a sequential procedure for window design which respects energy conservation priorities.

We would also mention the HyperLight PC-based tutorial package covering all aspects of lighting design. This package, widely used in United Kingdom schools of architecture, incorporates many of the lessons of the Building 2000 programme. Details are available from W. Burt, School of Architecture, University of Manchester, Manchester M13 9PL, England.

SIMULATION OF THERMAL ASPECTS OF THREE GERMAN PROJECTS BY ESP

Peter Voit, Universität Stuttgart, Institut für Thermodynamik und Wärmetechnik, Pfaffenwaldring 6, W-7000 Stuttgart 80, Germany

1. INTRODUCTION

The building simulation program ESP (Environmental Systems Performance) was used to carry out analyses of different designs for three German projects. At all project designs conservatories are incorporated as an important part of the building.
Two preparation phases preceded the analyses. First the architects got guidelines about principles and input needs for different kinds of thermal calcalation methods. Additionally comments to the first design were given. In the second phase improved designs were discussed during a workshop.

2. SURVEY OF METHODS

Various calculation methods are available to predict heating requirements. In general they can be subdivided into three classes:

1. Simplified determination of annual heating energy demand by a method, which is using degree days for the whole heating period (e.g. VDI 2067 /1/)

2. Extended method considering internal and solar gains by an utilization factor and an accumulated temperature difference (e.g. for each month) to predict heating requirements (e.g. ISO DP 9164 /2/).

3. Detailed model based on hourly weather data considering thermal capacity of the building. Ability to predict temperature processes. Can only carried out on computers (building simulation programs like ESP /3/, SUNCODE, TRNSYS or DEROB).

The accuracy of the predicted result depends not only on the method but also on the design of the calculated building or room.
It is quite easy to estimate the energy consumption of a not insulated building with small windows as can be concluded from its energy balance:

$$\dot{Q}_h = UA\ (T_i - T_e) - \dot{Q}_s - \dot{Q}_i - C_{bld}\ \Delta\dot{T}_i\ [W] \qquad (1)$$

where Q_h is the required heating power [W], UA the overall heat loss coefficient of the building [W/K], $(T_i - T_e)$ the difference between internal and external temperature [K], $\dot{Q}_s$ the solar and $\dot{Q}_i$

the internal gains [W] and C_{bld} the heat storage capability [J/K] of the considered building.
For the given example the thermal losses UA $(T_i - T_e)$ are the dominating part of the balance, whereas internal and solar gains are nearby neglectable (small window area) and the capacity has no effect for long term considerations. The prediction of the overall heat loss coefficient can be done quite accurate, if the wall constructions are well known. Solar radiation data can be extracted from handbooks (e.g. European passive solar handbook /4/, Atlas über Sonneneinstrahlung /5/)
In this case a simple method is quite sufficient. The difference to detailed simulation models is very small (about 5 %) for prediction of heating requirements.
But even in that simple case a prediction of overheating or freezing aspects is not possible, because of the used weather data, which are mean values for large time periods and the neglectance of short term capacity effects, which are very important for detailed investigations.
The thermal behaviour of a conservatory or atrium is much more difficult to predict in comparison to normal buildings like described before due to the very strong and very fast change of heat fluxes:
Without solar radiation there is a quite large heat loss of the conservatory because of the bad insulation effect of the glazing. But as soon as the sun is shining the gains will beat the thermal losses (see Eq.(2)):

$$\dot{Q}_{net} = UA\ (T_i - T_e) - gA\ I \quad [W] \tag{2}$$

where $\dot{Q}_{net}$ is the net heat flux [W], I the solar irradiation [W/m^2] and gA the solar aperture [m^2] of the glazing.
Hence the prediction of overheating and freezing aspects in conservatories is only meaningful by using detailed calculation models, which are based on hourly weather data.

3. INPUT NEEDS

Normally a lean-to conservatory is connected to three main environments:

- ambient air
- ground
- building

For the requested time period all environmental temperatures of the conservatory must be given.
Furthermore the following design characteristics are needed:

1. Location, orientation, geometry and size of the conservatory

2. Detailed description of all floor, roof and wall constructions and glazings. Thermo-physical properties of the material:

- conductivity	k	[W/(m K)]	for all layers
- density	ρ	[kg/m^3]	
- specific heat capacity	c	[J/(kg K)]	
- thickness	δ	[m]	
- solar absorptance	α		internal and external surfaces
- infrared emittance	ϵ		

If this detailed description is not available a minimum data set is mandatory consisting of:

- overall heat transfer coefficient U [$W/(m^2 K)$] of the whole construction (e.g. wall, roof, floor etc.),
- location of the insulating layer (inside, outside, center).

3. Thermal and optical properties of glazings:
   - heat transfer coefficient U [$W/(m^2 K)$]
   - transmittance τ
   - solar heat gain factor g

   Type and size of window frames.

4. Operational use of the considered conservatory: Air change rate [1/h], internal gains (occupancy, lights, moisture etc.), shading devices.

5. Controlling (temperature set down at night, preheating, freeze protection)

6. Meteorological data (solar radiation, temperature, wind speed and direction, humidity)

4. DESIGN GUIDELINES FOR ARCHITECTS

Because all listed input parameters has an influence, only a variation of the interesting input parameters can show their impact. Temperature processes as well as peak values can only be predicted by a detailed simulation program based on hourly weather data.
If energy saving is intended, a good insulation of the regular living space on which the glass house is leaned to has major priority. For a south facing glass house a well insulated opaque roof is recommended. This will reduce thermal losses in winter time and will have a self-shading effect in summer time, but will clearly reduce the illumination of the rooms behind the south facade. Therefore the use of transparent insulation material on roofs is a possibility to do two jobs in one, but on that case shading devices must be taken into account.
Finally, glazings with low U-values will diminish thermal losses during the heating period, which are nevertheless larger than solar gains in Central Europe (see Fisch et al. /6/). For energy saving aspects reducing the thermal losses is better than optimizing the solar gains. Considering Swedish insulation laws net gains are only possible if more than three transparent layers (e.g. low-e triple-glazing) are used /7/.

5. RESULTS OBTAINED BY ESP SIMULATIONS

For all simulations ESP release 6.14a was employed. For each run the German Test Reference Year Hannover is used as hourly meteorological input data. The geometrical input is carried out by using a cartesian coordinate system, where each corner of each surface has to be exactly defined. All projects are simulated for the whole year, which takes about 30 minutes CPU time on VAX 8300 for each.

For each project it is calculated by ESP:

- total annual heat consumption (also related to the floor area [kWh/m^2 a] or bounded volume [kWh/m^3 a])
- monthly heat consumption
- required hourly heating power in winter (including required peak power) for different operation modes (constant 20 deg.C, night temperature set back).
- room temperature evolution in winter
- room temperature evolution in summer (including peak overheating temperature)
- statistical analysis of comfort level

Exemplary results for the Permaculture Institute are presented in the following figures.

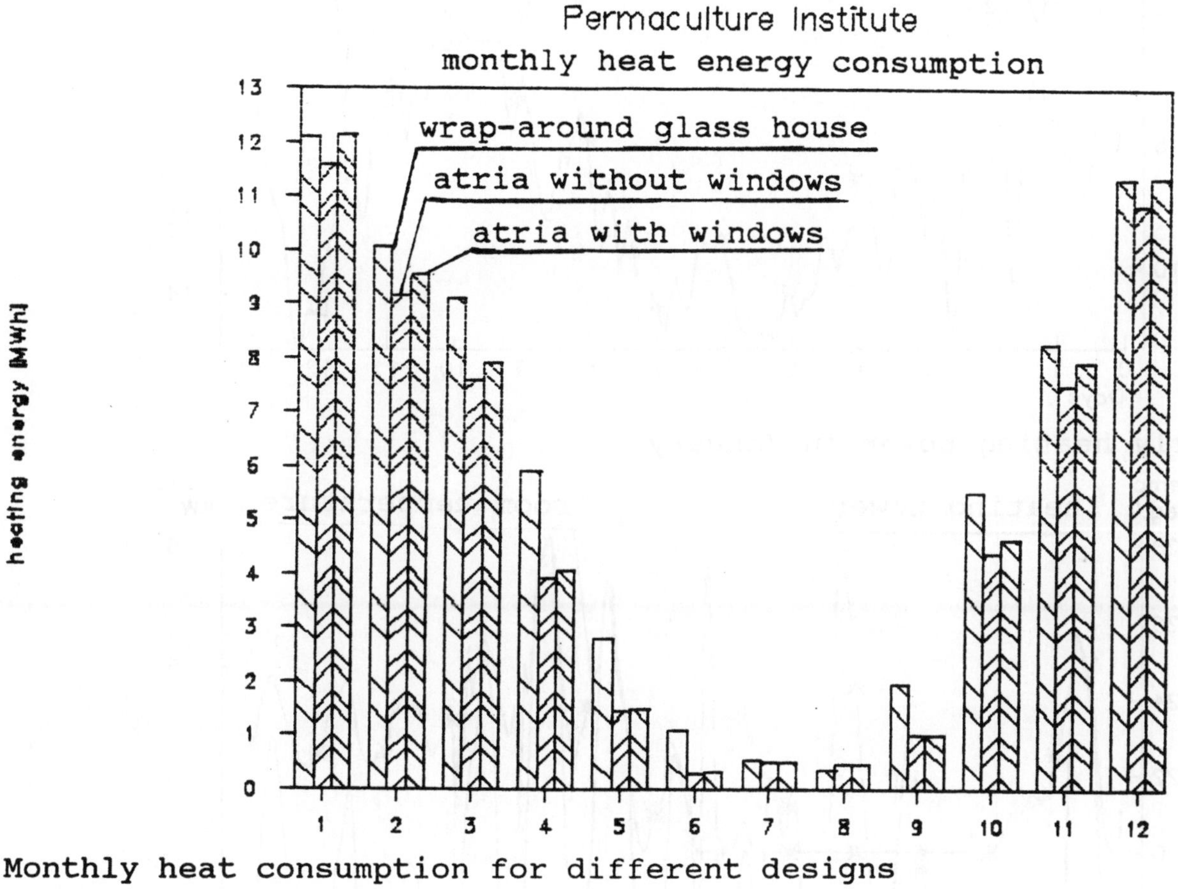

Monthly heat consumption for different designs

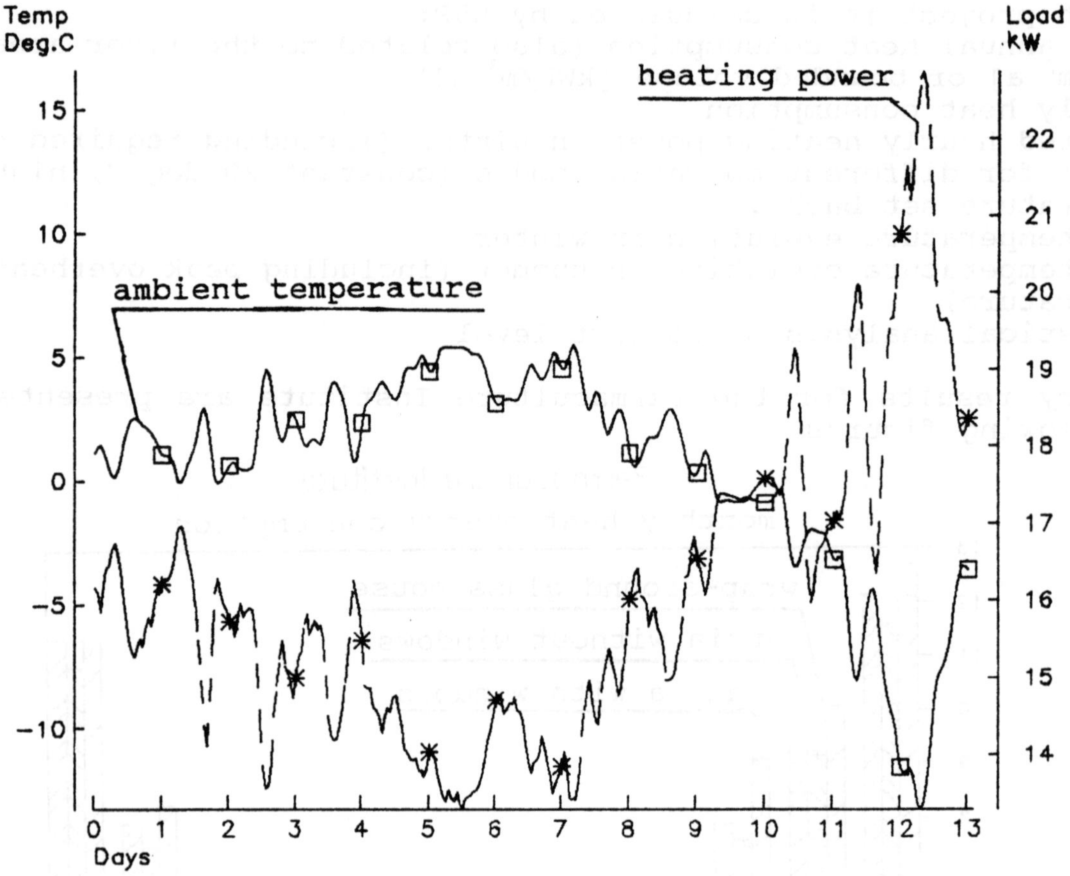

Hourly heating power in January

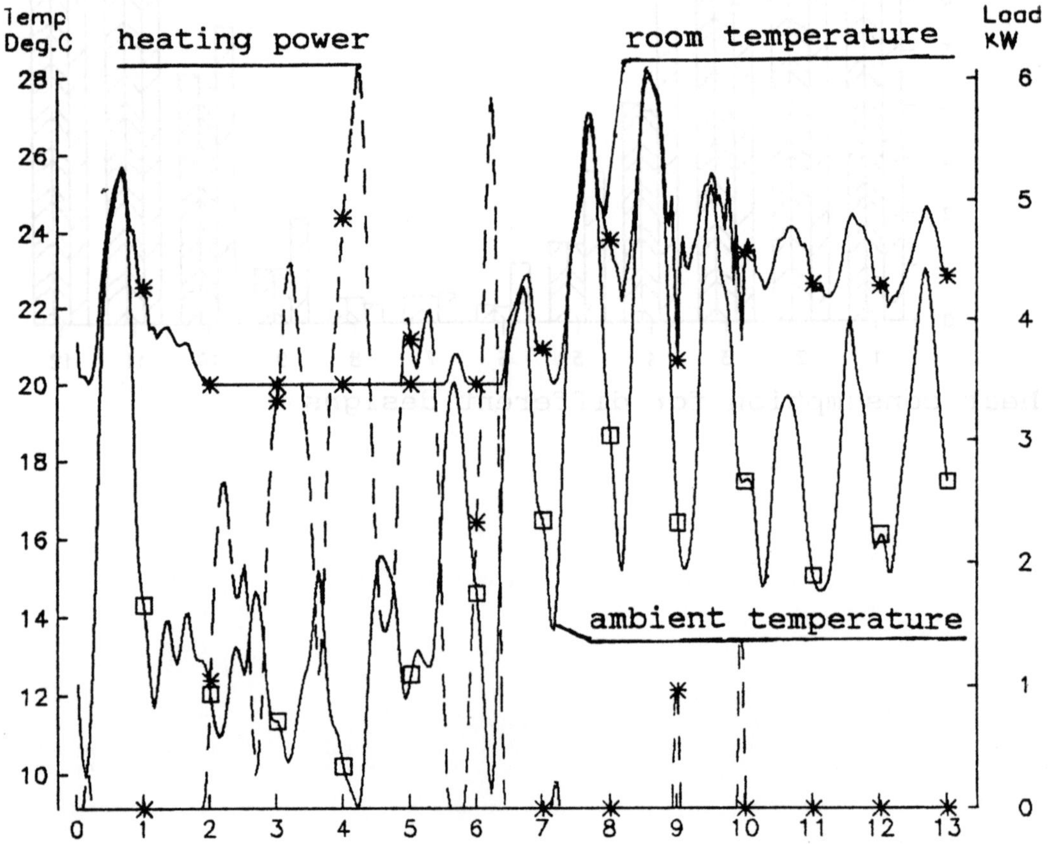

Hourly room temperature and required heating power in July

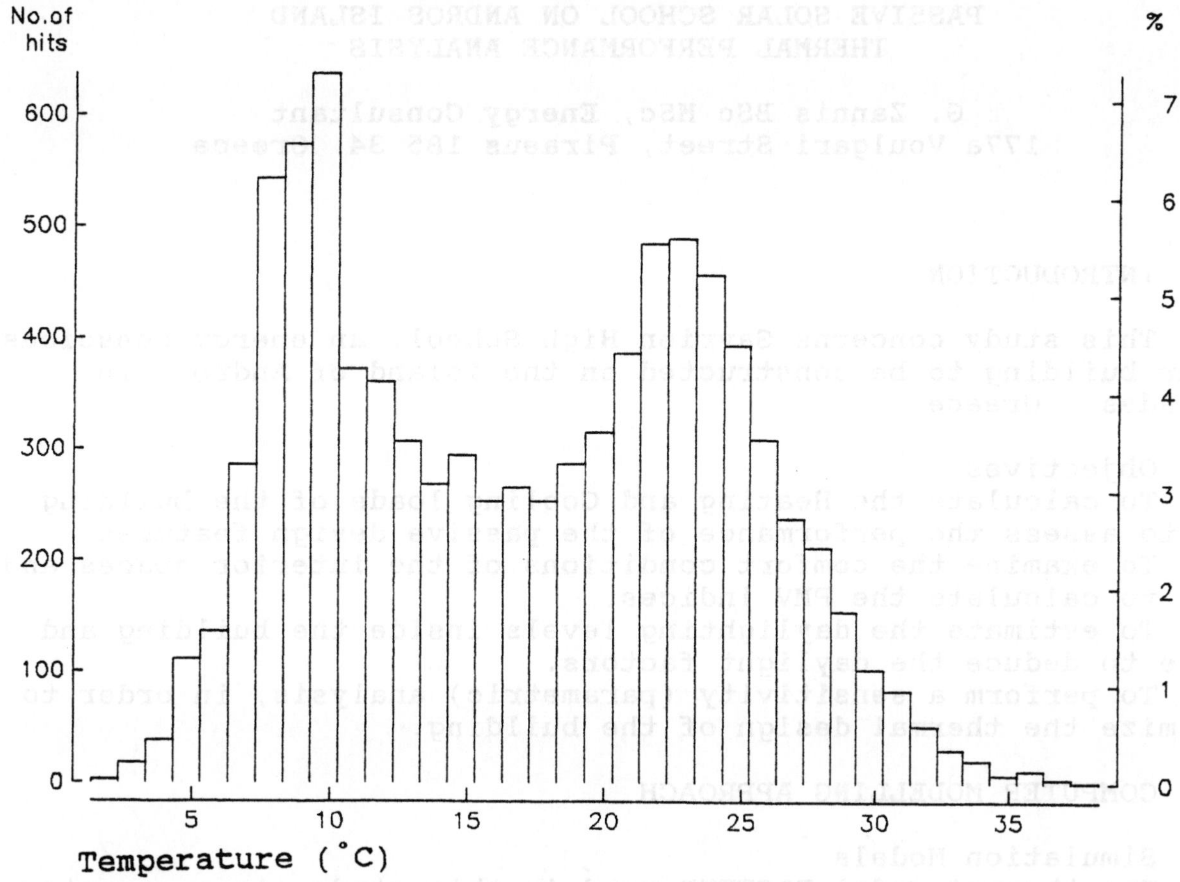

Annual frequency distribution of the temperature in the conservatory (operation mode: freeze protection)

6. REFERENCES

/1/ VDI 2067: Berechnung der Kosten von Wärmeversorgungsanlagen

/2/ ISO-TC-163-SC2-WG5: The calculation of space heating requirements for residential buildings DP 9164, 1985

/3/ ESP - A Building and Plant Simulation System, Version 5, Release 3, Glasgow 1986

/4/ Achard, P., R. Gicquel: European Passive Solar Handbook, Commission of the European Communities 1986

/5/ Kommission der Europäischen Gemeinschaften: Atlas über die Sonneneinstrahlung, TÜV Rheinland 1984

/6/ Fisch, N., J. Munder, E. Hahne: Untersuchungen über Maßnahmen zur Reduzierung des Wärmeverbrauchs eines Wohngebäudes, Bauphysik 2 (1984), S.55 - 62

/7/ Energy in the built environment, Swedish Council for Building Research, G 16, 1988

PASSIVE SOLAR SCHOOL ON ANDROS ISLAND
THERMAL PERFORMANCE ANALYSIS

G. Zannis BSc MSc, Energy Consultant
177a Voulgari Street, Piraeus 185 34, Greece

1. INTRODUCTION

This study concerns Gavrion High School, an energy conscious solar building to be constructed on the island of Andros, in Cyclades - Greece.

1.1 Objectives

- To calculate the Heating and Cooling loads of the building and to assess the performance of the passive design features.
- To examine the comfort conditions of the interior spaces and thus to calculate the PMV indices.
- To estimate the daylighting levels inside the building and hence to deduce the daylight factors.
- To perform a sensitivity (parametric) analysis, in order to optimize the thermal design of the building.

2. COMPUTER MODELLING APPROACH

2.1 Simulation Models

The thermal model PASTEMP used in this study, is a computer simulation package based on the Admittance method. The building is divided into several zones connected to each other and to the environment. The model calculates for a typical day in every month the heating and cooling loads, the indoor air temperatures (hourly), and taking into consideration any passive design features it predicts the solar contribution to the energy demand.

The Bruel & Kjaer demonstration programme COMFORT was used to compute the PMV indices inside the building, by integration of the indoor parameters such as air temperature, relative humidity, ventilation rate, clothing factor and activity level.

The daylighting analysis for overcast winter days, was performed with the MICROLITE V.1.0 computer model. This programme estimates, for a given internal zone geometry, the daylight factors, from the sky illumination data available for the region.

3. ANNUAL SIMULATION RESULTS

Because of the mild climate in the Cyclades region and the relatively short occupancy period (9am to 4pm - September to June), the passive features supply almost all of the heating demand, while no cooling is required during the warmer months.

3.1 Monthly Energy Demands and Gains

The monthly simulation results are presented in the following table and graph.

MONTH	HEATING LOAD	DIRECT GAIN	SOLAR FRACTION
OCT.	280 KWh	232 KWh	82.8 %
NOV.	1,055	934	88.5
DEC.	2,554	2,298	89.9
JAN.	3,602	3,096	85.9
FEB.	3,122	2,743	87.9
MAR.	2,403	2,136	88.9
APR.	793	761	95.9
YEAR	13,809 KWh	12,200 KWh	88.3 %

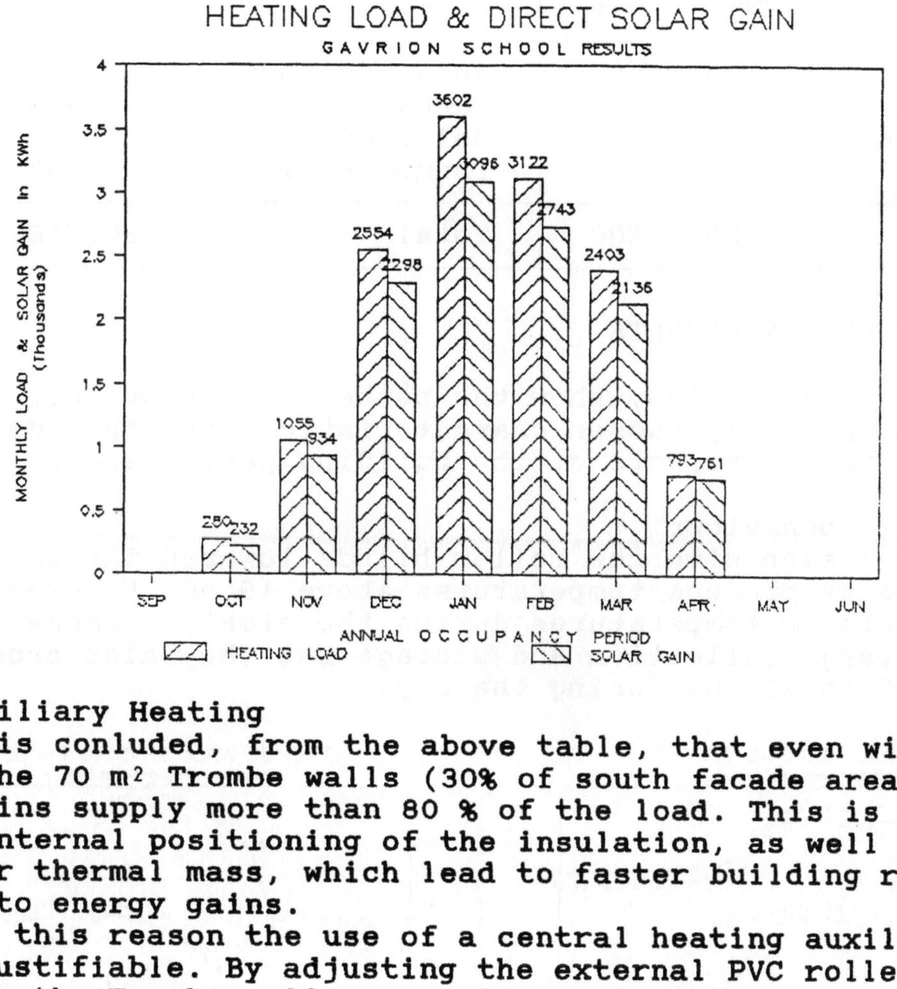

3.2 Auxiliary Heating

It is conluded, from the above table, that even without the use of the 70 m2 Trombe walls (30% of south facade area), the direct solar gains supply more than 80 % of the load. This is mainly due to the internal positioning of the insulation, as well as due to the low floor thermal mass, which lead to faster building response with respect to energy gains.

For this reason the use of a central heating auxiliary system is not justifiable. By adjusting the external PVC roller blinds and therefore the Trombe wall area, all of the demand can be met by the passive solar features. Under extreme weather conditions (i.e. on very cold and cloudy days), the individual wall mounted electric fan convector units can be used.

3.3 Air solar collectors

These will be placed on the roof of the auditorium in order to cope with its space heating load, which because of the low glazing area cannot be supplied by direct solar gains. This system, being an "active" one - since air is circulated through the collectors and ducts by electric fans, was analysed by a special version of the F-CHART model, developed for hybrid systems. The simulation was carried out with different solar collector areas and from the results obtained it was concluded that an optimum area of 15 m^2 (facing south with a 55 deg. tilt angle) will contribute 1060 KWh (91%) to the auditorium's space heating load, during the winter season.

3.4 Energy Balance

This table presents, the annual balance between all the sources contributing to the energy gains of the school building, during a typical heating season (October to April), and the corresponding losses.

L O S S E S	KWh	%	G A I N S	KWh	%
Fabric	17,120	56	Solar (direct)	12,200	40
Infiltration & Ventilation	13,590	44	Casual	16,900	54
			Solar collectors	1,060	4
			Trombe walls	550	2
Total	30,710	100	Total	30,710	100

4. DAILY SIMULATION RESULTS

The graphs that follow, show the pattern of indoor air temperature on specially chosen characteristic days, in order to illustrate particular aspects of the building performance.

4.1 Trombe Wall behaviour

In winter : even with the roller blinds lowered the casual and direct solar gains produce temperatures above 19 °C. Exposed Trombe walls lead to higher temperatures during the night, because of the higher solar energy collection and storage but they also produce overheating (27 to 29 °C) during the day.

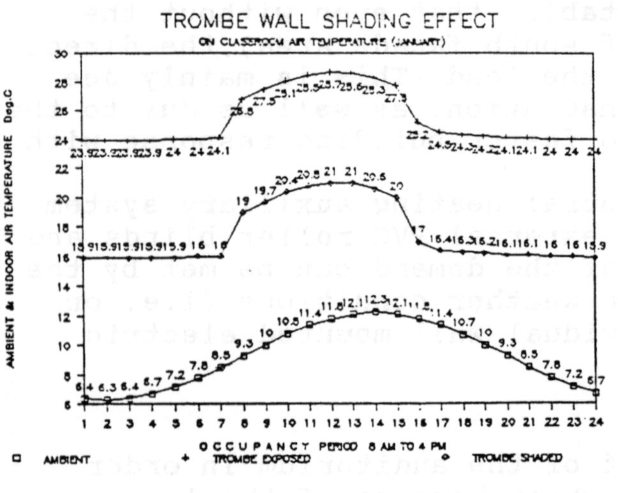

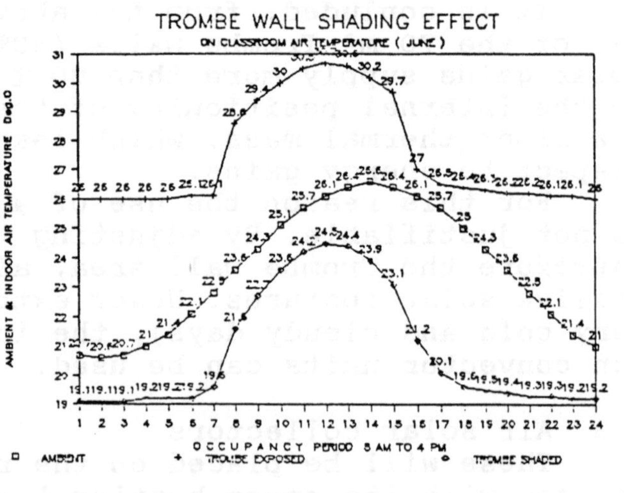

In summer: once again, an exposed Trombe wall leads to overheating (29 to 31 °C), mainly because although the ventilation rate has been increased (to 5 ACH from 2 ACH during the winter), there is considerable heat transfer by conduction through the solar wall, due to the lack of thermal insulation.

It should also be mentioned, that the above analysis was carried out assuming 100% utilizability of the direct solar gains. In practice, however, the use of venetian blinds (in order to alleviate glare problems,) would greatly affect these gains; therefore the existence of Trombe walls (with adjustable roller blinds) is considered essential for both heating and cooling the school building.

4.2 Performance under extreme conditions

To fully appreciate the potential of the passive features under various weather conditions, simulations were also carried out for a cold-cloudy and a cold-sunny day.

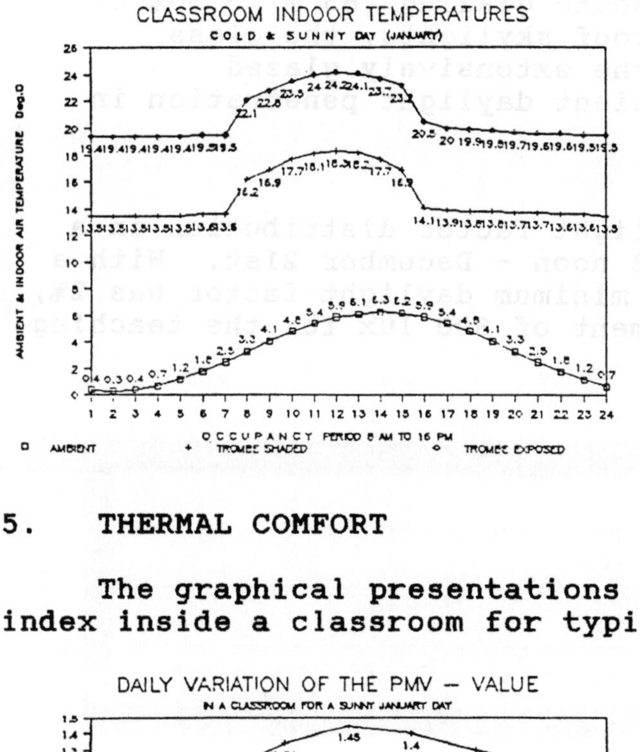

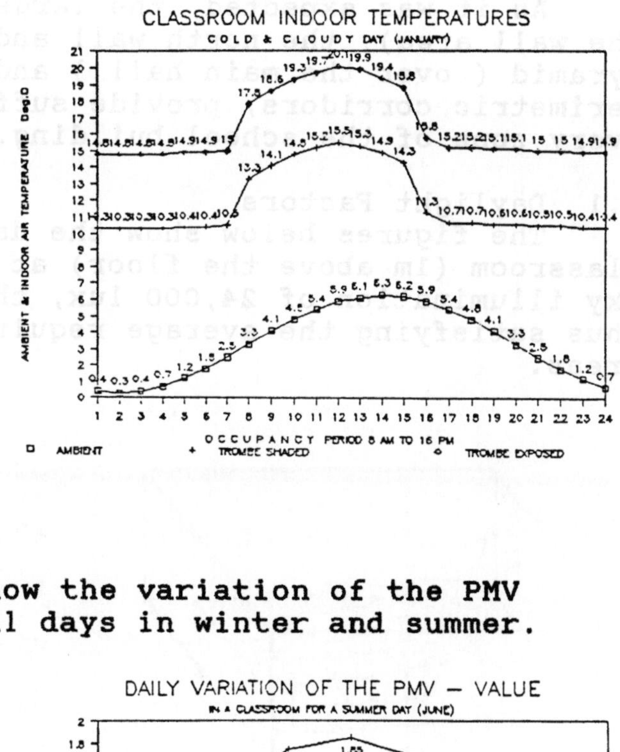

5. THERMAL COMFORT

The graphical presentations show the variation of the PMV index inside a classroom for typical days in winter and summer.

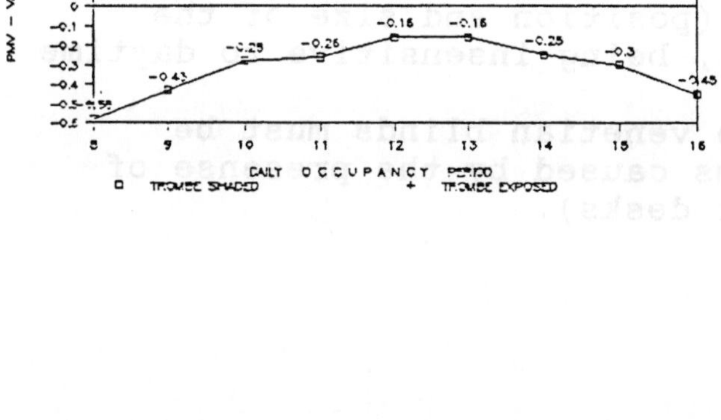

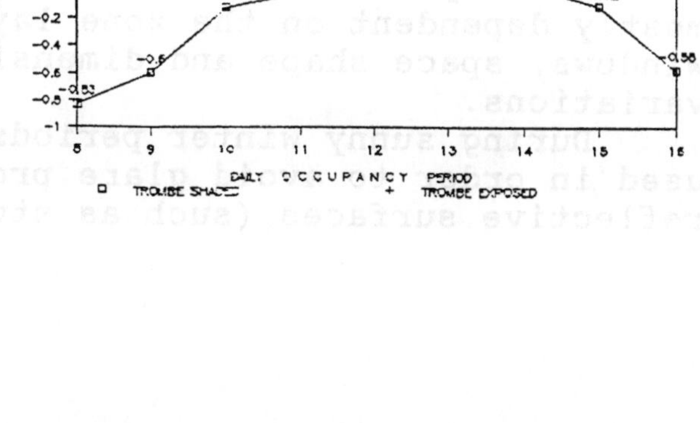

In both cases, it was assumed, that the mean radiant (wall) temperature was equal to the air temperature, since the building was well insulated, with the insulation installed internally.

The comfort analysis showed that the PMV (Predicted Mean Vote) and PPD (Percentage of People Dissatisfied) values, may vary considerably not only from month-to-month but also during the day, because of their dependence on many indoor variables such as air and mean radiant temperatures, relative humidity, ventilation rate, clothing factor and activity levels.

The results were satisfactory, both for winter and summer (PMV: -0.5 to +0.5 and PPD: 5% to 10%), depending on the use of the Trombe walls (solar gains and induced cross-ventilation).

An attempt to improve the comfort level by increasing the ventilation rate is not recommended, because of the high air velocities resulting in the classrooms.

6. DAYLIGHTING ANALYSIS

As it was expected, the large south glazing (85 m^2 - 30% of the wall area), the north wall and roof skylights, the glass pyramid (over the main hall) and the extensively glazed perimetric corridors, provide sufficient daylight penetration in every zone of the school building.

6.1 Daylight Factors

The figures below show the daylight factor distribution in a classroom (1m above the floor) at 12 noon - December 21st. With a sky illumination of 24,000 lux, the minimum daylight factor was 2%, thus satisfying the average requirement of 300 lux for the teaching areas.

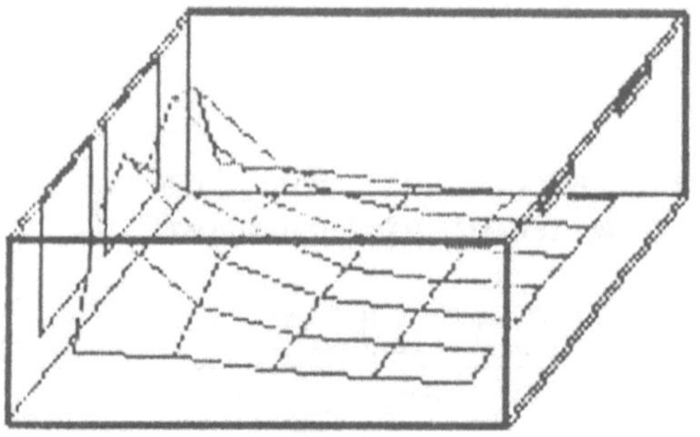

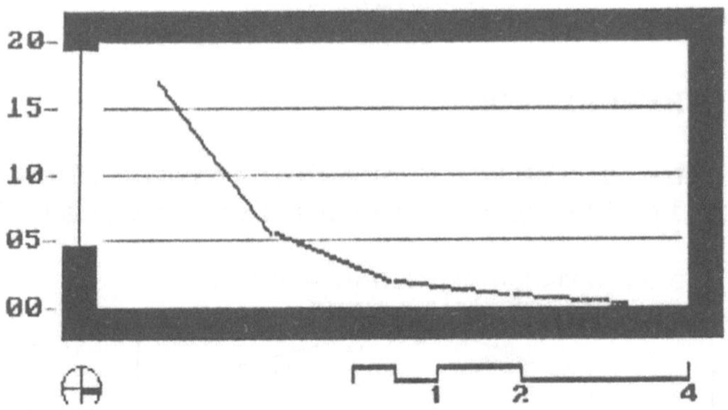

From the results it was also confirmed that during cloudy days when the sky radiation is mainly diffuse, the daylight factors are mostly dependent on the zone layout (position and size of the windows, space shape and dimensions), being insensitive to daytime variations.

During sunny winter periods the venetian blinds must be used in order to avoid glare problems caused by the presense of reflective surfaces (such as student desks).

7. PARAMETRIC ANALYSIS

Following the initial design simulations, a number of sensitivity analyses were carried out in order to explore opportunities for further improvements in thermal performance.

7.1 Glazing Area

A)Indoor air temperatures : With larger glazing areas (50%) the heating load was increased in winter, whereas at the same time higher indoor temperatures were experienced during sunny periods.

B)Daylight factors : A reduction in glazing area (20%), had a greater effect on the daylight factors than an increase and it is not recommended.

7.2 Ventilation Rate

Since the building was well insulated any changes in the ventilation rate affected the internal comfort conditions considerably. A decrease, had a greater effect on the air temperature due to the casual and direct solar gains.

7.3 Insulation Level

Increasing the insulation thickness had almost no effect on the indoor conditions, and it was concluded that the initial insulation level was near optimum (Law of Diminishing Returns).

7.4 Thermal Mass

The direct solar gain incident in a zone is mainly absorbed by the floor thermal mass. With a reduction in the floor thickness the resulting change was negligible, since the initial floor mass was relatively low.

8. RECOMMENDATIONS

From the analyses performed for the Gavrion passive solar school, it was concluded that the initial design options were quite correct for such a building, but a few improvements could be made as given below:

- Conversion of the Trombe walls to Rapid Response Solar walls, by lowering the thermal mass (thickness) and by introducing thermal insulation (80mm) internally, thus facilitating ambient energy collection and delivery within the occupational periods in winter (lower time-lag), together with faster cooling by cross-ventilation in summer.
- Increase of the floor Thermal Mass by up to 50 %, so that sudden fluctuations in the internal comfort conditions (caused by undesirable solar gains), will be avoided during the warmer months.
- Small reduction in the insulation level (20 % - as this would not produce a noticeable effect), but with parallel use of external shutters for night insulation of the large window areas.

In the future, monitoring and verification of the Gavrion high school thermal performance, may lead to a series of passive solar schools, that will be constructed on the Islands of Cyclades, thus contributing to the overall energy conservation in Greece.

METHILHILL SCHOOL, SCOTLAND

Nick Baker, Cambridge Architectural Research and
The Martin Centre for Architectural and Urban Studies
6 Chaucer Rd, Cambridge, UK

1. INTRODUCTION

Methilhill is in Fifeshire in mid-Scotland experiencing a fairly cold climate with 3567 degree days compared with 3178 for SE England at a latitude roughly the same as Copenhagen. Experience with earlier courtyard design schools had shown that the climate - rain, low temperatures and low angle winter sun, meant that the open courtyard was little used except for a relatively brief summer period. This suggested a covered courtyard or atrium design. The association of atrium designs with good energy performance prompted the design team to submit the scheme to the B2000 programme, requiring technical design support on all aspects of passive design, since the team had no previous experience in this field.

It was decided to provide under the B2000 programme and partly funded by the Energy Design Advice Service (EDAS) scheme, extensive simulation studies by ABACUS Simulations Ltd. Other input included an allowance of up to three days input from Cambridge Architectural Research (CAR) which is reported here.

2. CONSULTANTS COMMENTS ON EARLY PROPOSALS

An early proposal is shown in fig 1. Comments upon this proposal lead to a design meeting in December 1988, with the architect and engineer, and a consultant from ABACUS. The following is a summary of the advice given by CAR.

The north sloping glazing (1) of the central glazed roof would be a net looser - better thermal performance would result from an opaque insulated north slope. Transparent insulation could be considered for the north slope over the courtyards where there could be a daylighting function to surrounding rooms.

The function of the glazed roof over the plant rooms (2) is only to improve the COP of the heatpumps. The validity of this strategy should be checked by simulation. A similar scheme has been tried at Walton Gunfleet in Essex. The intention is to use the heatpump to pre-heat ventilation air, storing surplus heat in the courtyard floor(3). This does not seem to be a good choice because of the high losses to the courtyard glazing at night.

The existing proposal would suffer from summer overheating in the glazed courtyards. It is recommended that the ventilation openings be increased to between 5% and 10% of the glazed area, located partially in the ridge, and moveable shading blinds be incorporated on the south slope.

The general purpose hall(4) should be daylit and naturally ventilated. Openable rooflights with a ceiling in the plane of the roof pitches, as distinct from the horizontal ceiling in the proposal, would improve the quality of the space and permit daylighting and ventilation. Blackout blinds would be needed.

The depth of the classrooms(5) would result in insufficient daylighting at the back of the room. Daylighting from the courtyard is prevented by the low ceiling(6) beneath the service duct. This part of the section should be redesigned and the daylighting tested by model studies.

3. SIMULATION STUDIES

At the time of the meeting the simulation studies had not, in our opinion, been very useful in informing design decisions. The only parametric run was the comparison between two courtyard glazing profiles which gave a result which would have been predictable by much simpler analysis. "Solar performance" was a comparison between the existing proposal with and without "solar processes". This is misleading; we should have seen a comparison with a "non-solar" design with solar processes. At this stage, no annual simulations had been carried out - choosing a cold sunny sample day can also be misleading, since this will exaggerate solar performance. There seemed to be errors in some of the input - how can 90 children generate an internal gain of 23kW?

The high services content - mechanical tempered air plus perimeter radiators and underfloor heating, and high dependence on artificial lighting, meant that this proposal was some way away from the general aims of the B2000 programme. The designers were urged to make the reduction of services a design goal.

No further consultancy took place. However CAR were very pleased to see that the design developed in the general direction suggested.

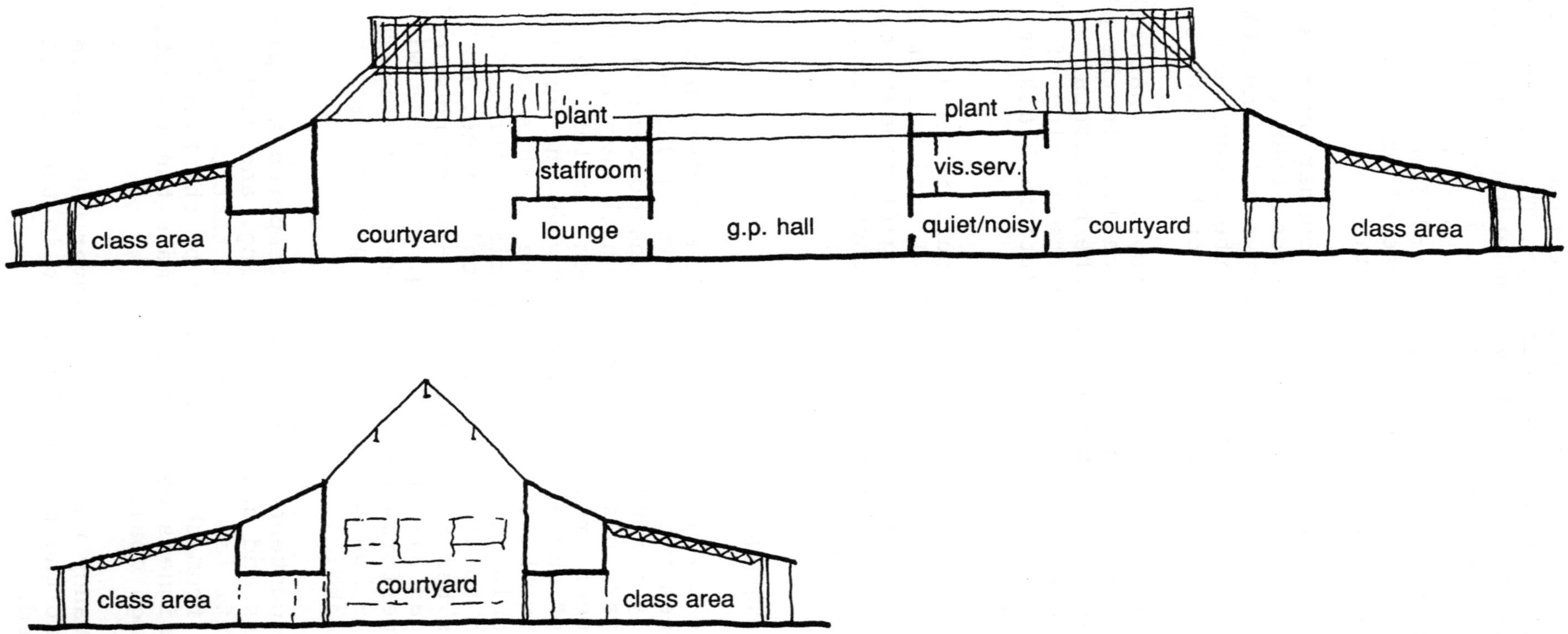

Figure 1 Simplified sections of Methilhil School
Early proposal

BUILDING 2000

Francisco de Oliveira Moita, Dipl. ing. Architekt (Ger)
Rue Luis de Freitas Branco, 42-13A, 1600 Lisboa-Portugal

1. INTRODUCTION

The present report makes a description of the author's work as thermal consultant of the Valongo do Vouga school plan. The study of the thermal componemte fell on the classrooms block, being analysed the thermal performance of three kinds of rooms.

2. PLAN DESCRIPTION

The village of Valongo do Vouga belongs to the municipality of Agueda, situated by the center of Portugal about 50 Km far from both the sea and Aveiro, district capital.
The Valongo do Vouga school is part of common programme carried through by both the Ministry of Public Instruction and the Municipality of Agueda under the technical support of Coimbra University.
This school meant for secondary education of childeren aged between ten and fifteen is equiped with 18 classrooms and all the requisite support areas such as laboratories, training rooms, amphitheatres, library, kitchen and refectory as well as all administrative rooms.
The school is located on a wel sunlighted land slightly inclined on south and grows on two floors with classrooms facing south.

3. CLIMATE DATA

Valongo do Vouga is affected by the average conditions of the soft weather typical of the most part of portugese land. The winds are mild and predominantely NW in summer and SW in winter (sea winds). The greatest temperature swings between extreme points are usually in summer with a variation of about 10°C, thereby enhancing cooling cross ventilation of the building. There are on the average 1540 degree days per annum and ca. 2520 sunshine hours (56% of the daylighting hours).

4. BUILDING SPECIFICATIONS

Being the design objective to create both in summer and in winter the most comfortable thermal conditions in the classrooms using auxiliary heating as little as possible, the design team tried to build a duly isolated shell to prevent heat losses and heavy enough to provide not only a high solar heat storage during the winter but also the appropriate radial temperature in summer improving cooling through night cross ventilation.
So, the whole opaque shell contains a thermal isolated tier (4 cm for external walls, 3 cm for groundfloor and 6 cm for roofing plate) to reduce transmission heat loss as much as possible.
The internal side of classroom walls is compact brick-build, thus formating the thermal mass.
For economical reasons and after concluding form thermal analyseis that double glazing was not absolutely requested, single transparent glazing was adopted as well as wood-framed windows.
This being so, the shell shows the following U values:

1) - External wall U value = 0,53 W/m2C.
2) - Beams and pillars U value = 0,65 W/m2C.
3) - Internal wall room/room U value = 1,75 W/m2C.
4) - Internal wall room/corridor U value = 1,12 W/m2C.
5) - Downstairs room floor U value = 0,58 W/m2C.
6) - Upstairs room ceiling U value = 0,52 W/m2C.

5. PASSIVE SOLAR COMPONENTS

The main passive solar components of this design, besides the above referred correct orientation of classrooms facing south and the adequate thermal isolation of the shell, could be briefly synthetized as follows:

* Solar Gains: exclusively obtained by direct gains through classroom-windows facing south and having a collecting glazed area of ca. 9,6 m2, wich corresponds to about 40% of the facade area and to ca. 13% of the room area. On the 1st. floor some classrooms provide direct gains through glazed rooflights with a collecting area of 3,45 m2 each, thus amouting to ca. 5% of the room floor area.

* Shading: glazed windows facing south are provided with overhang high as 2/3 which geometry ensures inside insulation during cool season and diffuse radiation all over warm season.

* Thermal mass: the design team was recommended to care about the use of heavy enough materials when of the classrooms interior construction (compact brick or concrete blocks) to avoid too high internal temperature swing during both winter and summer. In effect, partition walls are of compact brick thick of 22 cm (11 cm of brick as thermal mass for each classroom), the external walls are covered on the interior side with compact brick and both the roof and the floor plated are of concrete.

* Natural ventilation: windows facing both north and south as well as classrooms doors are supplied with upper tilting dumping wings to ensure cross ventilation when necessary without discomfortable draught effects. In winter the needful air renovations occur during the classes pauses but this does not reduce internal mass effect provide comfort temperature levels reducing internal overheating.

6. THERMAL PERFORMANCE

All the thermal calculations of this study were made by a computer programm using a thermal analysis based on a Method especially developed for portuguese conditions by Franciso Moita, published by the Portuguese General Directorate for Energy.

This analysis unfolds data considering the following figures:

* Exterior walling area	7,92 m2
* Window area (frame included)	11,78 m2
* Glazing area	9,6 m2
* Rooflight area (frame included)	5,95 m2
* Rooflight glazing area	4,2 m2
* Roofing area	78,48 m2
* Floor area	78,48 m2
* Structure thermal bridges	4,08 m2
* Buffer value - living room/corridor	0,42 m2
* Buffer value - floor/ground	0,70 m2
* Average air changes/hour	2 a/h
* Internal gains due to 30 childeren and 1 adult	33 kWh/day
* Comfort temperature	18°C

Thermal analysis took three kinds of classroom pointing out for each case besides common aspects, the following particularities:

1) Classroom in upper floor provided with rooflight:
   * Solar gains also through rooflight
   * Heat losses also through rooflight
   * Heat losses also through roof

Results:

- Space heating demand/year	ca. 5775 kWh
- Solar heating fraction/year	ca. 64 %
- Auxiliary load/year	ca. 2080 kWh

2) Classroom in upper floor without rooflight:
   * Heat losses through roof

Results:

- Space heating demand/year	ca. 4210 kWh
- Solar heating fraction/year	ca. 60 %
- Auxiliary load/year	ca. 1680 kWh

3) Classroom in groundfloor contiquos to the hall corridor:
   * Heat losses also to the corridor (buffer zone)
   * Heat losses through the floor

Results:

- Space heating demand/year	ca.	4280 kWh
- Solar heating fraction/year	ca.	60 %
- Auxiliary load/year	ca.	1710 kWh

It could be verified a posteriori that thermal performance of these three rooms didn't present but little differences, thus allowing to consider a rather satisfatory uniform performance for the whole classrooms.
It have also been calculated for each one of these three rooms the total thermal storage capacity and the inside temperature swing in a winter sunny day. The values obtained can be considered within quite normal parameter. So, following figures were reached:

1a) Classroom in upper floor with opened rooflight:

- Total heating storage load	ca.	7650 Wh/°C
- Inside temperature variation in a winter sunny day	ca.	6 °C

1b) Classroom in upper floor with closed rooflight:

- Total heating storage load	ca.	7650 Wh/°C
- Inside temperature variation in a winter sunny day	ca.	4,5 °C

2) Classroom in upper floor without rooflight:

- Total heating storage load	ca.	7650 Wh/°C
- Inside temperature variation in a winter sunny day	ca.	4,5 °C

3) Classroom in groundfloor contiquos to the hall corridor:

- Total heating storage load	ca.	7650 Wh/°C
- Inside temperature variation in a winter sunny day	ca.	4,5 °C

As it can be considered in case 1a) rooflight will be closed whenever direct radiation is excessive and it will be easy to reach 4,5 °C as highest variation - cases 1b), 2) and 3) - being this value quite normal.
Considering that determination of auxiliary annual heating needed to maintain a comfort level (on the basis of 18°C) was calculated as if the building is constantly occupied, this project must be counted as an excellente passive solar design.
As a matter of fact, on pretense of curiosity thermal analysis calculations were made on the basis of 15°C (usually considered for non-constantly occupied buildings) for the thermal less favoured classroom (classroom in groundfloor) and results were indeed spectacular!

Results:

- Space heating demand/year	ca.	1330 kWh
- Solar heating fraction/year	ca.	90 %
- Auxiliary load/year	ca.	130 kWh

DEPARTMENT OF MEDICINE, UNIVERSITY OF IOANNINA

Nick Baker, Cambridge Architectural Research, Cambridge, UK
Simos Yannas, Architectural Association, London, UK

1. INTRODUCTION

This project became part of the B2000 after winning a national competition. The original design had already undergone some revision following consultancy by others and is shown in the axonometric in figure 1. It can be seen that a major feature is the large atrium. The consultancy described here was primarily aimed at optimising the thermal performance of the atrium, both in energy saving to the parent building, and for comfort conditions in the atrium itself.

A characteristic of the climate is the relatively cold winters, due to elevation and the northerly location, but with good solar radiation, and the hot summers. Thus both summer and winter conditions were of special concern.

Consultancy took place in two phases, as represented in the two reports dated 9 May 89 and 25 Oct 89.

2. ENERGY PERFORMANCE IN WINTER

The first phase was mainly concerned with the overall morphology of the building. Use of the computer model ATRIUM indicated that by re-arranging the surrounding buildings to increase the area of contact with the atrium and reduce the external wall area, considerable improvements could be made in winter energy performance and winter comfort conditions. This can be seen as increasing the "protectivity ratio" - that is the ratio of the area of separating wall (between atrium and parent building) and the area of external wall. Atrium performance is very sensitive to this.

Another area of concern in this first phase was the design of the atrium roof. Two basic alternatives were considered - a cylindrical glazed roof, and a sawtooth form with vertical south-facing glazing and opaque sloping sections. The latter option represented an improved "solarity" ratio, that is the ratio of the area of south vertical glazing to non-south glazing area. The sawtooth option was shown to have better winter and summer performance. In summer it is partially "self-shading" due to the solar geometry. It was recommended that this was supplemented by strips of shading to prevent all direct radiation from reaching floor level in summer.The results of this first phase are shown in figs 2,3 and 4. Note "Revised 1 and 2" refer to the cylindrical roof, "Shed 1,2 and 3" to the sawtooth option, and "Recommendations" to the suggested modified atrium, based upon these findings. It must be pointed out that the relative performance of the options as a percentage of total thermal energy consumption is dependent upon assumed internal gains.

This modification, with increased protectivity ratio, also carried benefits by reducing built envelope and bringing circulation into the atrium, and improving the daylighting role of the atrium. The latter would have energy benefits which have not been assessed. The 'Recommended' form is shown in figs 5 and 6. Note that it also included the glazing over of the gaps between the laboratory blocks and the addition of rooflights on the top floor laboratories.

These recommendations were adopted by the design team.

3. COMFORT CONDITIONS IN WINTER

The second phase of consultancy was more concerned with the comfort conditions in the atrium in both summer and winter. The previous analysis had predicted mean mid-daytime (9am - 3pm) temperatures for January of 15.5°C for the recommended option. We estimated that these temperatures would drop to about 11.5°C for sunless days, where most of the heating effect is due to heat loss from the heated building. In this case the 'Recommended' option performs much better than the original proposals due to the more favorable "protectivity ratio".

These temperatures are a spatial average. An important question is how does stratification effect these.

Our view is that on sunless winter days there will be virtually no stratification because there are no high temperature heat sources. Thus the spatial average temperature is a good indicator. On sunny winter days, due to the shape and orientation of the atrium, the sunpatches tend to fall on the floor. This will minimise stratification and also have the benefit of increasing the mean radiant temperature by at least 3°C than predicted by the mean air temperature. Some heat will be stored in the massive floor which will result in better evening temperatures.

Experience of monitoring an unheated atrium in the UK supports the view that nighttime temperatures are unlikely to drop below about 9°C.

4. COMFORT CONDITIONS IN SUMMER

The earlier analysis showed peak spatial average temperatures of 3.2°C above ambient. This was predicted by a simple buoyancy model assuming moveable shading and top and bottom openable areas of 180 m^2 each. This increment drops to 2.3°C if the walkways are shaded.

In summer stratification will exist and will generally be in our favour. The temperature at the top of the atrium will be between 6 and 10°C higher than the spatial average, whilst the temperature at the shaded ground level will be close to ambient on average, and probably below ambient early in the day. The horizontal shading will reinforce the stratification and protect the walkways.

Fluid modelling carried out for the University of Seville Building, indicate that it may be possible to maintain cool air at the bottom of an atrium for some time, by allowing ventilation air to enter the atrium at first floor level as in fig 7.

5. SUMMARY

This consultancy was particularly interesting in that the atrium design had to take account of both summer and winter conditions. It was encouraging that analysis lead to a significant improvement in thermal energy performance and comfort performance of the atrium. It was also very good to see that the design team could react in a positive way to the technical input, and allow the design to respond without destroying the original concept.

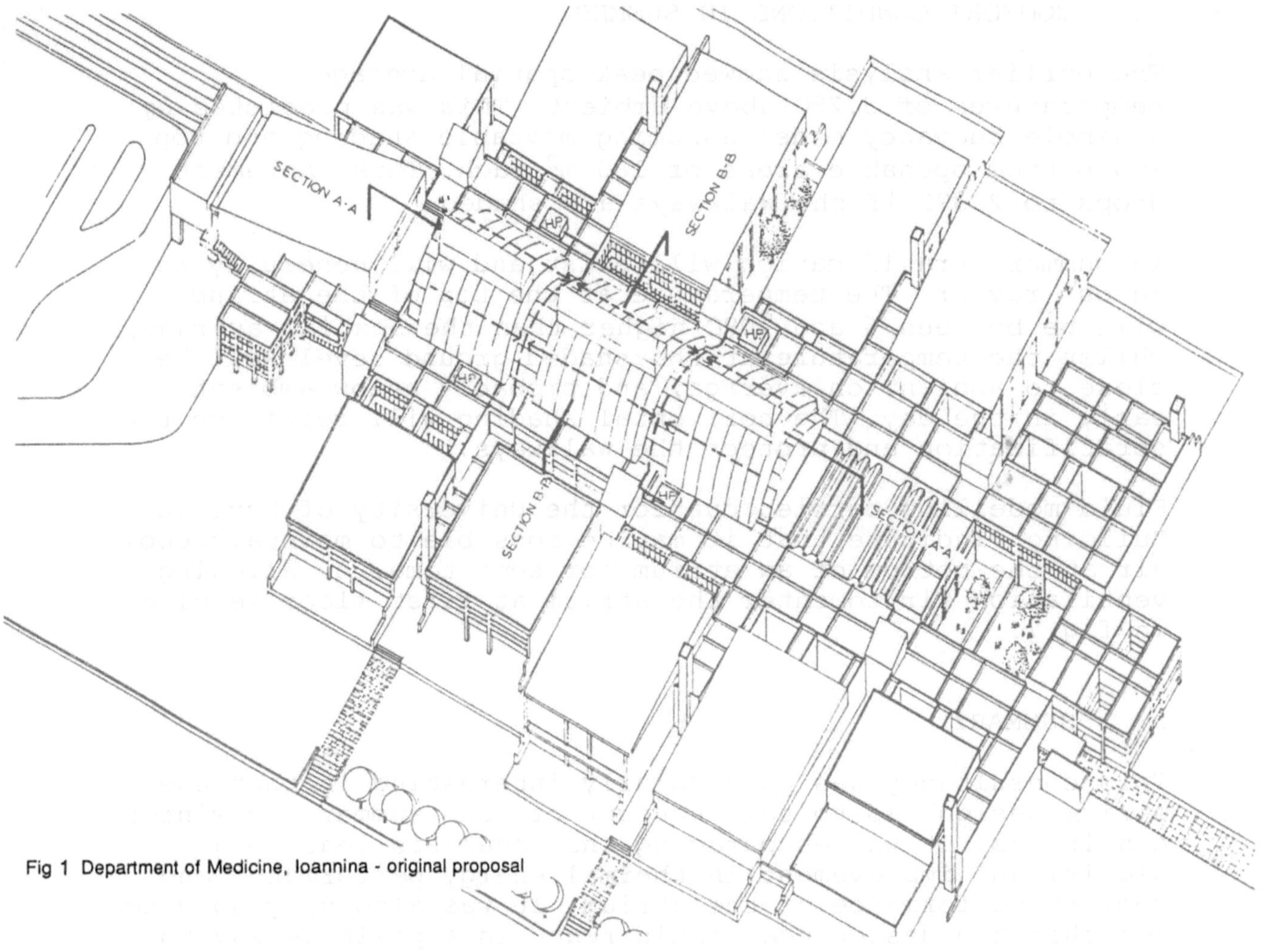

Fig 1 Department of Medicine, Ioannina - original proposal

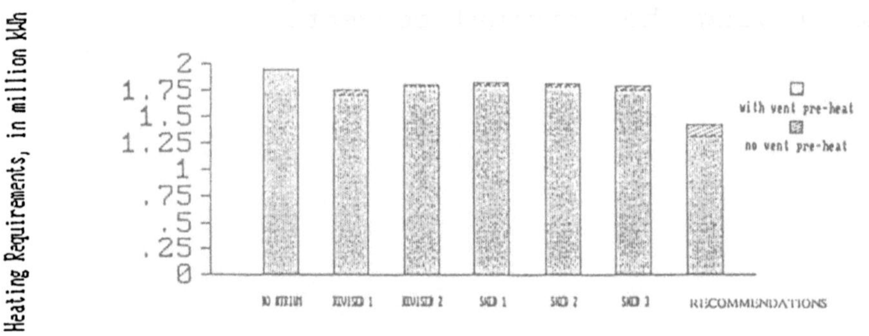

Fig 2 Effect of atrium on heating load

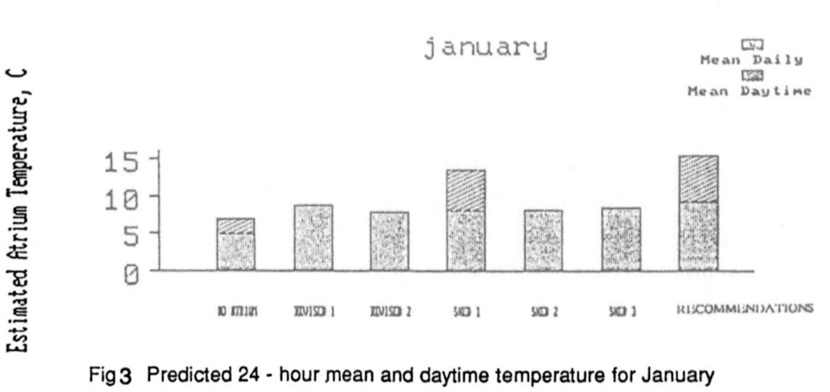

Fig 3 Predicted 24 - hour mean and daytime temperature for January

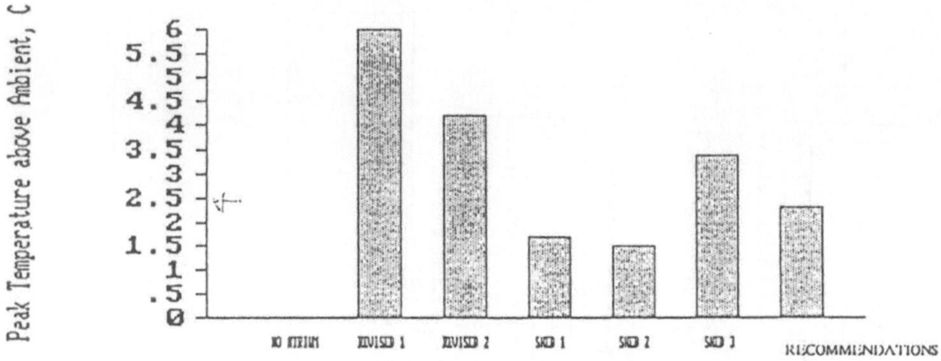

Fig 4 Predicted atrium temperature increment above ambient in summer (spatial mean, with lower and upper openable areas 100 - 180 m^2)

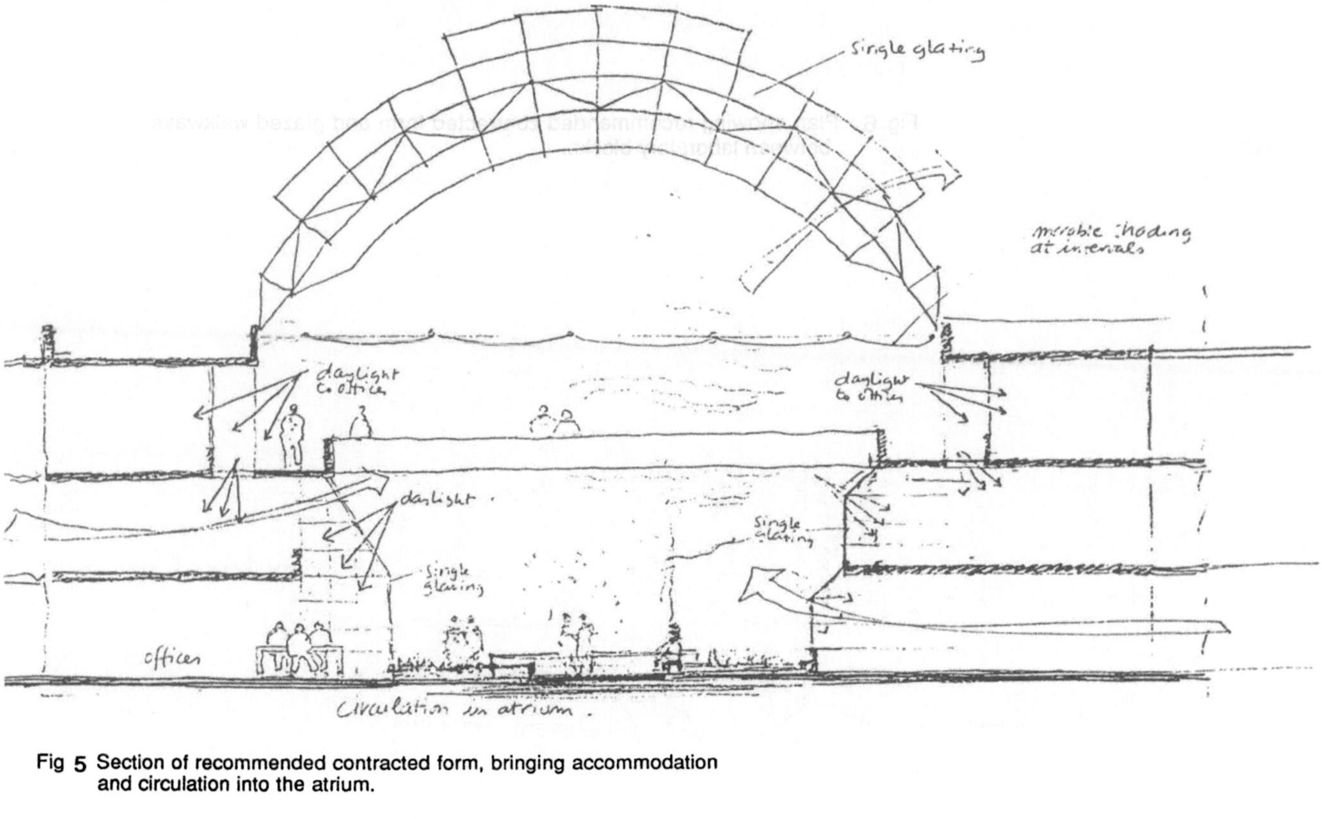

Fig 5 Section of recommended contracted form, bringing accommodation and circulation into the atrium.

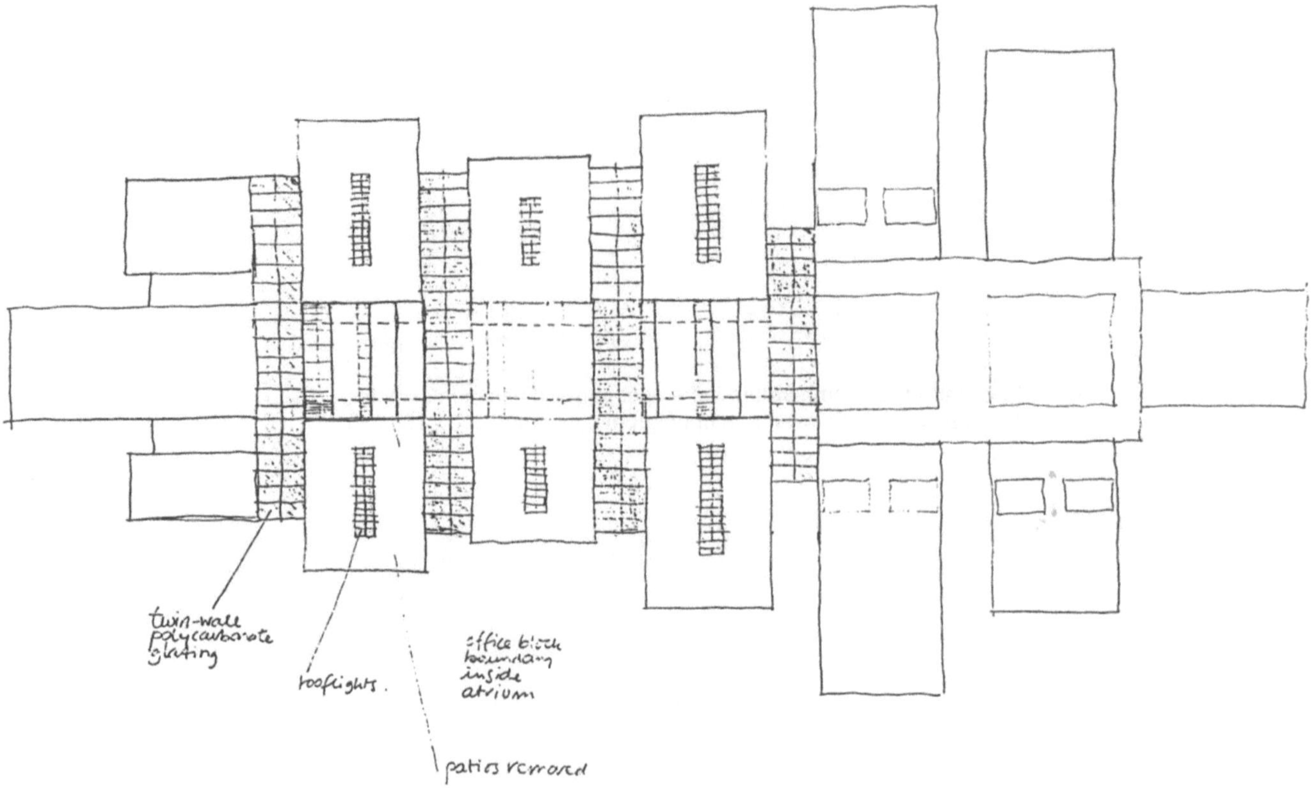

Fig 6. Plan showing recommended contracted form and glazed walkways between laboratory blocks.

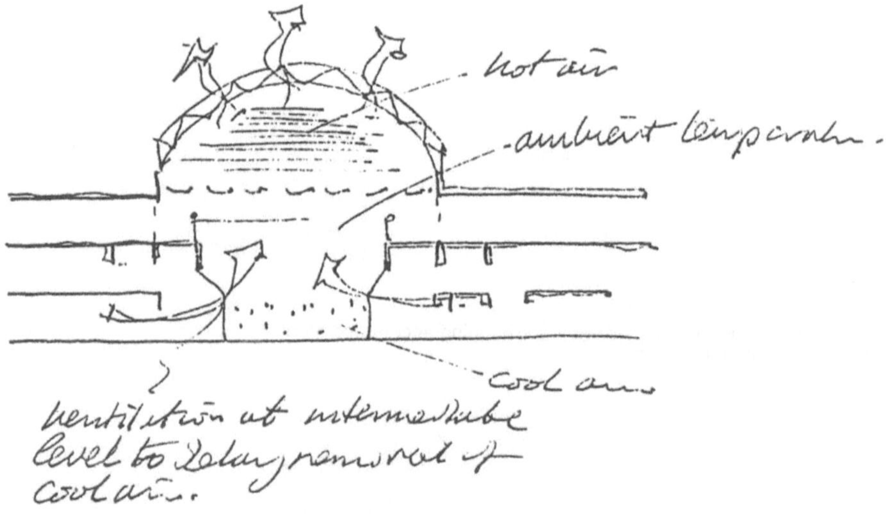

Fig 7 Useful temperature stratification for summer conditions.

EDUCATIONAL CENTRE WITH SOLAR ROOF MADRID - SPAIN

W.F.M. van der Vliet, EGM engineering bv
P.O. Box 1042, 3300 BA Dordrecht, The Netherlands

1. INTRODUCTION

One of the energy-saving systems of the Building 2000 project 'Educational centre at Madrid - Spain', is the solar roof.
The educational centre includes a school and a gymnasium.
The roof of the two buildings has been replaced by two solar roofs. The solar roof provides the buildings with heat, cold, light, ventilation, and gives protection from the cool weather and the strong sun in summer.
The water accumulator of the gymnasium is an enclosed reservoir at the ouside of the building. In July the temperature of the water will be about 10°C.

The two problems to be solved are:

- impact sound insulation of the roof in case of rain
- thermal insulation of the water reservoir, particularly to prevent heat losses to the ground

2. IMPACT SOUND

Impact sound is the term used for sound which is produced when energy is transmitted intermittently to a stucture, e.g. by foot-steps or by rain. Before it can be heard the energy must be radiated into the air by a wall or a ceiling. Once the energy is in the construction, large areas can be set into vibration giving rise to a high degree of radiated noise. At a school this phenomenon is undesirable.
Since impact noises are readily transmitted through building structures every effort should be made to prevent energy from the source from reaching the main structure.

3. IMPACT SOUND INSULATION

In case of rain on the roof of the school the two methods to reduce impact sound are:

damping materials
By the use of damping materials a part of the impact energy will be converted into heat. The damping factor indicates this ratio.
A 1 mm. steel plate has a damping factor of about 0.001 (this reduces about 11dB/sec.).

The same steel plate with a 1.5 mm. damping material has a damping factor of about 0.14 (this reduces up to 800 dB/sec.).
Damping materials have to be glued to a steel plate. The thickness has to be al least 1.5 times the thickness of the steel plate.
Dependent on the type and manufacture the best result will be reached between 15°C and 30°C.
In Holland the cost of damping materials (thickness 1.5 mm., mass 2.5 kg/m²) is about *f* 38,-/m² ex V.A.T. This is about 2,240 Ptas/m².

suspended ceilings
Suspended ceilings improve the insulation against both airborne and impact noise by an amount depending on the weight of the suspended ceiling and the degree of rigidity with which the ceiling is attached to the roof. Such ceilings reduce the noise level only in the room where they are used, by decreasing the radiation from the ceiling treated; no reduction is obtained in other rooms in the school having no suspended ceiling. Furthermore, they do not reduce the sound radiation from the side walls which results from flanking transmission.
The construction of a suspended ceiling is:

* dense ceiling of 2 x 12.5 mm. plasterboard. The seams between the two layers of plasterboard has to spring. The distance between plasterboard and steel roof has to be about 200 mm.
* 50 mm. absorbent blanket, mineral or glass wool with a mass of about 45 kg/m³

The number of points of suspension from the steel roof or steel stucture above should be kept to a minimum; resilient hangers have to be used.
Gaps between ceiling and surrounding structure must be sealed to avoid noise penetration through direct air paths. See also figure 1.

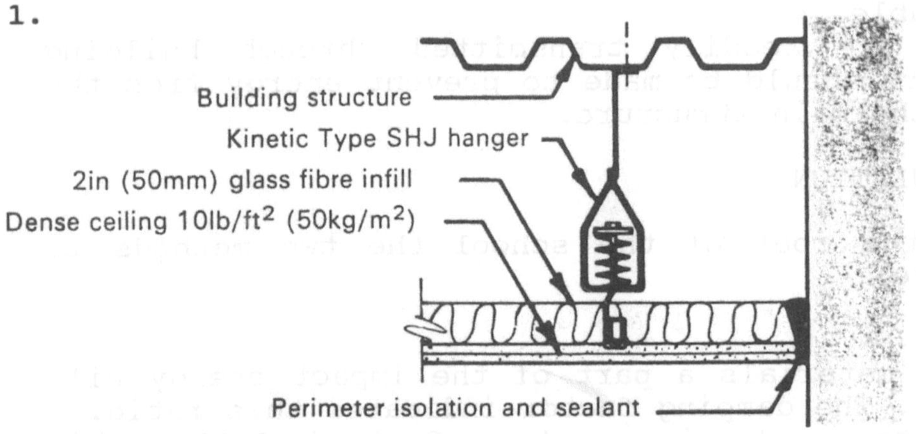

Figure 1 - Suspended ceiling

The sound reduction of this construction will be 20 - 25 dB(A). A greater reduction may be possible with a heavy suspended ceiling but this is in contrast with a steel roof.
Measurements of rainborne sound production on a steel roof are not known because in Holland this type of construction is not a applied. Every steel roof has a roofing and in most cases an insulation layer on top of the steel roof. We expect that the sound level is up to 75 dB(A).

The effect of rainborne impact sound can be heard in a car. On account of this effect we think that a reduction of 20 - 25 dB(A) is an acceptable solution.
A greater reduction is possible but it is a very expensive one.

4. THERMAL INSULATION OF THE WATER RESERVOIR

For refrigiration purposes during the night water at 10°C will be obtained.
Warming up of the water has to be minimized as much as possible.
Heat flow is possible through:

- the outside air temperature of 15 - 35°C
- the inside air temperature of 20°C
- the ground temperature of about 15°C

See also figure 2.
The water reservoir will be protected against direct solar radiation by the mass of the school.

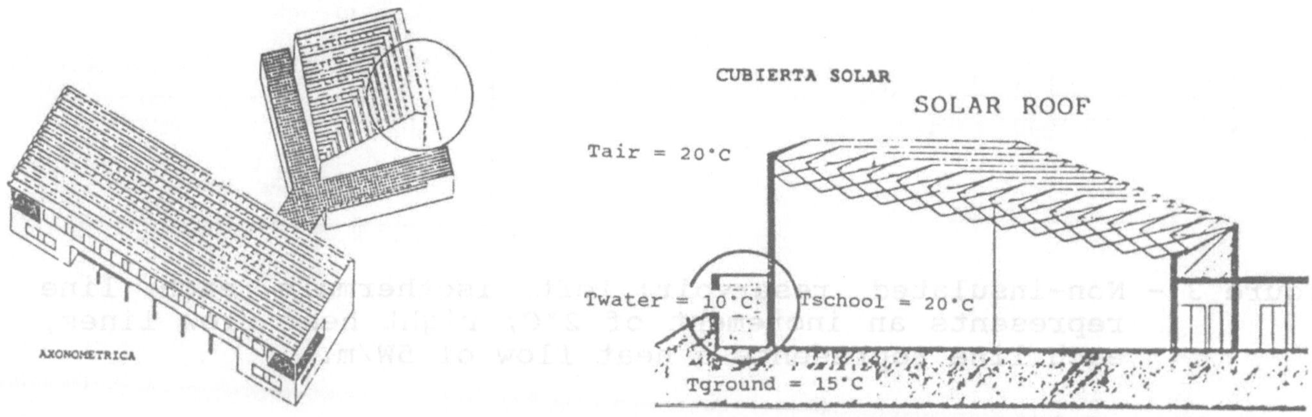

Figure 2 - Position of water reservoir

The distribution of the heat transfer over the elements has been calculated with KOBRU86, a computerprogram to calculate two-dimensional steady state heat transfer in objects using the energy balance technique.

In figure 3a the temperatures can be seen; in figure 3b the heat flow can be seen. Alse the distribution of the heat flow is given. The total flow is about 310 W/m.
The conclusion is that the main losses are to the air and to the school. The flow to the ground is rather small, even with a ground temperature of 20°C in stead of 15°C.
To reduce the flow the reservoir has to be insulated on the outside.
In figure 4 of the annex temperatures, heat flow and distribution of heat flows are given for an insulated reservoir. The insulation used in the calculation is 50 mm. extruded polystyrene, for instance Styrofoam SM-TG or Roofmate SL of the Dow Chemical Company. This material has to be protected against the influence of ultra-violet rays by a roofing felt. Another applicable insulating material is 50 mm. cellular glass, for instance Foamglass of Pittsburgh Corning.

Although in the calculation the air temperature is equal to the temperature in the school, the flow to the school is 2.5 times the flow to the air on the left side. The flow to the air via the top of the reservoir is about 1.5 times the flow to the air on the left side.
This is caused by so-called thermal bridges. A thermal bridge is a part of a construction with a relative great heat flow.
In figure 4 the thermal brigdes can be recognised by the encircled heat flow lines.

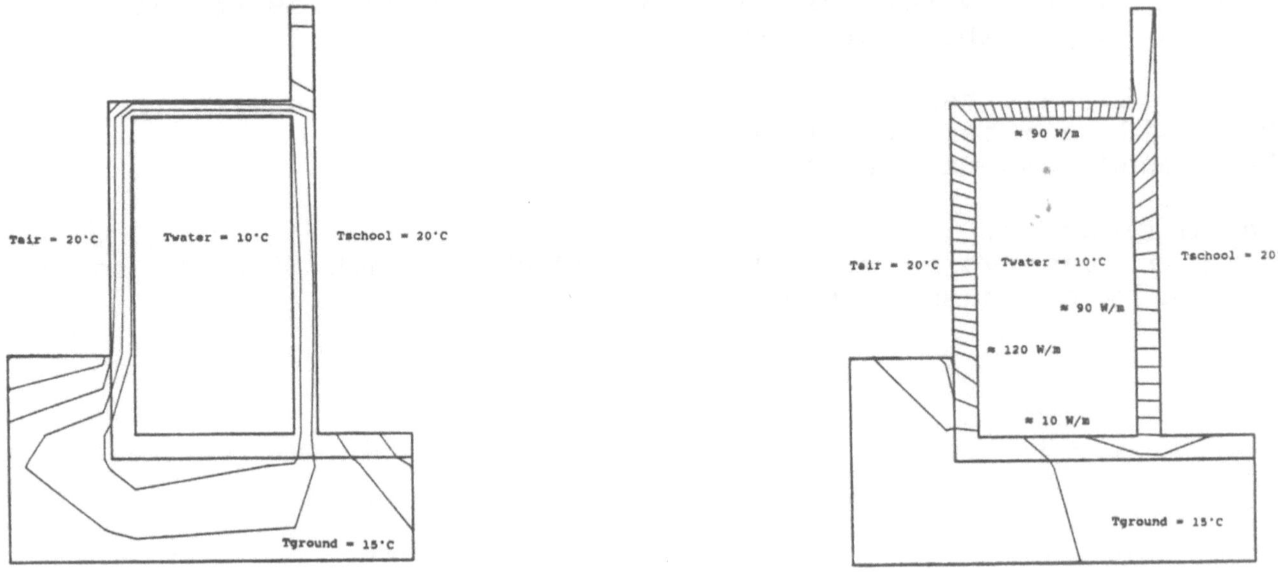

Figure 3 - Non-insulated reservoir; left isothermals, each line represents an increment of 2°C; right heat flow lines, each line represents a heat flow of 5W/m.

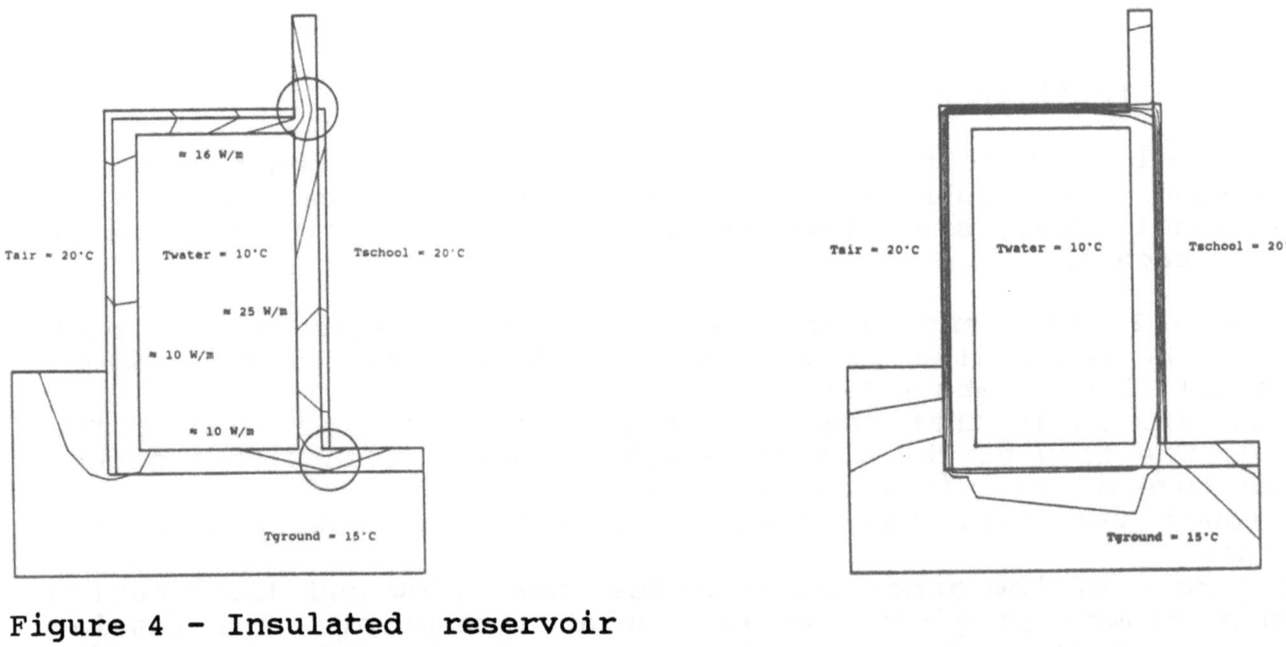

Figure 4 - Insulated reservoir

By insulating of the thermal bridges a further reduction is possible. This can be seen in figure 5. By insulating a strip of at least 30 cm. on the floor, and a strip on both sides of the wall above the reservoir of at least 30 cm. the heat flow will be reduced to 44 W/m.

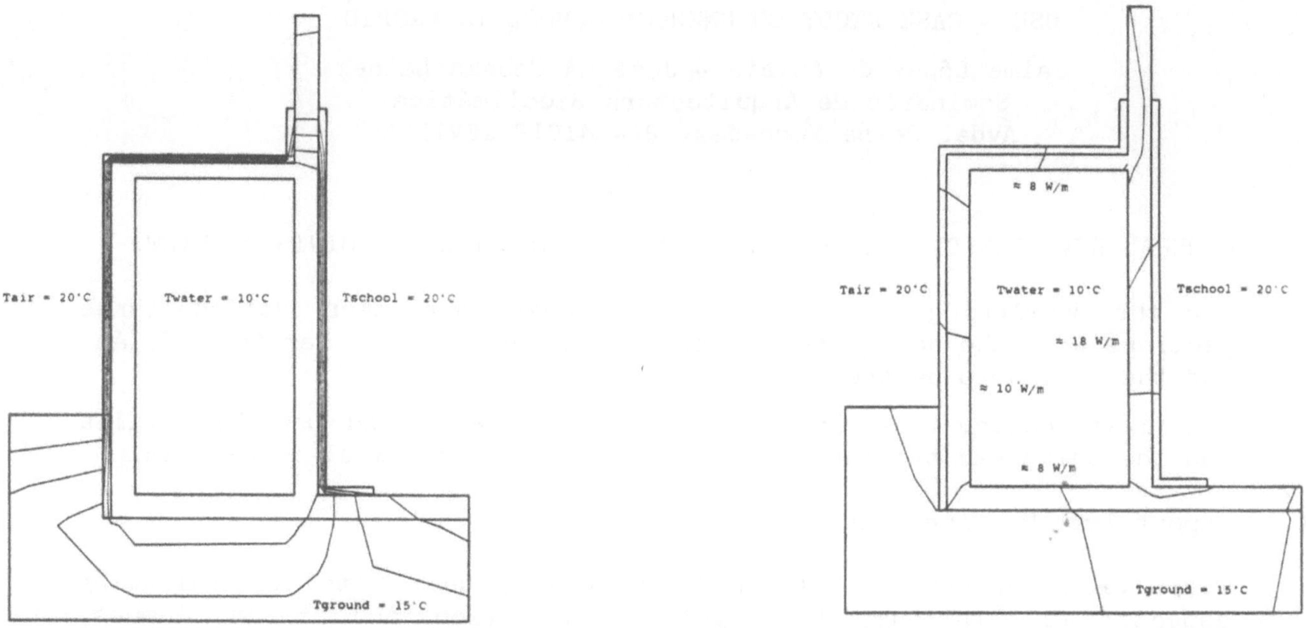

Figure 5 - Insulated reservoir, including part of the external wall and the floor

If insulating of the floor is impossible, the heat flow will be reduced to 46 W/m. See figure 6. The influence of the thermal bridge at the floor is relative a small one.

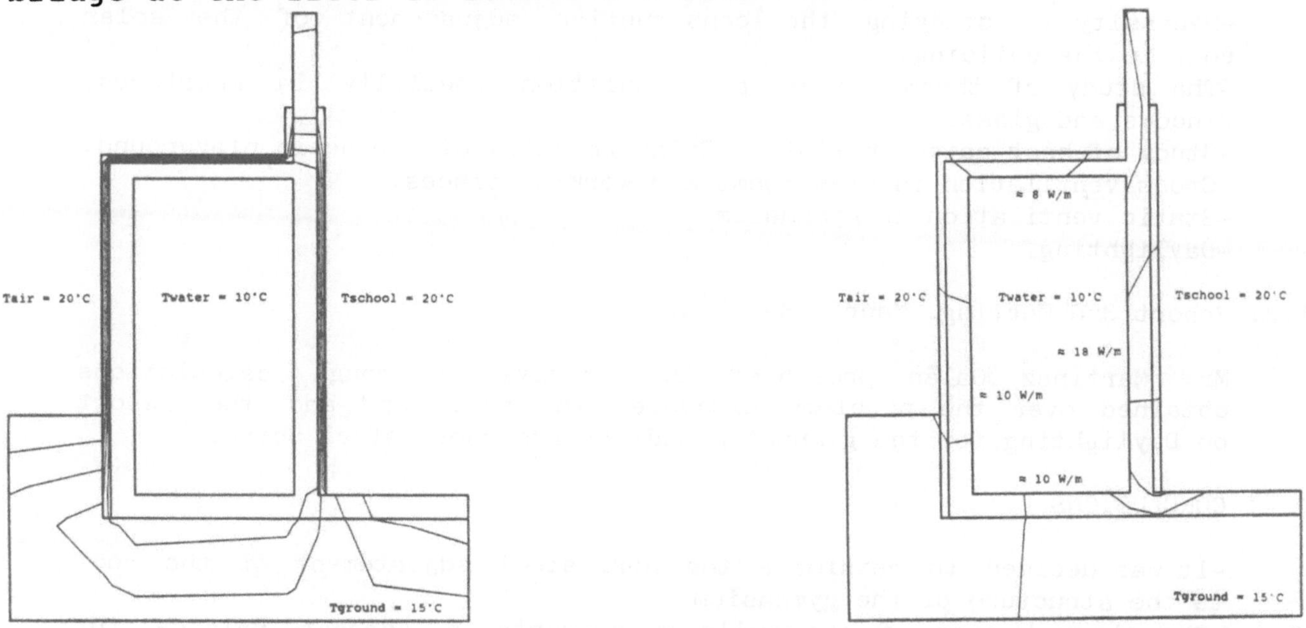

Figure 6 - Insulated reservoir, including part of the external wall

It is not necessary to insulate the bottom of the reservoir.

DSU - CASE STUDY ON ENERCOM SCHOOL IN MADRID

Jaime López de Asiain - José Mª Cabeza Laínez
Seminario de Arquitectura Bioclimática
Avda. Reina Mercedes, 2 - 41012 SEVILLA

1.- GENERAL REVISION OF THE PROJECT FROM THE BIOCLIMATIC POINTS OF VIEW.

For the developing of the DSU work there have been needed three meetings with the design team, represented by Mr. J. Martínez Galán, and the Seminario de Arquitectura Bioclimática.

The first meeting took place in Barcelone, the second one in Sevilla and the third had two parts, one in Madrid and the final in Sevilla.

1.1. Report 1rst Meeting. Barcelone

There was an agreement That the DSU work will deal with the following aspects: Orientations, heat gains, insulation and thermal mass, ventilation an refrigeration and that all these aspects from the point of view of design.

1.2. Report 2nd Meeting. Sevilla.

The main aspects treated were:
-Necessity of studying the constructive adjustement of the solar roof to the building.
-The study of thermal mass an insulation specially in apertures, windows and glass.
-Study of heat gains in winter. Solar geometry of porche in playground.
-Cross ventilation in classrooms and working places.
-Static ventilation in gymnasium.
-Daylighting.

1.3. Report 3rd Meeting. Madrid-Sevilla.

Mr. Martínez Galán presented the results of rough calculations obtained over the problems discussed in Meetin 2nd and the report on Daylighting for the gimnasium made by Profesor Rafael Serra.

2.- CONCLUSIONS

-It was decided to reinforce the perimetral adjustement of the roof to the structure of the gymnasium.
-The thermal mass of the building presents a correct balance for the thermal behaviour in general.

-It can be used conventional insulation in walls but in the roof it should be reinforced. The use of double glazing is recommended.
-It was accepted the possibility of a modification of the design of the porche of the playground for better solar gains.
-It was accepted the opening of complementary windows in the East and West facades for controled cross ventilation.
-It was also decide to open some apertures for fresh air in the gymnasium.
-Active refrigeration is not needed in general.
-Daylighting is correct in the gymnasium , according to the study made by Profesor Serra. It is also correct in classrooms.

3.- DESIGN RELATED ASPECTS

The general design of the building is correct from the point of view of bioclimatics.
Volumes and orientation N-S are addecuate.
The compactness of the building and the organization in two stories of the classrooms is convenient.

4.- PERFORMANCE EVALUATION OF THE SOLAR ROOF RELATED TO THE PROJECT

Considering that the adaptation of the roof to the building is solved, the performance of the whole construction will depend on the performance of the roof itself.

5.- DEFINITION OF CONSTRUCTIVE DETAILS. SIZING OF OPENINGS AND WINDOWS. DETERMINATION OF MASS AND INSULATION.

As it has been said along the report These aspects have been discussed and considered widely and the design team has made the convenient assumptions.

The present study and report has been made only on the basis of a pre-project, without detail constructive plans; now is the moment for the confectioning of the final constructive project and calculations to be done by the design team.

We Include the plans presented by the authors and the main aspects considdered.

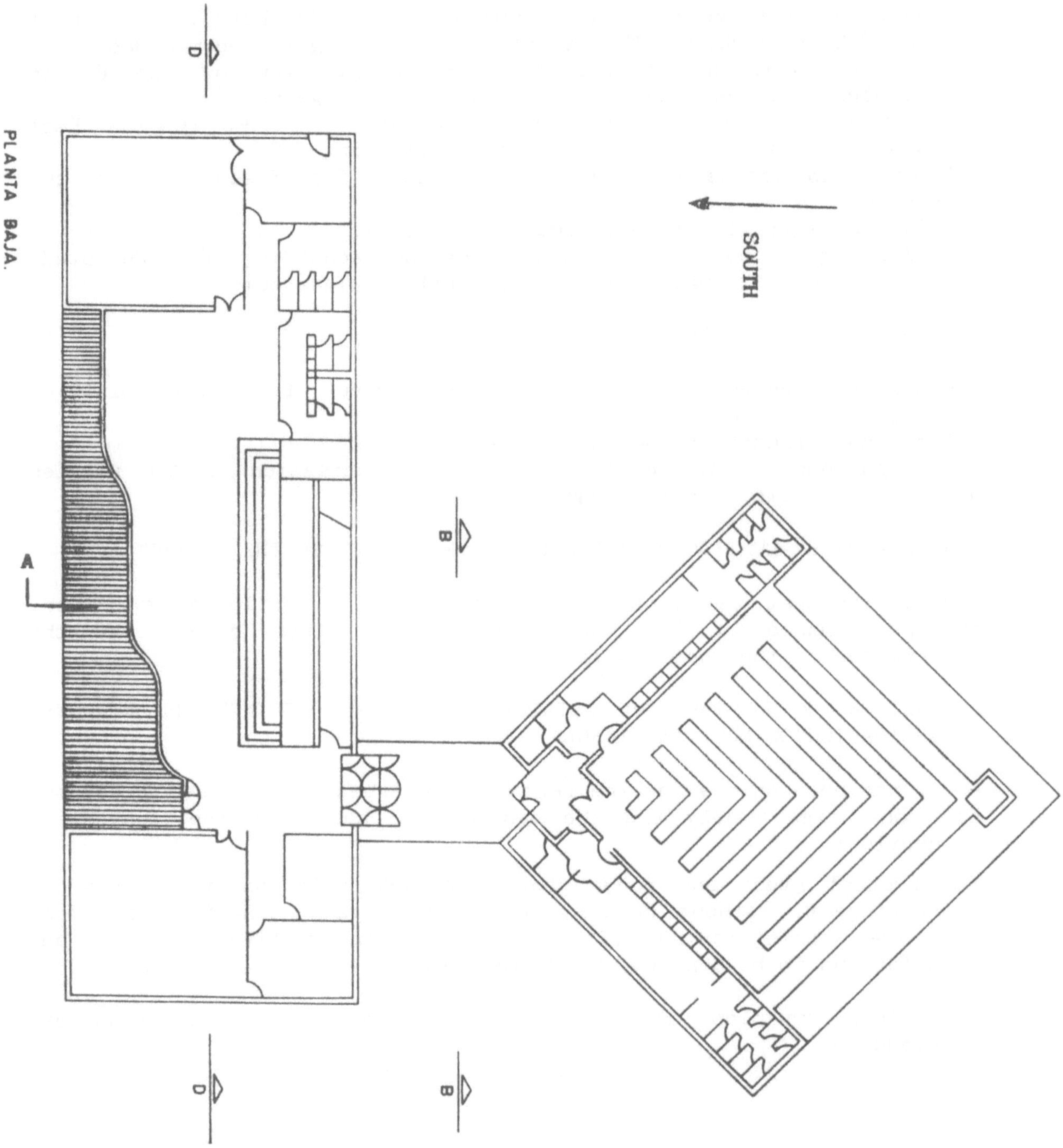

A- Study of heat gains in winter solar geometry of porche in playground.

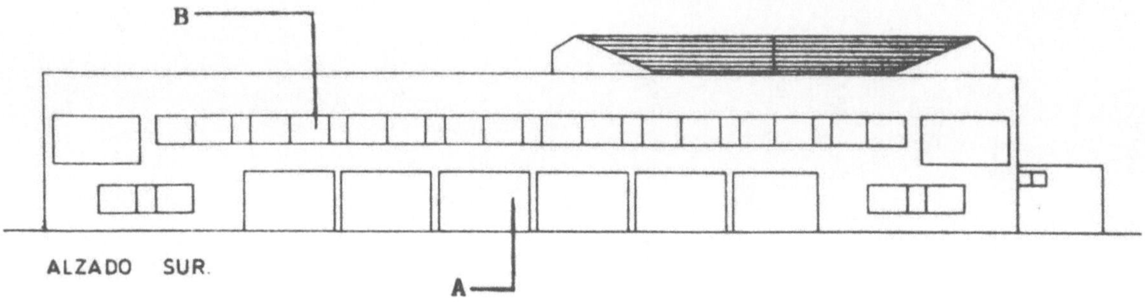

ALZADO SUR.

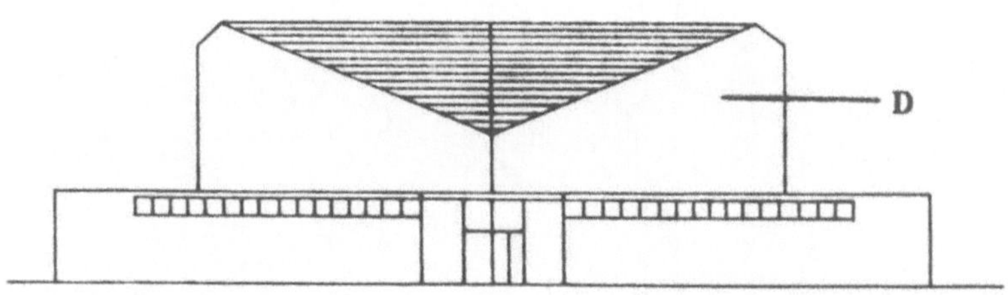

SECCION BB.

B- Study of solar protection and light diffusion -Horizontal movable louvers.

D- Possible complementary static ventilation and natural lighting.

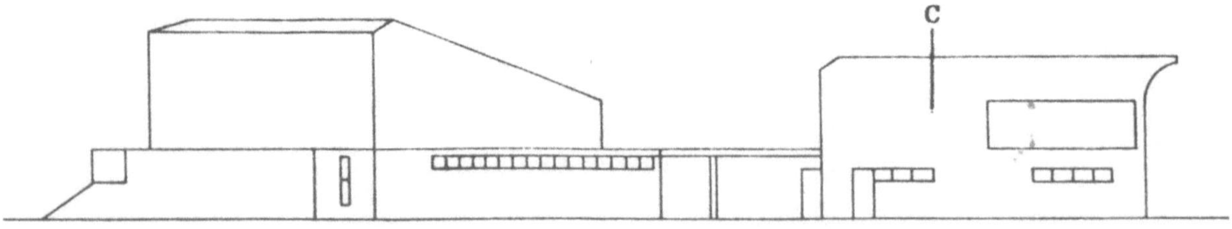

ALZADO OESTE.

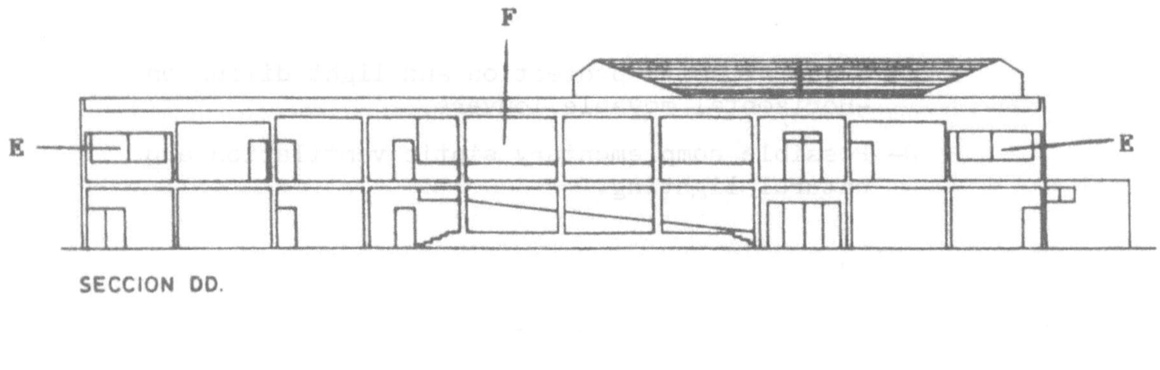

SECCION DD.

C- Possible new aperture for cross - ventilation.
Vertical louvers.

E- Cross ventilation.

F- Study of buffer space.

REPORT ON DAYLIGHTING THROUGH ROOF OPENINGS FOR A SCHOOL IN MADRID, J.A. de la Fuente and A. Sanz architects, J. Martinez-Galan and Eduard Morère engineers.

Research team directed by Prof. Rafael Serra
Escola T.S. d'Arquitectura de Barcelona (Univ. Polit. Cataluña)
Av. Diagonal, 649 08028 Barcelona (Spain)

1. INTRODUCTION

In this Building 2000 project, a study of daylighting has been carried out in order to facilitate the results to the authors of the project. This study has been carried out measuring a model of reduced dimensions under exterior real sky conditions.

Despite the size of the model the results are sufficient by accurate to give a global idea of the interior illumination conditions of the project and the interior problems and conditions that this design of roof can generate due to the particular climate of this geographic situation.

2. LUMINIC CONDITIONS

The model used represents a school gymnasium (scale:1/50). The daylighting level has been measured in this model with a unit of photovoltaic cells connected to a data acquisition system; the characteristics of IDAS are specified in the anex (at the end).

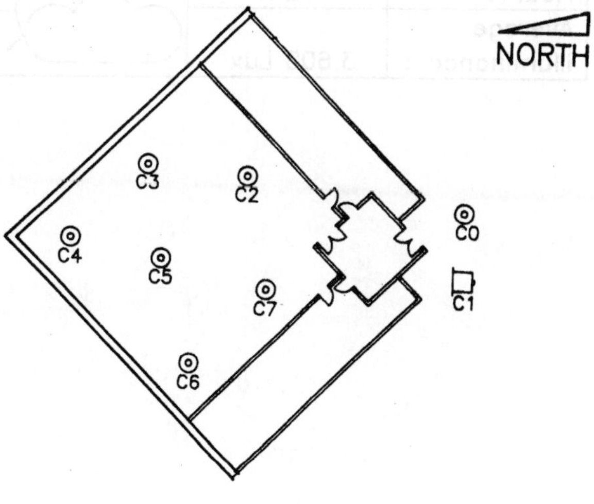

With this system the cells have been used in the following way:

- Cell 0: Exterior daylighting level on a horizontal plane.
- Cell 1: Exterior daylighting level on a vertical plane.
- Cell 2 to 7: Interior daylighting values on a horizontal plane (1.2 m above floor level).

- The measurements have been taken outside under local meteorological conditions. The cases that have been taken into account are the following:

a) Blue sky:

A simulation has been made of the different seasons by way of orientating the model according to sun position and latitude: (A simulation for every month of the year at 12 o'clock solar time has been made).

b) Overcast sky:

Measurements have been made under cloudy conditions at 12 o'clock solar time with most light coming from the South rather than from the North.

3. RESULTS

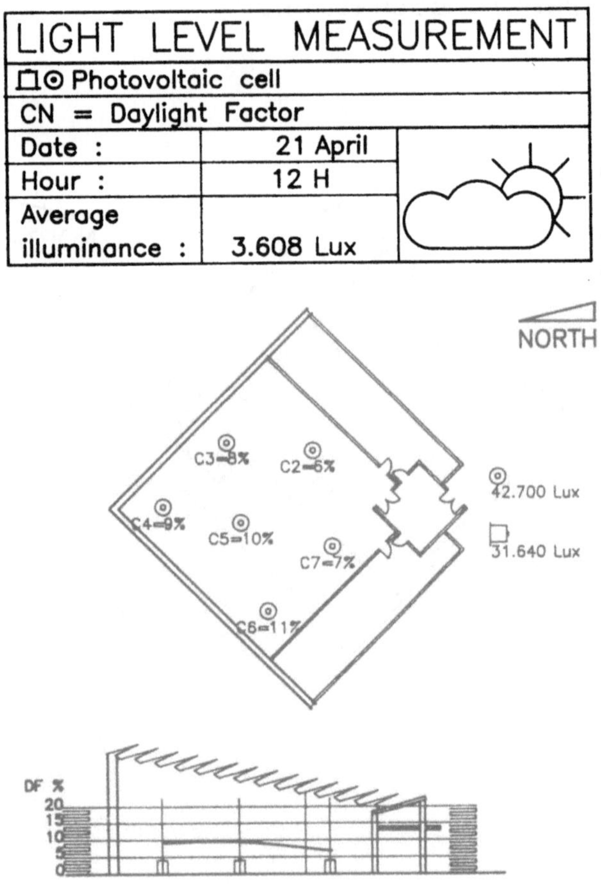

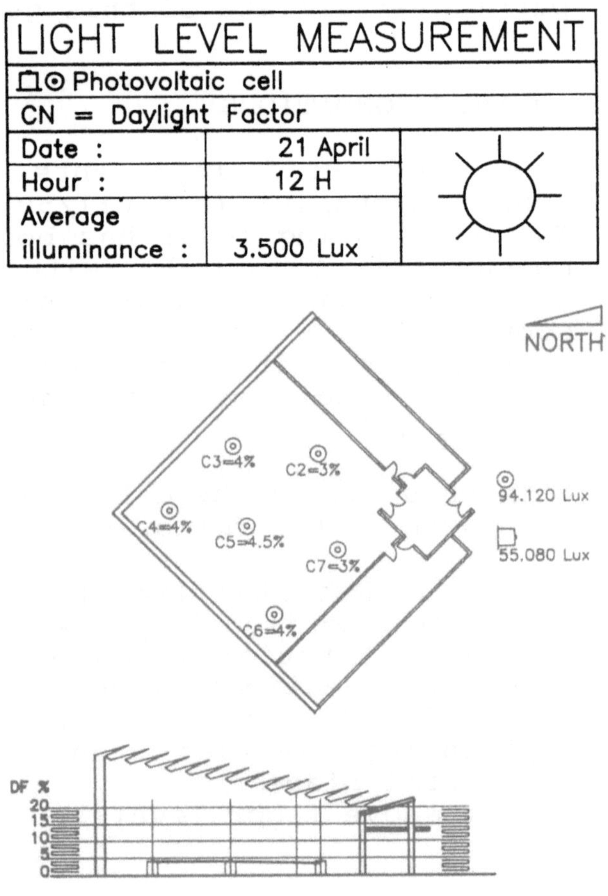

Date :	21 Feb.
Hour :	12 H
Average illuminance :	10.625 Lux

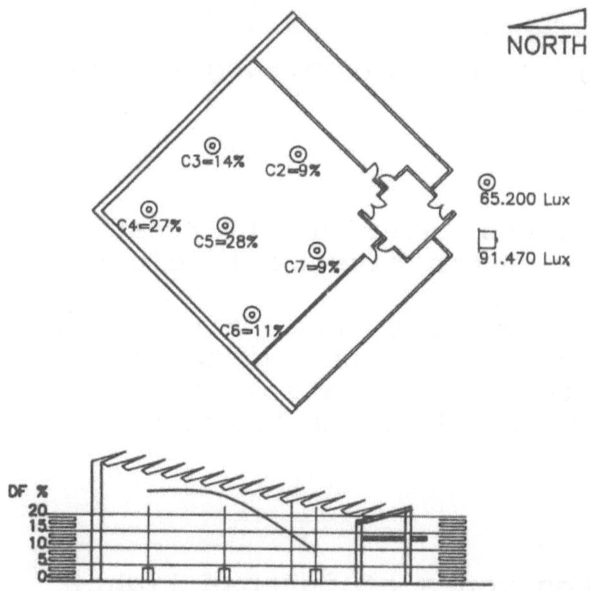

Date :	21 March
Hour :	12 H
Average illuminance :	10.015 Lux

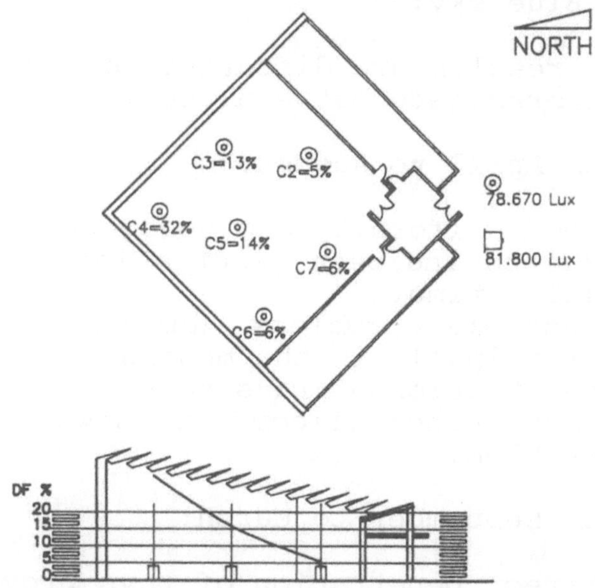

Date :	21 Dec.
Hour :	12 H
Average illuminance :	12.750 Lux

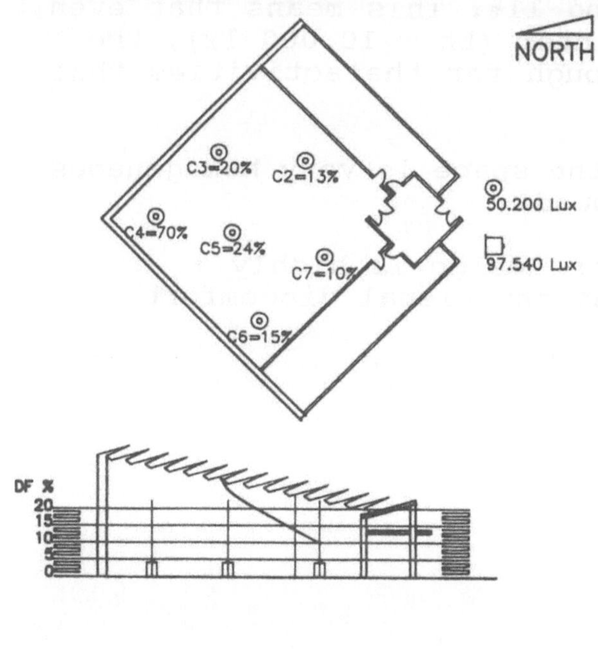

Date :	21 Jan.
Hour :	12 H
Average illuminance :	11.535 Lux

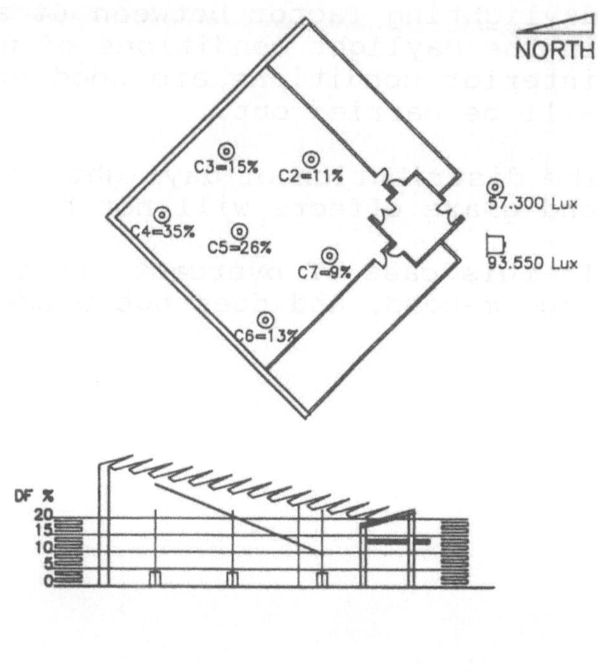

4. CONCLUSIONS

Having realised and analysed the different measurements we can note the following conclusions:

a) Blue sky:

The results are different according to the season, and we must differenciate between the following:

From April to August:

- Good uniformity and homogeneity in the interior illumination due to indirect penetration of the solar rays at 12 o'clock solar time.
- Possible problems during early morning and afternoon, principally in the months of spring and autumn due to direct penetration of rays because of lateral penetration which may cause minor discomfort. However this will be a limited problem.

From September to March:

- Direct penetration of rays provokes zones of intense light.
- Very high interior daylighting with high contrast.
- Considering the sport into activities which will be carried out in the interior, we will encounter important glare disturbances.

b) Overcast sky:

- Good general interior level of daylighting with values of daylighting factor between 6% and 11%. This means that even in the daylight conditions of winter (Eh = 10,000 lx), the interior conditions are good enough for the activities that will be carried out.

- The distribution of daylight in the space is very homogeneous and glare effects will not be found.

- In this case of overcast sky this design is highly recommended, and does not present any visual discomfort.

MIX
Papier aus verantwortungsvollen Quellen
Paper from responsible sources
FSC® C105338

If you have any concerns about our products,
you can contact us on
ProductSafety@springernature.com

In case Publisher is established outside the EU,
the EU authorized representative is:
Springer Nature Customer Service Center GmbH
Europaplatz 3, 69115 Heidelberg, Germany

Printed by Libri Plureos GmbH
in Hamburg, Germany